D1338820

Two week loan

Please return on or before the last
date stamped below.
Charges are made for late return.

04 DEC 2000

14 DEC 2000

CANCELLED
- MAY 2001

29 MAY 2001

11 JUN 2001

CANCELLED
14 JUN 2001

OCT 2001

CANCELLED

8 MAR 2002

CANCELLED

-9 MAY 2002

CANCELLED

IS.239/0799

INFORMATION SERVICES PO BOX 430, CARDIFF CF10 3XT

ABOUT ISLAND PRESS

Island Press is the only nonprofit organization in the United States whose principal purpose is the publication of books on environmental issues and natural resource management. We provide solutions-oriented information to professionals, public officials, business and community leaders, and concerned citizens who are shaping responses to environmental problems.

In 1994, Island Press celebrated its tenth anniversary as the leading provider of timely and practical books that take a multidisciplinary approach to critical environmental concerns. Our growing list of titles reflects our commitment to bringing the best of an expanding body of literature to the environmental community throughout North America and the world.

Support for Island Press is provided by Apple Computer, Inc., The Bullitt Foundation, The Geraldine R. Dodge Foundation, The Energy Foundation, The Ford Foundation, The W. Alton Jones Foundation, The Lyndhurst Foundation, The John D. and Catherine T. MacArthur Foundation, The Andrew W. Mellon Foundation, The Joyce Mertz-Gilmore Foundation, The National Fish and Wildlife Foundation, The Pew Charitable Trusts, The Pew Global Stewardship Initiative, The Philanthropic Collaborative, Inc., and individual donors.

COASTAL WATERS OF THE WORLD

Coastal Waters of the World

Trends, Threats, and Strategies

DON HINRICHSEN

Island Press WASHINGTON, D.C. • COVELO, CALIFORNIA

Library of Congress Cataloging-in-Publication Data
Hinrichsen, Don.
    Coastal waters of the world : trends, threats, and strategies / by
Don Hinrichsen.
        p.   cm.
    Includes bibliographical references and index.
    ISBN 1-55963-382-4 (cloth).—ISBN 1-55963-383-2(pbk.)
    1. Coasts.   2. Coastal zone management.   I. Title.
GB451.2.H56   1998
333.91'715—dc21   97-19856
                                                                     CIP

Printed on recycled, acid-free paper ∞ ⊛
Manufactured in the United States of America
10 9 8 7 6 5 4 3 2

To my parents, Dorothy and George Hinrichsen, with much gratitude

_____

CONTENTS

Human beings are altering coastal ecosystems at an accelerating pace. Much of this change is reducing the long-term capacity of these systems to provide an adequate quality of life and produce renewable wealth. Although the pace of degradation varies greatly and there are a few, small-scale instances of recovery and restoration, the planetary trends are downward. Coastal regions have emerged as the primary human habitat containing, on less than 10 percent of the nonpolar land space, half of the world's people. By 2030, these coastal lands are expected to contain three-quarters of a far larger human population. While the numbers of coastal people and the intensity of their activities spiral upward, in the vast majority of the world's coastal regions water quality is declining, fresh water flows to estuaries are being reduced, fish stocks are collapsing, and habitat critical to both people and fellow species is being destroyed. Conflicts among different competing groups and types of activities are becoming more intense. There is evidence that disease threats and toxic blooms of marine plankton are increasing. If human-induced climate change plays out as the great majority of today's scientists believe, by 2050 the impacts of the trends we can document today will be overlaid with region-wide shifts in rain-fall, a greater frequency of rising storms, and a rising sea level sufficient to threaten many low-lying, heavily populated coasts.

The rapidity of change and the knowledge that we are now as a species causing measurable changes to the planetary ecosystem has produced a concept that now peppers policy documents, mission statements, and speeches at international con-ferences. This is "sustainable development." The Bruntland Commission has sug-gested that sustainable development requires patterns and intensities of resource use today that do not compromise the ability of future generations to meet their needs. It means not eating our seed corn. It means reducing the gargantuan ap-petite for resource consumption in today's wealthy nations while reducing popula-tion growth in poor nations. Sustainable development is widely accepted as the central objective for programs of many stripes in the developing world, but, inter-estingly, it is a term and an idea that has little currency among the public or gov-ernment agencies in such wealthy nations as the United States.

This book suggests that we need to consider a global agenda for coasts as a re-sponse to their accelerating transformation. This requires that we examine the possible links between coastal management and the concept of sustainability. Is coastal management an endeavor that can, or should, lead to definition and pro-motion of sustainable levels of development? It is difficult to deny the favorability of sustainable modes of behavior and the creation of a sustainable balance be-tween humans and the coastal ecosystems most of us inhabit. But if we accept sus-tainable patterns of utilization as the objective for coastal management, then what does this in tangible terms imply for those who wish to improve how we govern coastal regions and thereby attempt to strike a sustainable balance among the need to develop and the need to conserve our own primary habitat?

First, we must recognize that thus far coastal management projects and programs have not been an attempt to define what is sustainable and then set about achieving it. The coastal zone management programs now being implemented by many coastal states in the United States have primarily been an attempt to bring order to a crowded strip of land and water. The objective has been to forestall needless degradation and negotiate conflicts, to set priorities among competing activities in specific places, and to mitigate some of the adverse impacts of development. Such efforts have had positive effects, and this model of coastal management has been adapted to some developing nations. Bringing order and a measure of predictability to the process of change can be an initial step toward sustainability, but it is no more than a first step in a long journey. More recently, many coastal management initiatives have adopted an ecosystem-based approach that addresses the full suite of social, economic, and environmental issues in watersheds and the adjoining coastal sea. These are much more complex undertakings that require reexamination of the nature of development itself and a long-term process of incremental planning and action. In these integrated coastal management programs, struggling to make the goal of sustainable development operational lies at the heart of the endeavor.

Why is it so difficult to think about how to define and how to achieve sustainable development for specific coastal places when so many are eager to endorse the concept? Part of the response is that those living in the so-called "developed" world assume that the big problems lie somewhere else—mostly in those hot developing countries. We too easily forget that the ones consuming the resources, be it rain, forest timber, or the fossil fuels that produce global warming, and creating the market demand that results in all those collapsing fish stocks and the bulldozing of wetlands to make shrimp ponds are in large part avid consumers of the "developed" world. Much of the problem also lies in the destructive practices of the poor—symbolized by those who fish with dynamite. These people feel that they have few alternatives and too often are marginalized by the contemporary development process.

There have been many instances in history when the idea that the great majority of people carried in their heads motivated societies to strive for sustainable agriculture, sustainable forestry, sustainable hunting and fishing, sustainable trade, and institutions that supported these objectives. Unfortunately, the dominant paradigm today creates a context in which there are overwhelming disincentives to seriously thinking about sustainable development. At present, few professionals can survive long if they seriously attempt to implement strategies by which sustainable patterns of behavior and resource utilization might be achieved. To a professional resource manager, particularly in developing nations where the pace of coastal transformation is most rapid, these disincentives manifest themselves as follows:

- The incentive structure in both business and government does not reward or encourage those who pursue sustainable levels of development; rewards in business are for quantitative growth and for profits, in government for quick and tangible results.

- The time horizons for the elected officials that so strongly influence development policy are short; in many developing nations the tenure of high-level officials is measured in months, not years, and priorities shift from administration to administration.

- There are no accepted techniques for calculating net long-term benefits.

- The people who make decisions affecting how natural resources are utilized in central government are seldom directly impacted by their decisions or the projects they launch. If the resource conservation or resource development actions that they design and fund fail or produce unperfected and undesired results, the bureaucrats involved are rarely held accountable.

- The development planning function in most nations is weak.

- Central government control over finances and decision-making means that local governments have little authority. Local initiative and action too often is perceived as a threat to centralized power and authority.

The climate for seriously thinking and actively pursuing sustainable levels of development is not positive.

If we take the objective of sustainable development seriously, how might we begin? The experience thus far in attempting to manage coastal ecosystems should have taught us to think strategically and to adopt an issue-driven, pragmatic approach to the problems we attempt to address. If we accept that the goal of coastal management is the sustained quality of life of the majority of humanity that lives along coasts, then an international agenda requires that we work to provide (1) a healthy life-support system, and (2) an effective, fair, transparent, and participatory system for making societal decisions. In other words, we should work to provide a healthy environment and a good government system. The emphasis is on the quality of development not the quantity of growth as illustrated by GNP. Experience in both the developed and developing world suggests that the goals of environmental quality and equitable, effective governance must be intimately linked if they are to earn popular support. Coastal managers should not focus only on creating preserves of nature in her undisturbed state, or programs that protect biodiversity. Nor should they only engage in the development of the infrastructure that provides clean drinking water, sewage disposal, roads, and employment. We need coherent programs that address the needs for both environmental protection and development simultaneously. The glue that should tie these packages together is a system of governance that provides the people of each place with a measure of effective and participatory governance. With this comes a sense of control, a sense of responsibility, and a reason for hope in the future.

We must also insert time into the equation. The chiefs of the Iroquis Indians were taught to make decisions that would benefit the future seventh generation. Such concepts are unfortunately beyond us. I would suggest that as a first step we attempt, if we're going to take on the sustainability concept, to adopt as a criteria the selection of those actions that will not reduce quality of life 50 years from now. This translates into our lifetimes and those of our children. Even this would be extremely radical.

The experience of the coastal management programs now being implemented in a number of developing nations suggests that we should bring greater order and equity to the process of development as we work toward more sustainable forms of coastal use. One promising strategy calls for working on two tracks simultaneously: with central government and at the community level. In the great majority of developing nations, central government holds unto itself most of the power and local government is therefore weak or nonexistent. Yet the demand for effective governance is often greatest, and the chances for tangible progress highest, at the local level. A focus on community-based coastal management strategies enables us to close the loop between planning and implementation frequently and early on in the process of formulating management strategies. When working at a small scale, it is often easier to resolve conflicts posed by overlapping agency jurisdictions. It promotes reality testing among planners and researchers. The problems become real and tangible and the people affected by the success or failure of the management strategies that emerge become individuals we know rather than a faceless mass. If such trials prove successful, we will accomplish several very important things. We will begin to build a constituency for the endeavor. Experience in developed and developing nations alike demonstrates that without an adequate constituency, development is not sustainable. To build constituencies we need places that we can point to and say, "Look, here improved governance is maintaining and even improving the health of the ecosystem and the peoples' quality of life. It can be done and it is worth the effort." Today, such examples are rare.

But coastal managers must also work with central government to bring order to the process of change. Degradation in environmental quality, and therefore in the quality of life of people in developing nations, is being caused by two major forces. One is the action of people living in poverty who have few choices and often cannot plan for the future, and the other is the looting of environmental assets by the rich and powerful. In many instances the latter is more important. Efforts with central government should be directed at changing the policies that encourage both sources of unsustainable behavior. Central government must also be persuaded to support local-level governance and the process of discovery at the community level that can teach us all in practical terms how to achieve an ethic of stewardship.

If coastal management is to become a vehicle for progressing toward more sustainable forms of coastal development and conservation, we must recognize that we have a great deal to learn and that learning must progress quickly. The need is urgent and we have a long way to go.

Stephen B. Olsen

ACKNOWLEDGMENTS

Like most books, this one would have been impossible to produce without the encouragement, assistance, and advice of many people. A full accounting would probably fill a volume of its own. I would, however, like to acknowledge those individuals and institutions who made significant contributions.

This book took nearly three years to put together. It would never have happened if the Swedish International Development Cooperation Agency (Sida) in Stockholm had not provided a generous research and travel grant, which allowed me to travel to all of the world's major seas and interview literally hundreds of people from all walks of life—from ministers of state to poor, artisanal fisherfolk. In particular, I want to thank Mats Segnestam, head of Sida's Environmental Policy Division, and Magnus Torell, from Sida's Natural Resources Department, for finding the funds in very trying economic times. In addition, a number of other experts within Sida provided me with valuable contacts in developing-country institutions working on coastal-area issues.

The United Nations Population Fund (UNFPA) in New York also provided a grant, to help cover the costs of my extensive demographic research. In particular, I want to thank Mr. Hiro Ando, deputy executive director for policy and administration at UNFPA; Sterling Scruggs, director of the Information and External Relations Division of UNFPA; and Alex Marshall, deputy to the director of the Information and External Relations Division of UNFPA. Their comments on the demographic sections were very helpful.

Three other institutions contributed significantly to this book: the Coastal Resources Center of the University of Rhode Island; the Center for Communication Programs at Johns Hopkins University's School of Hygiene and Public Health; and UNESCO in Paris. I am especially indebted to Stephen Olsen, director of the Coastal Resources Center, and Lynne Hale, associate director, for taking time from busy schedules to review the first ten chapters in this book. Their incisive comments improved the final product immeasureably. At the Johns Hopkins University's Center for Communication Programs, I am particularly grateful to Jose Rimon II, deputy director of the Center and project director of Population Communication Services, for arranging to administer the Sida funds under its Stewardship for Environment and Population initiatives. Jose also provided valuable comments on early drafts. I also wish to thank Susan Dugan and Barber Horton of the center's contract management staff for their support. At UNESCO I want to thank, in particular, two people for reviewing major sections of the text, including the concluding policy chapter: Jeanne Damlamian, senior programme specialist for the Project on Environment and Population, Education and Information for

Development, and Dr. Dirk Troost, chief of the Coastal Regions and Small Islands Unit.

Without logistics support and backup, extensive field trips to far corners of the planet would simply not have been possible, given the time constraints. The following staff members of UNFPA's country offices were instrumental in getting me into the field: George Walmsley, former country director of the Philippines, and his wonderful staff (especially Yong) for arranging my far-flung travels in the Philippines; Jay Parsons, former country director of Indonesia, for coordinating and planning my itinerary in Indonesia; Sterling Scruggs, former country director of China, for arranging my travels in China; Qu Geping, director of China's National Environment Protection Agency, for graciously arranging several field visits; and the staff of the Shanghai Municipal Bureau of Environmental Protection for hosting my stay in China's largest metropolis.

In addition, I must thank Dr. Rodney Salm, coordinator of IUCN's Marine and Coastal Conservation Program in East Africa, for his help in guiding me through the shark-infested waters of Nairobi and for getting me around East Africa in one piece.

Thanks also to Professor Chou Loke Ming and his colleagues at the National University of Singapore for hosting me and providing valuable information on the state of Southeast Asian seas; to Isao Mizohashi, former director of the Asian Urban Information Center in Kobe, Japan, for hosting my stay in his marvelous city; and to the staff of the Chesapeake Bay Foundation in Annapolis, Maryland, for hosting me and arranging a field trip on the bay.

In addition, a number of my colleagues have made substantive comments on various chapters. I would like to thank the following people: Anne Platt of the Worldwatch Institute for reviewing the first four chapters and the Caribbean chapter; Professor Gavin Jones, coordinator of the Demography Program at the Australian National University, and Dr. Andrew Smith, former coastal management officer of the South Pacific Regional Environment Program (SPREP), for their reviews of the South Pacific chapter; Philomene Verlaan, formerly of UNEP's Oceans and Coastal Areas Programme Activity Center in Nairobi, for helpful comments on the Northwest Pacific chapter; Reza Amini, coordinator of the Southeast Asian Seas Action Plan, Kim Looi Ch'ng, program officer for the Southeast Asian Seas Action Plan in Bangkok, and Drs. Ed Gomez and Liana McManus from the Marine Science Institute of the University of the Philippines in Metro Manila for their very thorough comments on the Southeast Asian Seas chapter; Emmanuel Weill-Halle, coordinator of the Caribbean Action Plan in Jamaica, and her staff for reviewing the Caribbean chapter; Mike Hirshfield of the Center for Marine Conservation for his detailed comments on the North American chapter; Clive Wilkinson of the Australian Institute of Marine Science and Wendy Craik, a fisheries biologist with the Great Barrier Reef Marine Park Authority, for very useful comments on chapter 4; Rodney Salm, coordinator of IUCN's Marine and Coastal Conservation Program in East Africa for reviewing the chapters on the Arabian Gulf and East Africa; Bob Engelman and Pamela LeRoy from Population Action International for reviewing chapter 2; Michel Batisse, former assistant director-general for science at UNESCO, and Peter Thacher, former deputy director of UNEP, for reviewing chapters 7 and 19; and Olof Lindén, professor of zoology at Stockholm University, for reviewing the Baltic and North Seas chapter.

I would also like to thank my friend and colleague Pieter Stemerding of the United Nations Development Programme for reviewing chapters 1–5 and providing valuable feedback from the layman's perspective, Uwe Deichmann, now with the U.N. Statistical Division, for his valuable population density maps, and my mother, Dorothy Hinrichsen, for wading through most of the manuscript and improving its readability.

Finally, I would like to thank my publisher, Island Press, for being patient, and my hardworking editors, Stacye White and Todd Baldwin, for doing such an excellent job of guiding the book through its various production stages.

Despite all the expert help, any errors in this book are entirely my own. The views expressed here are also my own and do not necessarily reflect the official views of any of the sponsoring agencies.

LIST OF ABBREVIATIONS

ASEAN: Association of Southeast Asian Nations
BOD: biological oxygen demand
COD: chemical oxygen demand
CRC: Coastal Resources Center (University of Rhode Island)
EBRD: European Bank for Reconstruction and Development
EEZ: Exclusive Economic Zone
EPA: Environmental Protection Agency (U.S.)
FAO: United Nations Food and Agriculture Organization
GEF: Global Environment Facility
GIS: Geographical Information Systems
ICLARM: International Center for Living Aquatic Resources Management
IDB: InterAmerican Development Bank
IMO: International Maritime Organization
IOC: Intergovernmental Oceanographic Commission
IUCN: World Conservation Union
NCGIA: National Center for Geographic Information and Analysis (U.S.)
NOAA: National Oceanic and Atmospheric Administration (U.S.)
NRDC: Natural Resources Defense Council (U.S.)
OAS: Organization of American States
OCA/PAC: Oceans and Coastal Areas Programme Activity Center (UNEP)
OECD: Organization for Economic Cooperation and Development
SAREC: Swedish Agency for Research Cooperation with Developing Countries
SPREP: South Pacific Regional Environment Program
UNCED: United Nations Conference on Environment and Development
UNDP: United Nations Development Programme
UNEP: United Nations Environment Programme
UNESCO: United Nations Educational, Scientific, and Cultural Organization
UNFPA: United Nations Population Fund
USAID: United States Agency for International Development
WMO: World Meteorological Organization
WRI: World Resources Institute
WWF: World Wide Fund for Nature

The Kuna Indians of Panama

Safeguarding the Resources of the San Blas

The flight from Panama City, on the Pacific side of Panama, to the San Blas Islands, on the Caribbean side, takes only twenty-five minutes. In that short time, the traveler can see amazing contrasts. The crowded Pacific side of the country shows all the classic symptoms of development stress: destruction of coastal mangrove forests for shellfish ponds, deforestation and conversion of the uplands into cattle ranches, and uncontrolled urban sprawl around Panama Bay. The Caribbean coast, east of the canal zone, is Kuna country. Here, a lush "green sea" of unbroken tropical forest undulates to the horizon. Looking like gigantic stalks of broccoli, huge tropical hardwoods pierce the emerald canopy.

Landing on narrow jungle airstrips is not easy. Only the most experienced pilots fly into the San Blas. The reason is soon apparent. As our twin-engine otter approaches the coast from the landward side, we slam into a vicious headwind. The plane groans and shakes as we make our approach, skimming just above the treeline. A flock of mealy amazons—crow-sized green parrots—explodes in a shrieking frenzy of feathers. Suddenly, the seam of green below opens up and the pilot dives for the runway, throwing the engines into reverse as we hit and braking hard. We stop just short of the sea.

A short boat ride away is Wichuala. This island boasts one of the few "guest houses" in this part of the San Blas featuring a generator, which is turned on at night to provide light and to power the village television set. Like virtually all of the islands in the San Blas, Wichuala is not connected to the electricity grid and has no source of drinking water. Freshwater must be brought in from mainland rivers in big barrels once or twice a week.

The San Blas Islands stretch from San Blas Point (east of the canal) halfway to the coast of Colombia. Here, laced over aquamarine waters clear as glass, is a complex of some 365 coral islands inhabited by roughly 35,000 Kuna Indians. Only around 40 of the islands sprout villages; many of the rest are uninhabitable because of disease-carrying sand fleas.

Although the outside world is beginning to intrude, many Kuna still live much as their ancestors did, relying on the bounty of the sea and on what they can grow in garden plots hacked out of the jungle. Kuna continue to exploit the forest for nearly all their basic medicines.

Traditional ways still prevail in village life. Kuna women wear multicolored, hand-embroidered dresses, sport beaded bracelets on their ankles and wrists, and wear nose rings—considered signs of beauty. The Kuna share whatever they have with other members of their village. No one goes hungry. Despite the fact that many marriages are still arranged by the parents, divorce is virtually unknown. Because the Kuna have their own rough system of justice, crime is almost nonexistent.

"We have been able to maintain our traditions because we took complete control over our own lands," states Eladio Garcia, one of the Kuna's elder statesmen. "But we had to fight for them, and we continue to do so."

In the 1920s, the Panamanian government decided to integrate the Kuna into Spanish society. The Indians revolted and fought a bush war with the Panamanians from 1922 to 1925, which they won with help from the United States. Ever since, the Kuna have controlled the islands of San Blas and the lands along the coast, thanks in part to continued American support.

Eladio Garcia, a robust man who looks much younger than his sixty-five years, is known affectionately around the islands as Mr. Blue. He is a man who has learned to straddle two worlds, Panamanian and Kuna. Until his semiretirement recently, he split his time between Panama City, where he ran two food concessions for the U.S. military in the Canal Zone, and the San Blas. Now he spends most of his time fishing and talking with old friends on the islands.

Garcia even attended junior college for one year, a level of education seldom achieved by the Kuna, who do not like living in "foreign lands," such as Panama City. A high school diploma is difficult to obtain, since there are no high schools on the islands. Kuna who want their children to be formally educated have the unpleasant, and costly, option of sending them to Panama City to attend a Spanish high school.

To say that the Kuna depend on fishing for their livelihood is an understatement. "Nearly all our protein comes from the sea," says Garcia. Fishing is left to the men, who usually go out alone or in small groups in dugout canoes. Kuna fishing techniques are very simple: a nylon fishing line is wrapped around one finger; the hook on the end is baited and tossed overboard. "This can be done by any Kuna with access to a dugout or motorboat," explains Garcia. "Usually we fish for snapper, jacks, barracuda, and sardines." But bigger fish are taken, such as tarpon, tuna, sailfish, even sharks. Of course, the technique differs. So that they do not risk losing any fingers, the Kuna catch larger game fish using stronger lines that they tie to the boat. Struggling fish are finished off with a harpoon thrust to the brain.

The Kuna also do some spearfishing. Garcia remembers vividly the time he was hunting for jack when he spotted a three-foot-long barracuda. "I aimed my speargun right at him, squeezed the trigger, and missed." The enraged fish attacked with lightning speed. "I was nearly gutted in the water," exclaims Garcia, rubbing his stomach where the barracuda missed him by centimeters. "After that I gave up spearfishing."

The Kuna have evolved a sustainable system of farming the jungle that has been used for centuries. Vegetables and fruit (except the coconut palm that thrives everywhere) are cultivated along the mainland coast. The Kuna have carved garden plots up to 20 acres out of the jungle. Here they grow rice, corn, squash, melons, yucca, and plantain, as well as bananas, coconuts, and mangos.

These plots are often tended by men and women from the same extended family, but entire villages also work collective gardens. When the soil begins to wear out after three to ten years, the Kuna allow the jungle to reclaim the land, and they clear a new garden in another part of the forest. This system of shifting (swidden) agriculture has worked for hundreds of years.

But the winds of change are blowing over the islands. Many young Kuna men now leave to work in Panama City or in the Canal Zone. A cash economy is rapidly evolving, and tourism is being developed. Young Kuna couples have fewer children, perhaps only two or three, compared to the families of ten or more that their parents fostered. The distant, outside world is no longer so far away.

Presently, the Kuna are fighting on two fronts. They must parry frequent attempts by the Panamanian national guard to annex some of their coastal lands; the government wants to log more of the Caribbean watershed and open it up to cattle ranching. And seaward, increasing numbers of foreign fishing boats ply the rich coral reefs of the San Blas. Such incursions worry Garcia and his friends. Recently, Colombian shrimp boats had to be driven away at gunpoint, and a party of crocodile poachers was apprehended. "We don't allow any kind of exploitation from outside," affirms Mr. Blue. "The Kuna's resources are used by the Kuna. And we are determined to hold on to our ancestral lands."

"We realize that once we lose our forests, we also will lose the sea," says Garcia. Where coastal forests are cleared, eroded sediment quickly buries valuable nearshore habitats such as seagrass beds and coral reefs, often precipitating the collapse of important fisheries. The Kuna are determined that this will not happen to the San Blas as it has to other parts of Panama.

The Kuna's management system works because it is based on sustainable practices that take account of ecosystem limitations while maximizing their productive capacities. Their system of management is also rigidly enforced. The Kuna's communal system is a form of "sustained coastal governance," a process that touches many aspects of their lives. From a very early age, Kuna understand the vital ecosystem linkages that tie together upland forests and near-shore coral reefs. Their process of governance offers a valuable model to other third world societies struggling against great odds to manage coastal areas.

Since fishing techniques are mostly traditional, stocks are not overfished, and edible varieties abound. However, some shellfish, such as lobsters, are now being overharvested in some areas and sold to restaurants in Panama City for dollars. If tourism is developed without sound limits, demands for fresh seafood may result in the overexploitation of certain stocks, such as groupers, coral trout, emperors, and a variety of shellfish.

The Panamian government does not have a coastal management plan for the country as a whole. So long as the Kuna can maintain control of the Caribbean watershed forests that protect their coastal waters, there is no reason why their sustainable management system cannot endure.

Even as outside forces close in on their way of life, Eladio Garcia is adamant about one thing: "We have been fishermen since the beginning of time, and we will continue to fish until it ends."

We Are All Coastal

The dilemma facing the Kuna of the San Blas is one that confronts much of humanity. Nearly everywhere, coastal resources are under tremendous development pressures, pressures that often result in the destruction of the resources coastal communities have traditionally relied on for their livelihood and long-term survival. The globalization of the economy has resulted in the globalization of nonsustainable practices for exploiting critical resources. We may be more efficient at utilizing natural resources, but the levels of exploitation and resource consumption have risen dramatically over the past five decades, in line with population growth and technological advances.

Modern societies, for the most part, increasingly are out of sync with their environments. Entire societies are mortgaging their futures by borrowing from finite natural resource stocks to pay for the present. Instead of living on the "interest" we could get by using resources sustainably, we are spending our principal.

Collectively, humankind is undergoing a crucial test: whether as a globe-shaping species we can design resource management systems that are ultimately sustainable, not only for humanity, but for the entire complement of plants and animals with which we share the earth.

Nowhere, perhaps, is this challenge more daunting than in the world's coastal areas, where increasingly the bulk of humanity and economic activity is concentrated. Over 50 percent of the entire population of the planet lives and works within 200 kilometers of a coast on about 10 percent of the earth's land area (Hinrichsen 1996), but two-thirds live within 400 km. And nothing more dramatically illustrates that fact than a satellite photograph of the earth taken at night. Viewed from space, an uninterrupted river of light flows around the world's coastlines. In some regions, population centers are packed so tightly along the coast that the outlines of continents are recognizable from 100 miles up.

Ultimately, of course, the whole of humankind is coastal. No matter where we live, we are connected to the world's oceans through an intricate drainage system of rivers and streams. This vast network of waterways drains continental and island watersheds and is responsible for dumping up to three-quarters of all the pollution found in coastal seas.

This watershed viewpoint was graphically illustrated within the Smithsonian Institution's exhibition "The Ocean Planet," which opened in Washington, D.C., in April 1995. Among its unique displays was one depicting America's watersheds, in which visitors were challenged to find their "watershed address" by tracing the drainage pattern from their home to the sea. Clearly, everyone has an impact on coastal environments, either directly or indirectly.

Coastal Areas Defined

By their very nature, coastal zones are mercurial places, formed and reformed constantly by the irresistible forces of the sea. Land and ocean are forever locked in conflict. Marine scientist Jens Sorensen defines a coastal area as "that part of the land affected by its proximity to the sea and that part of the ocean affected by its proximity to the land . . . an area in which processes depending on the interaction between land and sea are most intense" (Clark 1996).

Coastal zones can be wafer-thin strips of coastline not more than a few kilometers wide, extending from the low-tide mark inland; or they can extend inland so far as to include entire watersheds and may run seaward to the continental shelf, encompassing the full extent of a country's 200-nautical-mile Exclusive Economic Zone (EEZ) (OECD 1993).

Many estimates of coastal populations are based on an area within 60 to 100 kilometers of the shoreline. In this book, the definition of a coastal zone, or coastal area, is a strip of land 200 kilometers wide (roughly 120 miles) measured from the low-tide mark inland and extending seaward to include important near-shore ecosystems such as barrier islands, seagrass beds, and coral reefs. In many areas of the world, the sprawling suburbs and bedroom communities associated with very large urban areas, such as Los Angeles, Tokyo, and London extend up to 150 kilometers inland. These areas have tremendous impacts on coastal zones. It seems sensible to include them in any realistic analysis of coastal-area problems and proposed management plans.

No matter how they are defined, coastal areas always include intertidal zones and often incorporate coastal floodplains, estuaries, mangrove swamps, salt marshes, and tidal flats as well as beaches, dune complexes, barrier islands, near-shore seagrass beds, and coral reefs.

Safeguarding these complex, interlinked ecosystems is probably the planet's most challenging management job. Marine ecologist John Clark summarizes the dilemma: Coastal areas are the "place where agency authority changes abruptly, where storms hit, where waterfront development locates, where boats make their landfalls, and where some of the richest aquatic habitats are found. It is also the place where terrestrial-type planning and resource management programs are at their weakest" (Clark 1996).

Coastal Area Management

There are probably as many definitions of coastal zone management as there are coastal managers. One of the more succinct definitions comes from the Organization for Economic Cooperation and Development (OECD): "Integrated coastal zone management is most simply understood as management of the coastal zone *as a whole* in relation to local, regional, national and international goals. It implies a particular focus on the interactions between the various activities and resource demands that occur within the coastal zone and between coastal zone activities and activities in other regions" (OECD 1993).

Often, coastal management is a slow, incremental process that evolves over decades. The fact that it is a complicated process should not excuse policy makers and planners from taking the "wide-angle" view, however, from looking at our *total* impact on coastlines. With a comprehensive picture of the numerous stresses confronting coastal areas, responsible management strategies lead inevitably to the notion of managing entire watersheds. There are precedents: Watershed management is being tried out in the Chesapeake Bay, the largest brackish water estuary in the United States. In Europe, the Rhine and Danube rivers now have river-basin management plans in various stages of implementation.

A comprehensive approach to management, known as integrated coastal zone management, requires the institutional capacity to tackle problems that cut across academic disciplines, agency mandates, and institutional boundaries. A successful coastal area management strategy should provide a workable mechanism for analyzing and responding appropriately to the growing and varied threats facing coastlines. Above all, the process of coastal governance should include all major stakeholders.

Coastal management plans need to address several issues simultaneously. These include the accelerating impacts of rapid population growth and migration on coastal urban areas and the need for municipal governments to regulate urban and industrial growth along coastlines (this would greatly facilitate the creation of more livable cities). They must introduce clear zoning laws that prohibit or allow certain types of activities in coastal areas, both urban and rural; address local or regional concerns and involve the active participation of coastal communities; regulate the use of critical coastal ecosystems such as wetlands, barrier islands, mangrove swamps, seagrass beds, and coral reefs; and regulate coastal fisheries.

The world's collective failure to manage coastal resources sustainably and introduce rational, long-term development strategies—strategies that balance the needs and concerns of local communities with regional and national objectives—has undermined our ability to grapple with the conflicting and contradictory issues confronting the use of coastal areas.

Instead of integrated management plans and policies that are inclusive, we have opted for contained, vertical plans that are essentially exclusive. Instead of incremental growth and purposeful development for the common good, we have allowed management of coastal areas to degenerate into virtual anarchy. Instead of encouraging local communities to take the lead in regulating and managing coastal resources, we have allowed top-down management to dictate resource use. Instead of creating broad-based public constituencies in support of comprehensive coastal development plans, we have permitted special interests to set the agenda.

About This Book

It has taken more than three years to put this book together. During that time I covered over 100,000 kilometers, crossing every major sea in the world. In the course of my travels, I interviewed hundreds of people from all walks of life—from prime ministers to artisanal fisherfolk. The picture that emerged is not one I had expected to find when I began research in the spring of 1993. The world's coastal areas are now on the frontline of a battle—a battle that does not attract headlines or Pentagon-style budgets, but one whose outcome is nonetheless important for the future of human habitation on this planet. The threat humankind collectively faces in coastal areas is clearly one of our greatest challenges now and in the coming decades.

There are three broad themes throughout this book:

- The world's coastal areas are being overwhelmed with people and pollution. The pollution comes from coastal populations as well as inland agriculture and industry via rivers and estuaries.

- As a result of the concentration of economic activities along coastlines, especially coastal urban areas, critical coastal resources—such as wetlands, mangroves, seagrasses, and coral reefs—are being plundered in the name of development and lost through inertia and neglect.

- The inability of governments, with a few exceptions, to craft and implement rational coastal management plans is having far-reaching consequences, including the collapse of coastal and offshore fisheries, the continued impoverishment of poor coastal communities, runaway urbanization, and the loss of numerous amenities associated with healthy coastal environments.

Chapters 1 and 2 present an overview of population and resource pressures afflicting the world's coastlines and near-shore waters. Chapter 3 discusses the contentious area of coastal management and offers three successful case studies in coastal governance. It also looks at the prospects of global management embodied in the ratification of the Law of the Sea Convention in 1994.

In chapters 4 through 17, each of the world's regional seas is profiled. Each chapter contains a discussion of the main population, resource, and management challenges facing the sea. Subheadings have been standardized so that readers who want to browse through the book can find sections of interest easily. My hope is that this book will serve as a useful reference guide to which readers can turn again and again for specific information.

The only seas not included in this analysis are the polar seas, since few people live in polar regions, and those regions suffer from entirely different types of threats than the rest of the world's seas.

As noted above, I have in the course of my research for this book conducted innumerable interviews, and have quoted in the text many of the subjects of these interviews. With one or two exceptions, quotes for which no publication is referenced are from these interviews.

Another subject has also been left out deliberately: sea-level rise. Although this is a serious concern for coastal populations, particularly on low-lying islands, when compared to all the other immediate concerns and issues raised in this book, it is a more remote threat. In addition, a number of academic books have already been written that deal entirely with sea-level rise and its probable consequences (see, for example, *Submerging Coasts* by Eric Bird [John Wiley, 1993]).

The final chapter takes a critical look at some of the lessons learned from three decades of coastal management experiences and discusses elements common to successful management efforts. It also proposes an action agenda that could be implemented, given political commitment and funding.

The demographic data in this book come from a variety of sources, but much of the population data are from the Population Reference Bureau in Washington, D.C., which uses data from the UN Population Division in order to produce yearly population estimates based on current growth rates. All estimates are based on national censuses, which usually take place every decade; the last round was in 1990–1995.

Gus Speth, administrator of the United Nations Development Programme, sums up our main challenge for the coming millenium: "Since economic growth is the means and human development the end, the quality of growth is as important as its quantity. Otherwise, economic growth can be jobless rather than employment-creating, ruthless rather than equitable, voiceless rather than participatory, and futureless rather than environmentally sound. The script for human development in the twenty-first century will begin to be written by what choices we make today. We should not let it be said of our time that we, who had the power to do better, allowed the world to get worse."

It is my fervent wish that this book will help place coastal area concerns higher up on government agendas, and contribute constructively to the ongoing debate over how to manage these ever-changing areas, not only for our own benefit, but for the benefit of future generations.

Coastal Population Growth

The Ultimate Threat

Throughout much of the world, coastal areas are overdeveloped, overcrowded, and overexploited. Coastal waters and bays are often horribly polluted with untreated (or partially treated) municipal, industrial, and agricultural wastes. Rivers bring in more pollutants, including organic chemicals and heavy metals, along with increasing loads of sediment. Rich coastal ecosystems, such as estuaries, salt marshes, and mangrove swamps, have been decimated. Figure 1.1 summarizes some of the main threats to the world's coastlines by region, including population density. The dark lines along the coasts represent areas under stress from development-related activities. More than half of the world's coastlines suffer from severe development pressures (WRI 1995). Globally, little is being done to manage the crisis of our coasts.

Underlying the crisis are escalating human numbers and needs. Over 50 percent of the world's population—some 3.2 billion people—already live along a coastline or within 200 kilometers of one (Hinrichsen 1994; Deichmann 1996). Future population projections indicate that by 2025, 75 percent of the world's population, or 6.3 billion people, could reside in coastal areas—500 million more people than the current *global* population (Hinrichsen 1996).

Using new demographic techniques, including GIS (Geographical Information Systems) technology, demographer Uwe Deichmann and his colleagues at the National Center for Geographic Information and Analysis (NCGIA) at the University of California at Santa Barbara were able to calculate population densities for each major region of the world (Deichmann 1996). The population density maps reproduced in this and later chapters (see figure 1.2) graphically illustrate that the majority of the world's people are concentrated in coastal areas and along major river valleys (e.g., the Gangetic Plain in India).

Much has been made of the world's exploding population, but where people live and work is a more important demographic indicator than basic growth rates. Population distribution gives a clear picture of population stress on a country's resource base. It also allows for the development of more rational management plans, especially in terms of future infrastructure needs, crucial services, and the provision of jobs. The global database assembled at NCGIA contains over 19,000 administrative units for some 217 countries (Tobler et al. 1995). In most cases, data

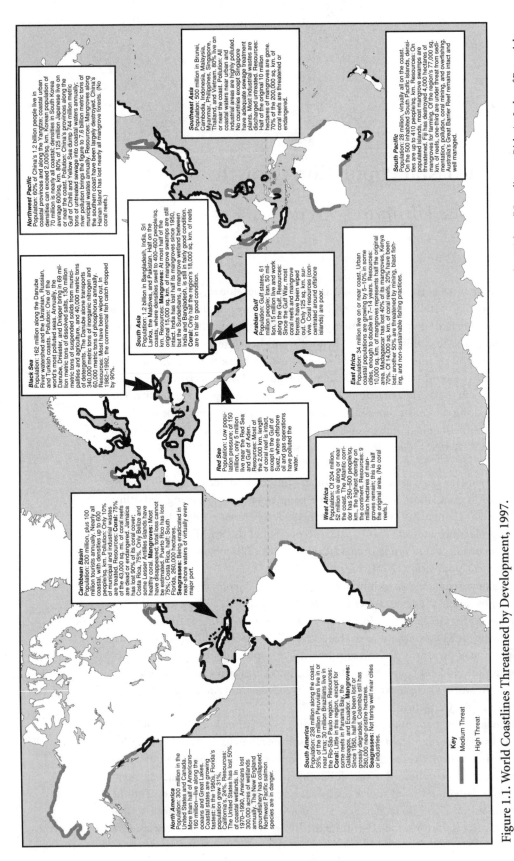

Figure 1.1. World Coastlines Threatened by Development, 1997.

Sources: D. Hinrichsen, "Pushing the Limits," *Amicus Journal* 18, no. 4 (winter 1997), pp. 18–19; D. Bryant, E. Rodenburg, T. Cox, and D. Nielsen, "Coastlines at Risk: An Index of Potential Development-Related Threats to Coastal Ecosystems," World Resources Institute, Washington, D.C., 1995, pp. 1–8.

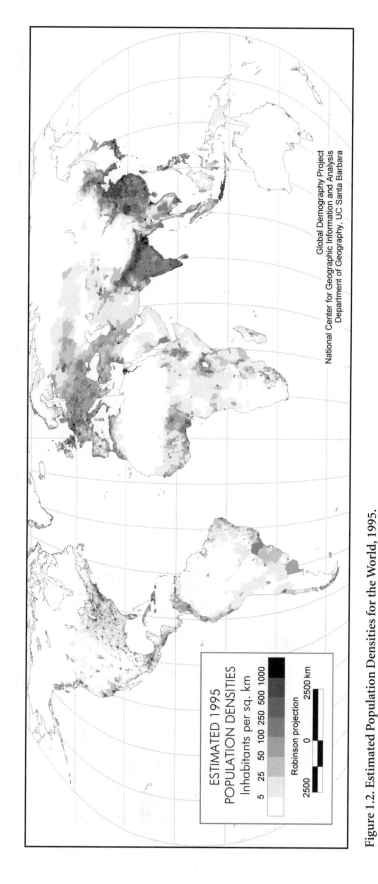

Within the image:

ESTIMATED 1995
POPULATION DENSITIES
Inhabitants per sq. km

5 25 50 100 250 500 1000

Robinson projection

2500   0   2500 km

Global Demography Project
National Center for Geographic Information and Analysis
Department of Geography, UC Santa Barbara

Figure 1.2. Estimated Population Densities for the World, 1995.

Source: Waldo Tobler, Uwe Deichmann, Jon Gottsegen, and Kelly Maloy, *Global Demography Project,* National Center for Geographic Information and Analysis, Department of Geography, University of California, Santa Barbara, 1995.

9

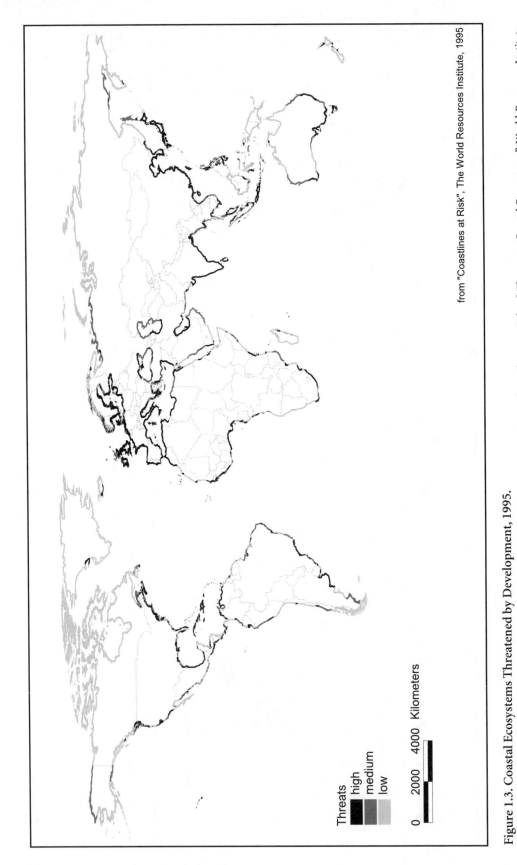

Threats
- high
- medium
- low

0   2000   4000   Kilometers

from "Coastlines at Risk", The World Resources Institute, 1995

Figure 1.3. Coastal Ecosystems Threatened by Development, 1995.
Source: D. Bryant, E. Rodenburg, T. Cox, and D. Nielsen, "Coastlines at Risk: An Index of Potential Development-Related Threats to Coastal Ecosystems," World Resources Institute, Washington, D.C., 1995.

are available for the smallest enumerated areas: local districts or counties. Such dis-aggregated data can be used to pinpoint areas of dense human populations. It should also motivate governments to design better urban planning systems, including zoning measures that regulate industrial, commercial, and residential development and minimize negative impacts on the environment.

In 1995, the World Resources Institute (WRI) in Washington, D.C., issued an indicator brief entitled *Coastlines at Risk: An Index of Potential Development-Related Threats to Coastal Ecosystems*. In addition to Deichmann's population density data, the WRI analysis includes four other basic indicators: cities, major ports, road density, and pipeline density (WRI 1995). The world map included in the WRI analysis and reprinted here (figure 1.3) indicates that half of the world's coastlines are already suffering from severe development impacts.

What the population data in figure 1.3 underscore is that in virtually every major region of the world, dramatic population shifts, mostly from in-migration, have fed the influx of people into coastal areas, especially municipalities. With coastal urbanization has come rapid industrial and commercial development. The rampant and often unplanned growth of coastal areas, in turn, has undermined the capacity of national governments to manage remaining resources on a sustainable basis.

The movement of people from the hinterlands to coastal areas is nothing new. It has been going on since the Middle Ages, when Europe's coastal cities became centers for international trade and commerce. As we approach the millennium, however, the mass movement of people from the interior to coastal urban areas has become one of the dominant demographic trends of the late twentieth century, clearly visible in developed and developing regions alike.

Since 1980, population growth rates have been dropping steadily throughout much of the developing world, with a few exceptions such as sub-Saharan Africa and the Middle East. Southeast Asia and Latin America had annual growth rates of 3 percent or above during the 1950s and 1960s. Today, both Southeast Asia's and Latin America's average population growth rates stand at 1.9 percent. In 1960, Thai women, on average, had nearly 6 children over the course of their reproductive lives. Today they average 2.2. Brazil's and Indonesia's total fertility rates—the average number of children a woman is likely to have over the course of her reproductive life—also dropped significantly: from an average of nearly 6 children per woman in 1960 to just under 3 in 1996 (Ehrlich 1990; Sadik 1993, 1995, 1996). Better maternal and child health care and access to reproductive health and family planning services have made the lower rates possible.

In most areas of the world, population growth and fertility levels continue to fall. But the world's population continues to increase because of the sheer momentum of human numbers. For example, although China's growth rate—now at 1.1 percent a year—is falling rapidly, the country's massive population base (1.2 billion) still translates into an extra 13 million people a year.

In the developing world, coastal areas harbor many of the most rapidly developing towns and cities. These cities are turning into economic hothouses, responsible for energizing economies and "growing" the bulk of new jobs. The 1970s and 1980s heralded the emergence of "primate cities," cities containing a preponderance of infrastructure, investment, services, and skilled workforce. Many of these primate cities are coastal, historic centers of trade and commerce that have experienced rapid economic development over the past four decades, especially as subsistence economies have been shoved aside by modern, interconnected market economies.

Rushing to the Coast: China

When Xiao Sun came to Shanghai in 1990, at the age of fifteen, she had only the clothes on her back. She came searching for a better life than the one she had left in a poor farming village in Jiangsu Province. An attractive girl, with limited education, she took a job as a nanny with a university professor's family.

Sun considers herself one of the lucky ones. She impressed her employer with her native intelligence and boundless capacity for work. After a few years, at nineteen, she married the professor's eldest son. She now lives a comfortable life in China's largest metropolis and vows to make sure her own child has advantages she did not. "I will never go back to my village to live," she affirms. "There is nothing for me or my family in rural China. There are few opportunities to make a better life. The future is here."

The overwhelming majority of Chinese—94 percent—live in the eastern third of that country. Of China's 1.2 billion people, over 677 million (56 percent) reside in thirteen southeast and coastal provinces and two coastal municipalities, Shanghai and Tianjin. Along much of China's 18,000 kilometers of continental coastline, population densities average over 600 people per square kilometer. In megacities like Shanghai they exceed 2,000 people per square kilometer (Tien et al. 1992).

Out of China's 467 cities with municipal status (as of 1990), 305 are coastal (*China's... Country Report* 1992). Many of those cities seem to be growing at more rapid rates than cities in the interior of the country. The country's official urban population jumped from 135 million in 1980 to 214 million by 1990, an increase of around 8 million a year. China's coastal cities are growing at an average rate of around 4.7 percent a year, enough to double their populations in fourteen years (Tien et al. 1992). Between 1982 and 1990, Shanghai's population increased by 13 percent. All but 5 percent of that growth was due to in-migration from the countryside. China's largest city now has around 3 million migrant workers, most of them living in the poorer areas of the city in makeshift housing.

According to Chinese demographer Tu Ping, most of the growth of coastal populations is due to in-migration from the interior of the country not natural increases. City dwellers in China tend to have fewer children (usually one) than their counterparts in the countryside (who average two to four, and even higher in some regions). But nearly 100 million Chinese are thought to have moved from the poorer provinces in the central and western regions to coastal areas in search of better economic opportunities for themselves and their families. At any given time, somewhere between 20 and 40 million Chinese are on the move, a population equal to that of Spain. The bulk of this large, "floating population" is concentrated in coastal provinces, precisely those areas with the highest economic growth rates.

In Guangzhou the city's floating population accounts for 45 percent of the total. "These migrants who move to cities without any planned arrangement have become an important force for urban development," claims Yukun Wang, an associate professor at the Development Research Centre of the State Council of China. "In Shenzhen, Guangzhou, and other cities and towns, authorities have adopted measures allowing peasants to settle in their cities after paying fees for urban infrastructure construction," he adds. "This practice is in fact an official recognition of population migration."

Since most of China's economic growth is concentrated in coastal provinces, migrants will continue to swell the population of coastal towns and cities. The country's demographic dilemma is that it can no longer keep up with such massive and rapid population shifts. Without comprehensive development and management strategies in place, China's coastal cities may choke on their own success. (See "China's Floating Population," p. 149, for more on this subject.)

Coastal Towns and Cities

High population growth rates in the countryside, poor living conditions, and limited economic opportunities fuel the migration of people from rural to urban areas. As hubs of economic activity, coastal towns and cities seem to be exploding everywhere. They continue to draw migrants out of the countryside like ants to sugar. This concentration of people and economic activity results in lopsided development. But few governments have been successful in promoting a more balanced distribution of population and resources.

In Latin America and the Caribbean, the transition from largely rural societies to largely urban ones took place over the course of three decades. In 1950, most Latin Americans lived in the countryside. By 1980, over half were living in towns and cities. Many urban areas grew by 4.5 percent a year during that period, doubling their populations every fifteen years or less (Gilbert 1900).

In some countries the pace of urban growth has been startling. By 1985, three out of every four residents of Caracas, Venezuela, over forty-five years of age, were born outside the city. Nearly 40 percent of Venezuela's population now lives on only 2 percent of its land area, the north-central coastal zone around Caracas. This area contains three-quarters of the country's industries and accounts for 61 percent of the gross national product and 40 percent of all fixed investments (Nieto 1993).

By the year 2010, the coastal zone from Rio de Janeiro to São Paulo, Brazil, is expected to be one large, contiguous urban area bulging at the seams with some 40 million people (Sadik 1993). A similar process is taking place along Chile's coast between Valparaiso and Concepción. This region already contains 75 percent of the country's population—10.5 million people—on only 15 percent of its land area.

Southeast Asia's coastal cities are also growing more rapidly than those in the interior. Jakarta, Manila, and Bangkok, for example, grew by around 4 percent a year during the 1980s. If present trends hold, by the year 2000 Jakarta will have over 13 million people, Manila over 11 million, and Bangkok 11 million—double their 1985 populations (Chapman 1992).

Bangkok, which already contains over 10 percent of Thailand's total population, has the greatest concentration of universities, hospitals, doctors, industries, banks, telephones, and cars in the country. It now generates 45 percent of the country's wealth, handles 95 percent of all imports and exports, and boasts an average per capita income over twice that of the rest of the country (Chapman and Baker 1992).

The Developed World

The forces at work in the developing world also account, in large measure, for the explosion of coastal towns and cities in the developed world. Historic patterns of economic development that fueled the first industrial revolution and transformed coastal cities into international centers of trade and commerce have been augmented since the end of the Second World War by a massive population shift from the hinterlands to coastal areas. Millions of middle-class families now have significantly more disposable income and more leisure time to enjoy the fruits of their labors; seacoasts, with their boundless economic opportunities and better quality of life, increasingly are viewed as preferred places to live, work, play, and retire.

In the United States, 55–60 percent of Americans (around 156 million) now live in 772 counties adjacent to the Atlantic and Pacific oceans, the Gulf of Mexico, and the Great Lakes (Population Reference Bureau 1993). The Washington, D.C.–based Population Reference Bureau reports that between 1960 and 1990 coastal population density in the United States increased from an average of 275 to nearly 400 people per square kilometer. In 1990, the most crowded coastline in the United States, stretching from Boston south through New York and Philadelphia, to Baltimore and the District of Columbia, had over 2,500 people per square kilometer. Another 101 coastal counties had population densities exceeding 1,250 per square kilometer (Culliton et al. 1990).

Florida, which is almost entirely coastal, is projected to have more than 16 million residents by 2010, an increase of over 200 percent from its 1960 level of 5 million. South Florida (the area south of Lake Okeechobee), which had a 1990 population of 6.3 million, is expected to have 15 to 30 million people by 2050. Similar dramatic increases are projected for California and Texas (Culliton et al. 1990).

The five states with the greatest rise in population are all coastal: California, Texas, Florida, Georgia, and Virginia (Population Reference Bureau 1993). By the year 2025, nearly 75 percent of Americans are expected to live in coastal counties. Coastal counties already contain fourteen of the country's twenty largest conurbations (see Table 1.1).

Japan transformed itself from a largely rural and noncoastal nation into an overwhelmingly urban and coastal one within two decades. In 1950, Japan's 83.2 million inhabitants were dispersed throughout the country, with nearly half living in farming households. By 1970 most Japanese were living in urban areas, the majority of them in the Pacific Coastal Belt, which extends from Tokyo southwest through the Seto Inland Sea to the northern part of the island of Kyushu. As early as 1970 the national census revealed that over 53 percent of the population lived in "densely inhabited districts" that occupy 1.7 percent of the country's land area, mostly in the Tokyo-Nagoya-Osaka region (Chapman and Baker 1992). Population densities in this crowded region average over 11,500 per square kilometer.

In 1997, Japan's total population amounted to 126 million. Of this, nearly 80 percent, or 100 million, are considered coastal. But no one in Japan lives more than

Table 1.1. Twenty Largest Conurbations in the United States, 1991

1. New York, Northern New Jersey, Long Island	19.6 million
2. Los Angeles, Riverside, Orange County	14.8 million
3. Chicago, Gary, Kenosha	8.4 million
4. Washington, D.C.; Baltimore	6.8 million
5. San Francisco, Oakland, San Jose	6.3 million
6. Philadelphia, Wilmington, Atlantic City	5.9 million
7. Boston, Worcester, Lawrence	5.4 million
8. Detroit, Ann Arbor, Flint	5.2 million
9. Dallas, Fort Worth	4.1 million
10. Houston, Galveston, Brazoria	3.9 million
11. Miami, Fort Lauderdale	3.3 million
12. Seattle, Tacoma, Bremerton	3.1 million
13. Atlanta	3.1 million
14. Cleveland, Akron	2.9 million
15. Minneapolis, St. Paul	2.6 million
16. San Diego	2.5 million
17. St. Louis	2.5 million
18. Pittsburgh	2.4 million
19. Phoenix, Mesa	2.2 million
20. Tampa, St. Petersburg, Clearwater	2.1 million

Source: Adapted from *U.S. Metro Data Sheet*, Population Reference Bureau, Washington, D.C., 1991.

120 kilometers from the sea. Furthermore, 77 percent of all Japanese now live in urban areas, predominately along or near the coast. The dramatic population shift has left much of the interior drained of workers. Nearly 47 percent of Japan's land area—mostly in the interior—is now designated as "depopulated" and eligible for special funding. By 1985, that area contained only 6.7 percent of the population, with an average density of 47 people per square kilometer.

The Mediterranean Blues

One of the most celebrated and threatened coastlines in the world is the Mediterranean. Here north and south meet, with all the tensions such a confluence cultivates. According to demographic projections worked out by the Mediterranean Blue Plan, the socioeconomic part of the Mediterranean Action Plan that links the protection of the environment with various levels of development, the Mediterranean Basin's resident population could go as high as 555 million by 2025. Also, according to Blue Plan projections, the urban population of coastal Mediterranean administrative regions could reach 176 million—30 million more people than the entire coastal population in 1990. Furthermore, depending on how tourism is developed in the future, the Med could be hosting up to 350 million seasonal tourists every year by 2025. At the same time, the number of the region's automobiles is expected to triple, causing serious air pollution problems in many urban areas (Grenon and Batisse 1989).

Michel Batisse, president and chief architect of the Mediterranean Blue Plan and former assistant director-general for science at UNESCO, is convinced that the future of the region is in jeopardy. "While northern populations with declining fertility rates will become progressively older, the southern and eastern regions will be dominated by young people," points out Batisse. "The numbers arriving on the labor market will largely exceed those leaving it, with a maximum gap around 2020, creating considerable unemployment and probably spawning waves of migrants heading to Europe in search of work."

Italy has already been flooded with more than a million illegal immigrants from North Africa, and more are arriving every year. One result is that most Italians now favor limiting the numbers allowed in from the poorer countries on the Mediterranean's south rim and from Asia. In both France and Spain, nationalist political parties are gaining momentum, due in large measure to their call for limiting immigration.

"In all the scenarios we developed for the southern- and eastern-rim countries, their development problems are aggravated by rapid, pell-mell urbanization," continues Batisse. "The greatest concentrations of people will continue to be in the narrow, mountain-lined coastal strips characteristic of the region."

Batisse argues that these trends are likely to generate enormous conflicts over dwindling resources in an increasingly polluted environment. "This is especially true for water availability, already critical in some areas," he insists, "as well as mounting land-use conflicts, traffic congestion, destruction of wetlands, soil erosion, and continued pollution of coastal waters."

Batisse is not sanguine about the future. "The protection of the marine environment was the starting point for pan-Mediterranean cooperative efforts, which began in 1975 with the launching of the Mediterranean Action Plan. Unless this expands to help the developing countries of the region meet the basic needs of their growing and destitute populations, it is not only the marine environment but the Mediterranean world that could face social and ecological disruption on a massive scale."

Coastal Gigantism

Coastal cities in many parts of the world especially in developing countries, are expected to continue growing until they simply run out of space, creating continuous concrete jungles. Already entire coastlines in Asia and Latin America are urbanized, or rapidly becoming so.

"Our coastal areas are suffering from gigantism," bemoans one Indian civil servant in Bombay. "They are becoming like Frankenstein monsters: They are too big and too out of control for us to manage."

As coastal areas continue to suffer the brunt of population and development pressures, the need to safeguard important coastal ecosystems takes on an added urgency. Rapid population growth combined with nonsustainable use of resources is reducing our timeframe for effective response.

CHAPTER 2

Coastal and Fisheries Resources in Danger

The tremendous population and development pressures that have been building in coastal areas for the past four decades have triggered widespread resource degradation. Throughout the world's coastal areas, valuable ecosystems such as estuaries, salt marshes, mangrove forests, seagrass beds, and coral reefs are being plundered in the name of development. This destruction, combined with massive overfishing, has precipitated a global fisheries crisis.

We may already have lost half the earth's coastal wetlands. Accurate estimates do not exist, since we don't know how extensive these ecosystems once were. What is known is that millions of hectares have disappeared since the end of the Second World War, mostly through a combination of urban expansion, drainage for agriculture, conversion to mariculture and aquaculture ponds, and land reclamation.

Some destruction goes back centuries. The Chinese have been draining coastal wetlands for the past 6,000 years, using the rich soils for rice production. In the United States, San Francisco Bay—the largest estuary in the West—has lost 60 percent of its water area to land reclamation projects over the past 140 years.

In many areas of the world, people do not comprehend the critical role these ecosystems play in providing them with food, fiber, building materials, and other necessities. In other cases, such as Japan, crowded coastal populations have nowhere to go but seaward as coastal economies explode.

The dual pressures of population growth and economic development have combined to undermine the health and viability of many coastal ecosystems. But three, in particular, have suffered disproportionate destruction: coastal wetlands, including salt marshes and mangrove forests; seagrass beds; and coral reefs. These three are among the most biologically productive ecosystems on the planet, besides providing a host of other benefits, including land protection and stabilization, water purification, climate modification, and food provision.

Coastal Wetlands

Too often, wetlands are viewed as wastelands to be converted into more "productive" uses. Coastal wetlands, however, including estuaries, salt marshes, and mangrove swamps, are highly fecund ecosystems. Hectare for hectare, wetlands produce more wildlife, both in numbers and in variety, and more primary plant growth than any other habitat on earth. They are also efficient natural filters. In the southeastern part of the United States, for instance, tidal marshes are used to filter wastewater from sewage treatment plants, a function that makes them worth around $123,000 per hectare (based on the cost of replacing them with artificial treatment facilities) (Dugan 1993). Coral reefs have been valued at $47,000 per square foot for their shore protection functions alone (Coral Reef Alliance 1996).

Coastal wetlands provide vital spawning, nursery, and feeding grounds for thousands of species of fish and shellfish; filter out pollutants washed off the land (e.g., heavy metals and nitrates); trap and stabilize sediments; and serve as buffers between land and sea.

As no-cost fish farms, these ecosystems are unrivaled. Globally, nearly two-thirds of all fish harvested depend on the health of wetlands, seagrasses, and coral reefs for various stages in their life cycles. Roughly 90 percent of all commercial species of fish and shellfish caught in the Gulf of Mexico and the Caribbean are dependent on estuaries, tidal flats, mangroves, seagrasses, and coral reefs for all or part of their lives (Lindén 1990). Similarly, around 80 percent of the Indian fish catch from the lower delta region of the Ganges and Brahmaputra rivers comes from the extensive mangrove swamps of the Sundarbans, which cover roughly 20,000 square kilometers. Half the current 500,000 metric tons of fish taken in the lower Mekong Delta are of wetland origin. In Malaysia, at least 65 percent of the fish and shellfish harvested every year are associated with mangrove swamps. And tropical Pacific islanders rely on coastal marine resources for around 90 percent of their total animal protein intake.

Fish from Trees: Mangrove Forests

Biologists consider mangrove forests to be one of the most productive and biologically diverse wetlands on earth, supplying habitats for over 2,000 species of fish, shellfish, invertebrates, and epiphytic plants. Their root zones provide sanctuary for sponges, crested worms, crustaceans, and mollusks, as well as green, red, and brown algae. Intertidal zones create habitats for a variety of crabs, mud skippers, and small animals. Mangrove canopies harbor hundreds of species of birds, while mangrove estuaries shelter marine mammals such as dugongs, manatees, and otters, as well as endangered reptiles like the South American caiman and the Indo-Pacific crocodile. Figure 2.1 lists some of the products from mangrove forests that benefit humankind.

All told, some eighty species of salt-tolerant mangrove trees and shrubs cover roughly 182,000 square kilometers (18.2 million hectares) of intertidal, lagoonal, and riverine flatlands throughout the world. Most species are found in a wide tropical belt, reaching their greatest concentrations along the coasts of South and Southeast Asia, South America, and Africa. The largest expanse of mangrove forests—around 20 percent of the world's total—borders the Sunda Shelf, a region in Southeast Asia encompassing Vietnam, Kampuchea, Thailand, Malaysia, and the Indonesian islands of Sumatra, Java, and Borneo.

Indonesia's mangrove forests remain the most extensive in the region, covering some 4.25 million hectares, with about 75 percent of the total amount concentrated

Fuel	Textiles, Leathers	• Matchsticks
• Firewood	• Synthetic fibers (e.g., rayon)	• Incense
• Charcoal	• Dye for cloth	**Agriculture**
• Alcohol	• Tannins for leather preservation	• Fodder and "green manure"
Construction Materials	**Food, Drugs, Beverages**	**Fauna and Flora**
• Timber, scaffolds	• Sugar	• Fish
• Railroad ties	• Alcohol	• Crustaceans
• Mining pit props	• Cooking oil	• Shellfish
• Boat building	• Vinegar	• Honey
• Dock pilings	• Tea substitutes	• Wax
• Beams and poles for buildings	• Fermented drinks	• Birds
• Flooring, paneling	• Condiments from bark	• Mammals
• Thatch or matting	• Sweetmeats from propagules	• Reptiles and reptile skins
• Fence posts, water pipes, chipboards, glues	• Vegetables from propagules, fruit, and leaves	• Other fauna (amphibians, insects)
Fishing	**Household**	**Other Products**
• Poles for fish traps	• Furniture	• Packing boxes
• Fishing floats	• Glue	• Wood for smoking sheet rubber
• Wood for smoking fish	• Hairdressing oil	• Wood for burning bricks
• Fish poison	• Tool handles	• Medicines from bark, leaves, and fruit
• Tannins for net and line preservation	• Rice mortar	• Paper of various kinds
• Fish shelters	• Toys	

Figure 2.1. Products from Mangrove Forests.

Source: Adapted from John Clark, *Coastal Zone Management Handbook,* New York: Lewis Publishers, 1996, pp. 348–349.

on Irian Jaya, the Indonesian half of the large island of New Guinea, and nearly 400,000 hectares along the south and west coasts of Kalimantan (the Indonesian part of Borneo). The two Malaysian states of Sarawak and Sabah, which share the island of Borneo with Indonesia, contain an estimated 173,000 hectares and 365,000 hectares, respectively (see Spalding, Blasco, and Field 1996).

Mangrove communities manufacture a nutrient-rich broth for sustaining a wealth of marine life. Concocted from the decomposition of mangrove leaves and twigs, this broth is the first link in a long food chain that extends through seagrass meadows to coral reefs and ends in open ocean fisheries (figure 2.2). The abundance of offshore shrimp, for example, is directly related to the amount of mangrove nursery available. On the Fiji Islands, about half of all fish and shellfish caught by commercial and artisanal fishermen are dependent on mangrove swamps for at least one stage in their life development.

The Matang mangrove area in Malaysia—where mangroves are exploited sustainably—has a gross aquatic productivity of 1–3.5 tons of carbon per hectare per year. An area of 10,000 hectares was found to produce 67,000 tons of carbon a year. The same area also generated 400,000 tons of organic matter per annum from leaves, twigs, and bark.

U.S. scientists have estimated that 1 hectare of mangrove forest in the Philippines, if properly managed, could produce an annual yield of 100 kg of fish, 25 kg of

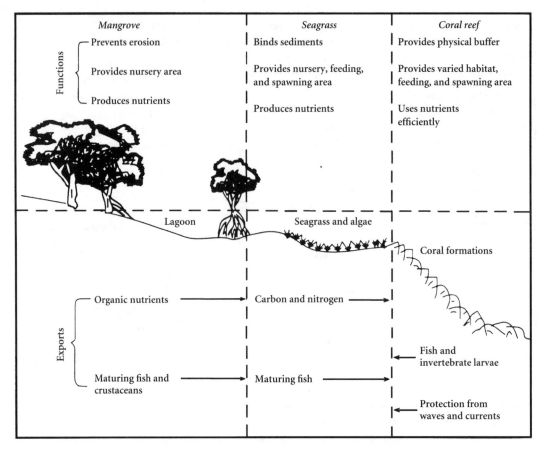

Mangrove	Seagrass	Coral reef
Prevents erosion	Binds sediments	Provides physical buffer
Provides nursery area	Provides nursery, feeding, and spawning area	Provides varied habitat, feeding, and spawning area
Produces nutrients	Produces nutrients	Uses nutrients efficiently

Functions

Lagoon — Seagrass and algae — Coral formations

Exports

Organic nutrients → Carbon and nitrogen →

Fish and invertebrate larvae ←

Maturing fish and crustaceans → Maturing fish →

Protection from waves and currents ←

Figure 2.2. Linkages Between Mangrove, Seagrass, and Coral Reef Ecosystems.

Source: Alan White et al., *Collaborative and Community-Based Management of Coral Reefs,* West Hartford, Conn: Kumarian Press, 1994, p. 8.

shrimp, 15 kg of crabmeat, 200 kg of mollusks, and 40 kg of sea cucumber. The same area could theoretically supply an indirect harvest of up to 400 kg of fish and 75 kg of shrimp that mature elsewhere (Porter 1988).

Marine biologist John Clark puts it this way: "Natural systems, like mangrove forests, should be managed as an investment, where the interest earned by the investment is analogous to the sustained productivity of the system." (Clark 1996).

Managed mangroves are a rarity, however. In many parts of the third world, mangroves are completely disregarded. Despite their value, they are in retreat throughout their range. Clear-cutting for timber, fuelwood, and woodchips; conversion to brackish water fish and shellfish ponds; and expansion of urban areas and agricultural lands have claimed perhaps half of their original area. Indirect threats include runoff from pesticides used on agricultural fields and orchards and eroded sediment brought in by the deforestation of upland watersheds and large-scale mining operations.

The total amount of mangrove area lost or grossly degraded over the past one hundred years is difficult to estimate but is thought to be roughly 25 million hectares. If that is true, mangrove forests originally covered around 43 million hectares in the tropics and subtropics. Of the Philippine's original mangrove area, estimated to have been between 500,000 and 1,000,000 hectares, only a little more than 100,000 hectares remain, according to recent remote-sensing data. And most of the surviving stands are found in small, isolated patches. Only two islands, Palawan and Mindanao, still have some pristine (undisturbed) areas of mangrove forest.

Indonesia has not fared much better. Over the past two and a half decades, over 70,000 square kilometers of mangroves have been converted into rice paddies and brackish water ponds (called *tambaks*) for the cultivation of shrimp and fish. Most of Java's and Bali's mangroves have been destroyed, replaced with tambaks, rice fields, or tourist facilities. Over 2,000 square kilometers have been converted into woodchips and exported to Japan. Another 50,000 square kilometers of virgin mangrove forest—mostly along the coast of Kalimantan on Borneo and on the west coast of Irian Jaya—are under license to Japanese and Indonesian logging companies bent on turning them into woodchips and plywood. Once cleared, the land is slated for agricultural development.

Thailand, now in the fast lane of economic development, has devastated its mangrove resources over the past three decades. About half of the country's mangrove forests have been destroyed since 1960; only around 196,000 hectares remain; though some estimates go as low as 160,000 hectares (McNeely and Dobias 1991). By the end of the 1980s, conversion to shrimp and prawn farms had replaced illegal cutting for charcoal production as the leading cause of mangrove loss. In 1989, Thai authorities estimated that shrimp and prawn farms had replaced one-quarter of the country's mangroves. In two years, 1986–88, Chanthaburi Province, on the Gulf of Thailand, lost nearly 90 percent of its mangrove forests to shrimp ponds.

In the entire East African region, only around 1 million hectares of mangrove forests remain, less than half the original area. The last major stand of mangroves in West Africa—the 500,000 hectares in the delta of the Niger River in Nigeria—is threatened. Seaward, these mangroves are victims of chronic oil pollution and coastal erosion. On land they are exploited for timber, tannin, fuelwood, fodder, and fish ponds.

In 1920, the island states of the Caribbean were 50 percent forested, with extensive mangrove swamps. By 1990, coastal forest cover had been slashed to just 15 percent. Mangrove forests and tidal marshes have been reduced by 40–50 percent in Tampa Bay, Florida, and by 75 percent in Puerto Rico since pre-Columbian times. In Tabasco State, on the Caribbean coast of Mexico, around 65 percent of all mangrove stands have been degraded due to near-shore oil production (Lindén 1990; Lundin and Lindén 1993). Ecuador has lost around 86,000 hectares of mangroves and salt marshes since 1969—most of them replaced with commercial shrimp ponds catering to the export market.

The replacement of mangrove forests with shrimp ponds is particularly senseless. In Malaysia alone, over 30 percent of all fish and shellfish landed by commercial operators every year (around 200,000 metric tons) is mangrove dependent. The juveniles of most commercially important shrimp species depend on healthy mangroves as nurseries and feeding grounds. With the mangroves gone, shrimp cannot survive. As the restocking of shrimp ponds with postlarvae from near-shore waters becomes more and more difficult, pond owners cut down more mangroves to raise more shrimp, initiating a downward cycle of destruction. The result is the complete collapse of wild shrimp fisheries and their substitution with mariculture for export. Local fishing communities become impoverished and are forced off the sea.

As of 1990, only around 1 percent of the world's mangroves received any form of protection. The figure may be higher today, since countries such as Thailand, Bangladesh, and Ecuador have placed their remaining mangrove forests under the protection of law. In many cases, however, enforcement is lax, and many areas continue to be exploited at nonsustainable rates.

Seagrasses: An Undervalued Resource

Seagrasses are the only land plants that have returned to the sea. Their underwater meadows are found in a wide swath around the world, in both temperate and tropical seas. All fifty species of seagrass grow close to shore in shallow water. Like mangroves, they trap and consolidate sediment, slowing down erosion and improving water quality not only for themselves but also for associated communities of filter feeders (clams, oysters, mussels) and nearby coral reefs. They also provide nurseries, shelter, and food for a host of marine life, including commercially important fish and shellfish, and interact with mangroves and coral reefs to reduce wave energy and regulate water flow. However, being close to shore makes them particularly vulnerable to land-based pollution and to overloads of sediment that can bury them.

Submarine meadows provide a host of organisms with shelter, nurseries, and food. Many fish and invertebrates, as well as sea turtles and dugongs, graze the meadows, eating algae and other plant matter that flourish on the surface of seagrasses or in their litter. In addition, because of their strategic position between mangroves and coral reefs, tropical seagrass communities act as effective buffers, modifying wave action and ferrying nutrients to and from these other ecosystems.

The global extent of seagrasses has never been adequately estimated, but they are found in virtually all coastal waters except polar seas and highly saline inland seas. Their current status is unknown except in those few areas where biodiversity inventories and similar detailed studies have been carried out.

It is known that seagrasses, like mangroves and coral reefs, are extemely fertile ecosystems, with productivity levels comparable to that of agricultural croplands. The crude protein levels found in tropical eelgrass and dugong grass, for example, reach as high as 23 percent of dry weight, higher values than those for terrestrial forage grasses.

What's more, seagrasses can grow as fast as cultivated corn, rice, and tallgrass prairies without the benefit of fertilizers. Notes Miguel Fortes, a seagrass specialist at the Marine Science Institute in Manila, "Per area production can be higher than phytoplankton production off the coast of Peru, one of the most productive areas in the world's oceans" (Fortes 1990).

Extensive studies carried out in Bolinao Bay, in the Lingayen Gulf on the west coast of the big Philippine island of Luzon, revealed that seagrass beds are extremely prolific, producing up to 18,900 kg of carbon per day. "This suggests that a square meter area of the bed produces 8,635 calories daily, or roughly 20 percent of the daily caloric requirements per kilogram of an ordinary individual," observes Fortes. "Thus, the daily caloric need of an adult weighing 70 kg is equivalent to that which is naturally processed daily by seagrass tissues within a 350-square-meter area of the bed" (Fortes 1990).

What this means, concludes Fortes, is that if productivity data for all of Southeast Asia were considered, "seagrass beds as nutrient providers might very well be the most important ecosystem in the marine environment for the entire region."

Seagrass beds in the Philippines harbor at least 123 species of fish from 51 families. In Bais Bay, on Negros in the southern part of the Philippine archipelago, researchers found 49 species of fish belonging to 21 families, along with 19 species of crustaceans from 9 families and 27 edible species of bivalves (clams, mussels, and oysters). Many of them, including cardinal fish, rabbitfish, snappers, sardines, parrotfish, Caridean shrimp, and prawns (*Penaeus*), are commercially important. In general, there are five times more fish over seagrass beds than over mud, shells, and sand (Dolar 1991).

Monetary values have been assigned to seagrass beds based primarily on the fisheries they support. At Cairns, North Queensland, Australia, fisheries supported by seagrass meadows are worth about $540,000 a year. Studies done in Puget Sound, Washington, discovered that a little over a third of a hectare of eelgrass has a value of over $412,000 annually, based on the amount of energy derived from the system as well as the nutrition generated for oyster culture, commercial and sport fisheries, and waterfowl (Dugan 1993).

Like mangroves, seagrasses are under increasing threat from coastal development, deforestation, and pollution. Mining for sand, coral, and minerals, coupled with dredge-and-fill operations in ship channels and harbors, has smothered millions of hectares of seagrass beds throughout the world. There are no accurate estimates, or even crude "guesstimates," of the number of seagrass beds lost through human activities.

What is known is that even the loss of a relatively small area of seagrass can have serious and costly consequences. An extensive study carried out by Dr. Anitra Thorhaug, a professor of biological sciences at Florida International University in Miami, revealed that one dredge-and-fill operation in Boca Ciega Bay, Florida, to enlarge a boat harbor, destroyed only a fifth of the seagrass bed but reduced the number of fish by four-fifths and cost nearly $1.4 million in lost catches for local fishermen (Thorhaug 1981).

Coral Reefs: Underwater Rainforests

There are roughly 600,000 square kilometers of coral reefs throughout the world's tropical and subtropical seas. Reef colonies range as far north as the Ryukyu archipelago off southern Japan and as far south as the southwestern coast of Australia.

Like mangroves and seagrasses, coral reefs yield multiple benefits. They protect coastlines from storm damage and beach erosion; provide homes, breeding areas, nurseries, and food for tens of thousands of species of fish, shellfish, and invertebrates; and form an important link in cycling nutrients from the land to the open ocean.

Coral reefs are among the oldest living communities of plants and animals on earth, having evolved between 200 and 450 million years ago. Today, most coral reefs are between 5,000 and 10,000 years old, many of them forming thin veneers over older, much thicker reef structures. Most of the reef colony is actually dead. Only the upper layer is covered by a thin, changeable "skin" of living coral. Coral polyps (the tiny animals that build the reefs) are the master bricklayers of the sea, cementing their homes on the remains of their predecessors (Hinrichsen 1996).

As marine biologists have discovered, coral reefs rival tropical rainforests in species diversity. The barrier reef that surrounds the main island group in the South Pacific Republic of Belau (also known as Palau), for example, has 9 species of seagrass, more than 300 species of coral, and 2,000 varieties of fish and shellfish. The Great Barrier Reef of Australia shelters 400 species of coral, providing habitat for over 1,500 species of fish, 4,000 different kinds of mollusks, and 400 species of sponge. Some marine biologists think there are upwards of 1,000,000 species of plants and animals on tropical coral reefs. If that estimate is correct, marine science has identified only a fraction of the species found on reefs.

It is known that nearly a third of all fish species live on coral reefs, while others are dependent on reefs and nearby seagrass beds and mangrove swamps for critical stages in their life cycles. Somewhere between 70 and 90 percent of all fish caught by coastal fishermen in tropical Asia are reef dependent at one time or another.

The high productivity of coral reefs is due to a number of factors, namely their efficient biological recycling and retention of nutrients and their capacity to provide shelter and feeding areas for tens of thousands of species of fish, shellfish, and invertebrates.

Rodney Salm, coordinator of the Marine and Coastal Conservation Programme for the World Conservation Union (IUCN) in East Africa, explains how their remarkable recycling system works:

> The coral animals have tiny algal cells, called *zooxanthellae*, in their tissues. These process the coral polyp's wastes before they are excreted, thereby retaining valuable nutrients. Nutrients such as nitrates, phosphates and carbon dioxide produced in the polyp are used by the *zooxanthellae* during photosynthesis to generate oxygen and organic compounds which in turn may be used by the coral polyp. In this way, the *zooxanthellae* recycle waste products to form nutrients within the tissues of the coral animal, saving the polyp the energy it would have expended on these activities. In addition, movement of water over reefs by waves and currents constantly washes the corals, and eliminates the need for the polyps to clean themselves. The energy freed up in these ways can go to damage control and new growth (Salm 1983).

Efficient as they are, reef-building animals still depend on sunlight. Most coral growth stops at a depth of between 20 and 40 meters. And they cannot reproduce in polluted or murky water—the main reason they are so vulnerable to human impacts.

The secret to the reef's richness of species lies in its complex multistory architecture, like an apartment building under the sea. Resident reef fishes that forage during the day share their living space with other, nocturnal species. While daytime fishes are feeding, their living space is occupied by another organism, say an octopus, that is only active at night. This sharing of quarters allows a reef to shelter two populations of marine organisms.

Reefs can support five to fifteen times the number of species of fish found in the north Atlantic. Some highly productive reef fisheries in the southern Philippines yield sustainable harvests of 30 metric tons per square kilometer per year and produce up to 70 percent of the total fish catch in a given area. Small-scale fishermen in such areas can harvest up to 5,000 kg of fish a year (White, Hale, Renard, and Cortesi 1994).

"But putting yields aside," says Salm, "the fundamental thing to remember is that coral reefs are self-perpetuating fish farms which produce high-quality protein from essentially empty sea water."

The high species diversity on coral reefs gives rise to another, often overlooked, benefit: their potential as a source of new drugs. In order to cope with competition, many reef organisms have developed substances harmful to other organisms. Researchers have discovered that a number of these highly active compounds have useful medical applications. For example, certain reef-dwelling seafans and anemones possess compounds with antimicrobial, antileukemic, anticoagulant, and cardioactive properties. Such species may prove invaluable in developing more powerful anticancer drugs or other pharmaceuticals. The Australian Institute of Marine Science has even isolated a compound that protects the coral from sunburn, giving it great potential for application in more effective sunscreen products.

Despite all the benefits reefs bring to coastal areas, they are suffering widespread decline. Clive Wilkinson, a coral reef specialist working at the Australian Institute of Marine Science in Townsville, has estimated that 10 percent of the world's reefs have already been degraded "beyond recognition." Thirty percent are in critical condition and will be lost completely in ten to twenty years, while another 30 percent are threatened and may be gone within twenty to forty years unless effective management programs are implemented. Only 30 percent of the world's reefs are in stable condition—those removed from inhabited areas or otherwise too remote to be exploited (Wilkinson 1992; Jameson, McManus, and Spalding 1995; see figure 2.3).

According to a survey carried out under the auspices of the United Nations Environment Programme (UNEP) and the World Conservation Union (IUCN) in the late 1980s, of the 109 countries with significant coral communities, 93 were damaging them. In over 50 countries coral is being killed by eroded sediment washed off the land, largely a result of massive deforestation and poor agricultural practices. In nearly 70 countries reefs have been badly affected by dredging and land reclamation, and the building of harbors, airports, and tourist resorts. Blast fishing (with dynamite) and coral mining are problems in 41 countries (Wells 1988).

In the Philippines, Indonesia, Malaysia, Thailand, and East Africa, subsistence fishermen have resorted to using dynamite on reefs in a desperate attempt to put food on the table. Even a small bottle bomb can pulverize all coral within a radius of 1.5 meters; a gallon jug kills everything within 5 meters, and the shock waves from the blast kills or incapacitates many more organisms farther away. Few dynamited reefs ever fully recover.

Other fishermen use poisons, like sodium cynanide, in small amounts to stun and capture fish for the lucrative aquarium trade or for restaurants specializing in live fish. The poisons often kill many other kinds of fish and impoverish the reef further by killing off the coral polyps.

Coral mining and blast fishing are particularly destructive, since most coral species grow very slowly. Like the trees of rainforests, coral is not a readily renewable resource. It takes about twenty years for a brain coral to grow to the size of a

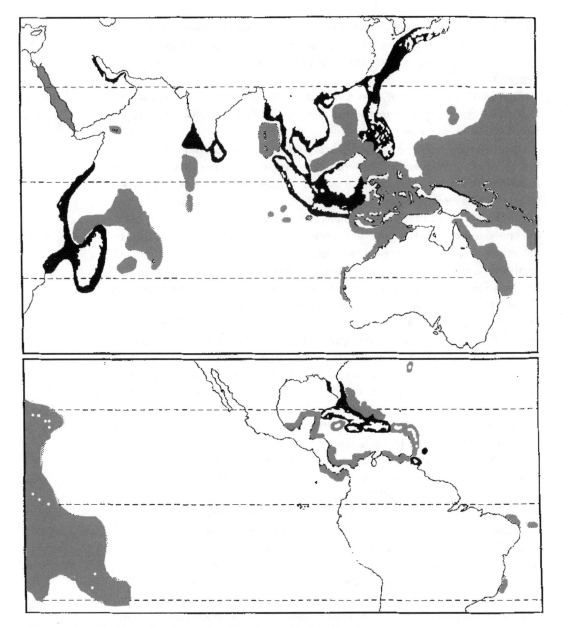

Figure 2.3. Global Distribution of Coral Reefs and Their Status, 1992.
Source: Clive Wilkinson, "Coral Reefs of the World Are Facing Widespread Devastation: Can We Prevent This Through Sustainable Management Practices?" in *Proceedings of the Seventh International Coral Reef Symposium*, Vol. 1, Guam, 1992.
Note: critical, those reefs under severe threat and likely to collapse within 10 to 20 years; *threatened,* reefs under increasing anthropogenic stress that will collapse within 20 to 40 years unless management and conservation are implemented; and *stable,* reefs remote from population stress or under effective management, which should suffer minor impacts in the next 100 years. Indo-Pacific center of Southeast Asia; Atlantic and Caribbean region.

■ critical   ▨ threatened   □ stable

Many reefs in the mid- and eastern Pacific are not shown, as they are classified as *stable.* Although some individual reefs are in the *critical* and *threatened* categories, the scale of these maps prohibits illustrating them.

man's head. It may take several hundred years before it is large enough to provide suitable habitat for fish.

A more mysterious problem is the bleaching of corals throughout the Pacific and in parts of the Caribbean. Bleaching occurs when polyps expel the colorful algae that live in their tissues, causing the corals to turn white. Sometimes bleached corals recover, and sometimes they do not. Scientists have reported bleaching at sixty sites around the world. In 1983, bleaching wiped out a significant number of corals around the Galapagos Islands, off Ecuador.

The cause of bleaching is uncertain. Various studies have linked it to unusually warm surface water, to infestations of newly discovered bacteria, to pollution, and to an interaction of all those factors. Coral experts are particularly concerned that if higher water temperatures are a factor, global warming could kill off a significant portion of the world's reef-building corals.

"We need to advocate the role of coral reefs and mangroves as natural fish farms, wave breakers, shore defenders, leisure areas, and storehouses of many other goods and services, whether we see them or not," says Rodney Salm. "Not just those who dive over reefs or daily glean their livelihood there, but all of us need to recognize the hive of activity hidden beneath the sea's reflective surface" (Salm 1994).

Fishing the Limits

During the last three decades, expansion of the world's fishing fleet, coupled with quantum leaps in fishing technologies and a dim understanding of fish population dynamics, habitats, and interactions, has led to overexploitation of nearly every commercial fishery in the world's seas.

By 1995, according to the United Nations Food and Agriculture Organization (FAO), nearly 70 percent of the world's marine fish stocks were either "fully-to-heavily exploited, overexploited, depleted, or slowly recovering" (FAO 1995). Of the earth's fifteen major fishing regions or areas (as defined by FAO), productivity in all but two has fallen over the last few years. In fact, in the four hardest hit areas—the northwest, the west-central, and the southeast Atlantic, and the east-central Pacific—catches have dropped by more than 30 percent since the peak year of 1989.

According to former Worldwatch Institute researcher Peter Weber, eighteen commercial fisheries, including those of herrings, cods, haddocks, redfishes, mackerels, pilchards, and hakes, have seen their productivity drop by more than 100,000 tons since 1989. "Together, these drops represent a fall of nearly 30 million tons—more than one-third of the 1992 marine catch," observes Weber (Weber 1994).

Over the past few years, the collapse of major fisheries has attracted headlines and political attention. Years of chronic overfishing and mismanagement virtually wiped out one of the world's most productive fisheries, the 4 million tons of cod resources on Canada's Grand Banks, off the coast of Newfoundland on the Atlantic coast. In 1991, commercial fishermen landed some 1.15 million tons of cod; two years later, in 1993, the Canadian government was forced to close the fishery, pushing some 50,000 fishermen and their families off the sea and onto the dole. The central government in Ottawa authorized a $400 million emergency rescue package, including unemployment compensation and retraining (Weber 1993, 1994).

The catch of cod has fallen dramatically over the past two decades throughout the North Atlantic, not just in Canada. U.S. catches have also suffered—plummeting from a grand total of 3.1 million tons in 1970 to 916,000 tons in 1991. The drop occurred despite enormous increases in fishing technology and effort over the same period. Atlantic mackerel catches also declined, from 418,000 tons in 1973 to 56,000 tons in 1989.

Overall, New England fisheries are in disarray. Between 1982 and 1991, fishing efforts in New England waters rose 13 percent while the total catch slumped by 43 percent. The total groundfish population—cod, haddock, redfish, hake, pollock, and flounder—fell by 65 percent in the decade between 1977 and 1987. Moreover, chronic overfishing on Georges Bank altered fish populations; skates and dogfish have replaced cod and haddock as the most dominant species. According to one study, even if fishing were to stop completely on Georges Bank, stocks would still take five to twenty years to recover (Weber 1994).

The giant bluefin tuna is now so rare that the big, tasty fish, which can weigh up to 680 kg (1,500 pounds), sells for an average of $20,000 *each* in wholesale Tokyo fish markets. According to the International Commission for the Conservation of Altantic Tuna, bluefin were thought to number only around 33,000 adults in the entire western Atlantic in 1993, a mere tenth of their 1970 numbers, which were assessed at close to 300,000. Faced with such intense fishing pressures and huge profits, the bluefin tuna fishery is unlikely ever to recover (Craft 1994).

But gross overfishing of commercially important stocks is only part of the picture. The widespread use of destructive fishing gear, such as bottom trawls, has ruined habitats and killed many more types of fish and shellfish than target species. In the Dutch part of the North Sea, researchers discovered that nearly every square meter of the seafloor had been churned up by bottom-trawling operations at least once a year. In some areas of the North and Barents seas and the northwest Atlantic, trawlers have destroyed all bottom habitats and associated populations of urchins, starfish, worms, crustaceans, and other life, creating what one scientist described as "furrowed deserts under the sea."

Bottom trawlers, which drag huge socklike nets through the water, not only turn productive seafloor into sterile desert, but also wreak havoc on many other species of fish and shellfish. Shrimp trawlers are particularly destructive of other marine life because of the small mesh size of the net necessary to capture tiny shrimp. In the process they routinely haul in tons of bycatch, or nontarget species. In tropical waters, the by-catch of "trash fish" can comprise up to 90 percent of the contents of the net. "Worldwide," notes Peter Weber, "shrimp fishers are estimated to jettison up to 15 million tons of unwanted fish each year, and other fishers are thought to discard at least another 5 million tons" (Weber 1994).

Throughout the 1970s and 1980s, drift nets, some up to 70 kilometers long (40 miles), were strung out across wide swaths of the Pacific, Indian, and Atlantic oceans. Drift nets, known as "walls of death," are indiscriminate killers. In one year (1990), drift netters are thought to have been responsible for the death of up to 42 million marine mammals, seabirds, turtles, and other nontarget species in their pursuit of tuna and squid. That same year, the U.S. Marine Mammal Commission estimated the total length of these lethal nets in the North Pacific alone at around 40,000 kilometers, enough to girdle the earth. Despite a general United Nations ban on the use of drift nets over 3.2 kilometers (2 miles) in length at the end of 1992, some fishing nations such as Japan, South Korea, Taiwan, Spain, and Italy, are still suspected of using longer nets.

As commercial operations continue to vacuum the sea of valuable species, fishing communities throughout the world suffer the consequences of unplanned coastal development and depleted resources. Many third world fishing communities, once able to support themselves, are sinking deeper into poverty as coastal waters become fished out and increasingly polluted. Burgeoning coastal populations throughout the third world also have placed intense pressure on fisheries. In many coastal developing countries, the sea has become the last resort for many destitute people pushed off the land. In the Philippines, slash-and-burn cultivators, no longer able to scratch a living from eroded uplands, have turned to the sea in an effort to feed their families. With little knowledge of fishing, they are often the first to use dynamite and poisons to put food on the table, initiating a destructive cycle that ruins the resource base for everyone.

The other problem facing poor fishing communities is that the momentum of marine fisheries is going in the wrong direction. Over the past decade, prices have risen, while the largest increases in supply have come either from trash fish used primarily as animal feed, or from high-priced species such as tuna, squid, and prawns. "Neither of these extremes benefit low-income consumers," points out Peter Weber. In Southeast Asia fish constitute the sole source of animal protein for the majority of the region's population, especially the poor.

As marine catches have leveled off or fallen, mariculture and aquaculture operations have surged ahead, producing 16 million metric tons in 1993. Fish farming for the export market is now big business in many parts of the developing world. By 1989 shrimp farming was producing 500,000 tons a year, about a quarter of the total world shrimp supply (FAO 1995). Often, large-scale mariculture operations are built at the expense of mangrove swamps, contributing to the depletion of wild stocks and further impoverishing traditional fishing communities.

Other than generating a few low-paid jobs, large-scale fish and shellfish farming usually does not benefit poor coastal communities. Instead, it is well-off consumers in developed countries who reap the benefits. The rich world—North America, Europe, and Japan—account for 84 percent of world fishery imports, by value. This protein drain from south to north is a particularly worrisome trend, especially for those poor developing countries with degraded coastal environments where seafood is the major source of animal protein.

"The undeniable fact is that national fishing fleets have grown too big for existing stocks," points out Peter Weber. "The UN Food and Agriculture Organization conservatively estimates that globally, annual expenditures on fishing amount to $124 billion, in order to catch just $70 billion worth of fish. Governments apparently make up most of the $54 billion difference with low-interest loans, paying access fees for foreign fishing grounds, and direct subsidies for boats and operations. These government subsidies keep more people fishing than the oceans can support" (Weber 1994).

Managed fisheries would not only save jobs, such plans would also save taxpayers billions of dollars a year. "Governments could potentially save $54 billion per year by eliminating those subsidies," reasons Weber, "and earn another $25 billion per year in rents, with a net budgetary benefit of more than the current gross value of the entire marine catch."

The UN Food and Agriculture Organization estimates that, if stocks were allowed to recover, fishermen could increase their annual catch by as much as 20 million tons, worth around $16 billion (at 1994 prices). Although FAO's estimate does not take into consideration the broader adjustments that societies will have to make in an effort to redirect former fishermen into other occupations or turn them into fish farmers, it does convey a sense of the economic mismanagement behind the ecological mismanagement of the world's seas.

Managing Coastal Areas Sustainably

At the beginning of 1997, there were 177 sovereign coastal states scattered around the globe, 80 percent of the world's 218 sovereign nations. As of 1996, 55 countries (up from 13 in 1975) had initiated integrated coastal management plans, but only 16 had implemented those plans by 1996 (20 if protectorates, dependencies, and territories are included) (Sorensen 1993).

There are no easy solutions to the human and resource crises afflicting the world's coastlines. Land and sea must be managed in a way that permits economic development yet sustains the resource base. This involves the balancing of a multitude of human uses, as well as managing resources in such a way that the needs of tomorrow are not sacrificed on the altar of expediency today. Well integrated coastal management strategies take account of population growth and distribution, urbanization trends, consumption patterns, industrial and agricultural development, generation of wastes, and the availability and use (or abuse) of critical resources, among other factors.

The development of coastal areas without a management plan can have disastrous results. When Dumaguete City on the island of Negros in the Philippines decided to lengthen the runway at the local airport in an effort to accommodate bigger jets with more tourists, local planners failed to take account of environmental side effects that might result. A few months after the runway was completed, the resort hotels along the bay—the main beneficiaries of the scheme—noticed that their beaches were disappearing. Research carried out by Silliman University solved the mystery: extending the runway into the bay had altered tidal currents, causing them to sweep along the shore and erode the beaches. A concrete wall had to be erected so that the hotels themselves would not be devoured by the advancing sea.

Such examples of counter development are not limited to the third world. Miami Beach, Florida, is one of the most heavily developed and commercially valuable coastlines in the world. Yet over a forty-year period, its highly prized white sand beaches all but eroded away. The cause was found to be poor management practices, particularly the unplanned dredging of inlets for recreational boating. Faced with the annihilation of its beaches, the city had no choice but to approve a massive beach nourishment project at a total cost of over $65 million (UNESCO 1993). A proper environmental impact assessment, costing as little as $100,000, might have forecast the consequences of dredging in the city's near-shore waters.

However, unless management efforts have clear and achievable objectives and are worked out in cooperation with local communities, they often end up missing the mark or even worsening the problems they were meant to address. In the late 1960s, for example, the government of Tanzania decided to establish some marine reserves in the clear, coralline waters south of Dar es Salaam, but it neglected to build support for the reserves among the local fishing communities. Several years later when researchers visited the proposed sites to take an inventory, they found that six of the coral reefs slated for protection no longer existed. Villagers had blasted them apart for building material.

When the Organization of American States (OAS) carried out a review of protected marine areas in the Caribbean in 1986, it discovered that there were 112, accounting for just under a quarter of the protected marine areas in the world. Of those 112 parks and reserves, only 28 had a budget, a staff, a management plan, and institutional support. The rest turned out to be nothing more than "paper parks." In most cases, the boundaries of the parks were not known or respected by local populations, so they continued to be exploited (OAS 1986).

Many management schemes are not integrated with land management or coastal resource use. The management of fisheries, for instance, usually is concerned with enforcing catch quotas, limiting the number of fishing vessels, regulating the type of fishing gear, and imposing bans on harvesting certain stocks during all or part of the year. Even though over 90 percent of the world marine fisheries harvest—which averaged 83 million tons between 1989 and 1991—is taken in near-shore waters, little effort is put into managing critical ecosystems such as estuaries, salt marshes, mangroves, seagrasses, and coral reefs on which many commercially important fisheries depend.

To make matters more confusing, many countries have no institutional mechanisms that would allow government departments or ministries to cooperate on implementing integrated management plans. Mangroves may be cleared for agricultural use under the jurisdiction of an agriculture ministry completely ignorant of the decline in fisheries that the loss of mangroves entails. Overlapping jurisdictions and political infighting blur lines of authority and delay actions that often need the cooperation of government agencies. Unless there is a broad-based constituency supporting national coastal management efforts or high-level political support from the executive branch of government, the best coastal management plans can remain mired in turf battles as ministries squabble over jurisdiction and money.

Even in developed countries with strong institutional support, the implementation of coastal management plans can involve dozens of provincial, state, county, and municipal authorities. Plans often end up colliding with hundreds of existing laws and regulations that can impinge on, even negate, their intended benefits. In the United Kingdom, for example, some forty-eight subnational units of government—from the parliamentary level down to town councils—have the authority to create an autonomous or semi-autonomous integrated coastal zone management strategy. In Brazil, coastal zone planners have to wade through twenty levels of government.

It may be that the only way to introduce sensible coastal governance in many financially strapped countries is community by community, later combining community plans into regional programs. A number of countries have scaled back their expectations, settling for more workable "special area management programs," which regulate only a portion of their coastal zone or focus on pilot project sites.

"It's not the quantity of the coverage that matters so much as the quality of the management plan and its implementation with local community support," maintains Stephen Olsen, director of the Coastal Resources Center (CRC) at the University of Rhode Island. "A comprehensive national coastal zone management plan is useless if it cannot be enacted and enforced" (Olsen 1993).

Ecuador provides an example. Alarmed by the rampant and uncontrolled proliferation of shrimp farms constructed at the expense of coastal mangrove wetlands, the government first enacted a series of tough laws and regulations in the early 1980s protecting mangroves from any form of exploitation. In the absence of enforcement, however, the pace of mangrove destruction actually accelerated. "The problem was not the absence of a database or adopted policies, laws, and regulations," notes Olsen, "but the absence of effective implementation."

The government of Ecuador, in cooperation with the CRC and the U.S. Agency for International Development (USAID), recognized the problems and developed a community-based coastal management process that could be built up province by province. At an early stage, CRC staff and their Ecuadoran counterparts prepared a detailed profile of each of the country's four mainland coastal provinces that traced the origins of its problems. Once the profiles were disseminated throughout the country and discussed at a series of public workshops, the program was able to create broad-based constituencies of support involving resource managers, local fishing communities, journalists, and political leaders, among other key groups.

Although Ecuador's current coastal management plan covers only 8 percent of its coastline, each of the four coastal provinces has at least one special-area management plan in place and working (see chapter 9, "Ecuador's Experience," p. 130).

Australia's Great Barrier Reef Marine Park

Australia's Great Barrier Reef Marine Park is by far the largest in the world, covering nearly 350,000 square kilometers, an area bigger than the United Kingdom. It is also the longest reef system in the world, stretching for more than 2,000 km along the Queensland coast. This complex system contains 2,900 separate reef formations, including 760 fringing reefs ranging in size from 1 hectare to more than 30,000 hectares. In addition, there are some 300 reef islands or cays, 87 of them permanently covered with vegetation, along with about 600 continental or high islands close to shore, many with fringing reefs around their margins. Technically, the region is not a park but a multiple-use management system (Kelleher 1996; Craik 1992).

The Great Barrier Reef Marine Region, in which the national park would be established, was annexed in 1975 by an act of Parliament that had the support of all political parties. Public outrage at the prospect of offshore oil drilling and mining on the reef was the driving force behind its passage. The Great Barrier Reef Marine Park Authority was established to administer the park, with broad powers to regulate and prohibit activities within its borders. However, the day-to-day management of the park is carried out by the Queensland Department of Environment and Heritage. The commonwealth government and the state of Queensland worked out an agreement whereby areas adjacent to the park are planned and managed in a way that does not compromise the integrity of the park.

Setting up a marine park was only the first step in a long process. Managing the reef system in a sustainable way required comprehensive planning. Before a management strategy could be worked out, a complete resource inventory was carried out and highly detailed maps of the entire reef system were drawn up. At the same time the Marine Park Authority sent out thousands of questionnaires to individuals and organizations that used the reef on a regular basis, in an effort to learn how the reef was used, where, and by whom. The authority followed up this process with public meetings and information campaigns designed to enlist the support of public and special interests—e.g., recreational and commercial fishermen and divers—for the park's proposed zoning plans.

Once the background information was processed, the authority began the painstaking task of dividing the reef into management zones. In order to facilitate the management of such a vast area of sea and coast, the barrier reef itself was divided up into four huge management sections, running from north to south—the Far Northern Section, the Cairns Section, the Central Section, and the Mackay-Capricorn Section. Within each section, the authority designated three broad types of management zones, each permitting or prohibiting certain kinds of activities:

- *General use zones.* These are more or less all-purpose zones in which most human activities are allowed, with the exception of mining and oil drilling.

- *National park zones.* These zones are more or less "look but don't touch" areas. Human activities are permitted that do not remove living marine resources or that remove only very small quantities of such resources.

- *Preservation or scientific research zones.* In these areas the only activity permitted is scientific research.

"The only activities that are not permitted in any part of the barrier reef are oil exploration, mining, littering, spear-fishing with SCUBA gear, and the taking of large specimens of certain species of fish," explains Graeme Kelleher, chairman of the Great Barrier Reef Marine Park Authority since 1979.

Most of the sections have large areas of "general use" zoning, where multiple activities are permitted. However, the success of the entire management concept rests on the voluntary compliance of the general public and specialized user groups like divers. "We cannot possibly patrol the entire reef with a field staff of only ninety-five people and a few airplanes," points out Kelleher. "Some reefs are so remote that we visit them only once a year, even though we spend close to a million dollars per annum on aerial surveillance."

Despite the Marine Park Authority's enormous responsibilities, it is able to manage an area the size of Great Britain on a budget of around $20 million a year (1992 U.S. dollars). "Because of the park's vast size, we invest a lot of time and effort in extensive information and education programs," says Kelleher. "These programs involve meeting with interested members of the public and user groups, the publication of information and educational booklets and brochures, and the production of special educational materials for primary and secondary schools."

One of the park's showcases is the Great Barrier Reef Aquarium in Townsville. It contains the world's largest captive reef in a huge tank that even simulates wave action, vital for cleansing the coral animals and keeping the entire reef healthy. The aquarium has enormous conservation, educational, and scientific value. It not only serves as a living laboratory for research scientists, but also affords millions of ordinary people, who may never get a chance to dive over a real reef, an opportunity see how one functions.

"An educated public is a vital element in our management strategy," admits Kelleher. "People who understand the importance of coral reefs will act to protect them." So far, relying on self-policing has paid off.

The park's main source of income is from domestic and international tourism, which is developing rapidly. At present, nearly half a million visitors spend 2.5 million visitor-nights and pump over $1 billion into the region's economy every year.

Since all tourist facilities within the park's boundaries require a permit, the authority is able to review each request in terms of its overall likely impact on reef ecosystems. This review system allows for a more rational and planned development of tourist infrastructure. Fears that the barrier reef would be overrun with trophy-hunting tourists have not proven valid. Of the region's 3,000 reefs and islands, only 24 contain tourist resorts, while another 19 have permanently moored tourist platforms for accommodating day trips.

The other major activity within the park is fishing. Two commercial fisheries— one bottom-trawling for a variety of prawns and scallops, the other a reef line fishery—haul in about $300 million worth of fish and shellfish a year. Some reefs have been closed to fishing in order to assess stocks. And a number of seagrass beds, vital as nurseries for shrimp and prawns, have been closed to bottom-trawling—a move supported by most commercial operators.

Threats to the Reef

Over the past decade there has been a noticeable deterioration in the quality of near-shore reef ecosystems. Increasing loads of nutrient pollution, especially nitrogen and phosphorus from agricultural activities, pose serious problems in some areas. According to Australia's Coral Reef Research Institute, by the mid-1990s, some 77,000 metric tons of nitrogen and 11,000 metric tons of phosphorus were finding their way into coastal waters. High levels of phosphorus are known to increase the porosity of coral skeletons, while elevated levels of nitrogen can lead to increases in the amount of plankton and attached algae, which can out-compete coral animals for light and nutrients needed for growth.

Increasing loads of eroded sediment—15 million metric tons in 1995—have also reduced the viability of coral reefs in near-shore waters. "However, it is the synergistic effects of multiple impacts that may be the most difficult to comprehend and remedy," observes Wendy Craik, a fisheries biologist with the park authority, "and yet the most insidious and serious" (Craik 1992).

Nevertheless, those coral reefs affected by pollution and sedimentation constitute only a small portion of the entire reef system. Most of the Great Barrier Reef lies between 40 and 100 kilometers offshore, far enough away that most land-based pollution will not pose much of a danger. A more serious concern is the prospect of an oil tanker accident: Even a small oil spill could have devastating consequences for offshore reefs.

The other major management problem for the park is the crown-of-thorns starfish (*Acanthaster planci*). The crown-of-thorns is an efficient predator with a taste for the coral polyps that build reefs. Under normal conditions, a few starfish feeding on coral polyps make little difference to the health of the reef. However, the crown-of-thorns starfish has a habit of descending on reefs by the millions. Since each starfish can consume its own area of coral in a day, a large infestation can soon destroy an entire reef. Worse, each female crown-of-thorns starfish is capable of laying 20–100 million eggs. The offspring of just one individual, if they survive, are enough to trigger a dangerous infestation. By the late-1980s, nearly 17 percent of the barrier reef had been infiltrated by this organism, but only 5 percent was seriously affected.

By 1991, the number of reefs affected by crown-of-thorns invasions had fallen to 3 percent, and in 1992 the outbreaks ceased. Recent research has found that tidal currents are the main reason for the southward movement of crown-of-thorns starfish populations, explaining why some reefs are more susceptible than others. Pollution, after all, may not play much of a role in their spread.

The Future of the Reef

Barring natural catastrophes, the multi-use management concept seems to be a promising way to preserve coastal resources from overexploitation. The management system that has evolved for the Great Barrier Reef has proved flexible enough, so far, to respond to the growing challenges facing the park, though, at some point, the authority, working with the state of Queensland, may have to start regulating land-based sources of pollution. One thing everyone agrees on: Whatever else happens, community involvement in the management of the park is crucial to its long-term survival.

Kobe, Japan: Planned for People

Most visitors to Japan are probably struck by at least one noticeable fact: All of the country's big cities, compared to their American and European counterparts, are almost antiseptically clean. No litter clogs drainage gutters, no garbage piles up on sidewalks. Japan's coastal waters, however, are often badly polluted with untreated or partially treated municipal and industrial wastes. Only about 50 percent of the population is connected to a sewage treatment plant, one of the lowest percentages in the industrialized world.

The Seto Inland Sea covers 22,000 square kilometers of shallow coastal waters along the south coast of the big island of Honshu, west to the northern part of the island of Kyushu. Eleven prefectures border the sea: Osaka, Hyogo, Wakayama, Okayama, Hiroshima, Yamaguchi (on the Honshu side); Tokushima, Kagawa, and Ehime (on the Shikoku side); and Fukuoka and Oita (on the Kyushu side). Over 30 million people live and work along its shores, a quarter of the population. Population density along the entire coastline *averages* 626 people per square kilometer, nearly twice the national average (Environment Agency of Japan 1993).

Although water quality in the Seto Inland Sea has been improving gradually over the past twenty years, it is still far from clean. Only 208 of the region's 605 cities and towns have sewage treatment plants working or under construction. Industrial pollution has dropped but not significantly: from 1,000 tons per day in 1979, as measured by chemical oxygen demand (COD), to 760 tons of COD per day in 1994. The incidence of killer "red tides" in the Inland Sea has been reduced from 300 a year in 1976 to 162 in 1986, a reduction of nearly half in a decade. Reductions since 1986 have not been as impressive. By 1993 there were still about 100 incidences of red tides every year, roughly two a week. In 1991 they cost the fishing industry over 1.5 billion yen in lost production (Environment Agency of Japan 1993).

Kobe, situated on the Seto Inland Sea, about 40 kilometers southwest of Osaka, is an exception, in more ways than one. This bustling port city of 1.5 million inhabitants on Osaka Bay has been planned so thoroughly that it is now a model of environmentally sound development for medium-sized port cities everywhere.

At first glance, Kobe seems to suffer from Japan's claustrophobic lack of space. Situated along a strip of coast only 3 kilometers wide, the city is hemmed in by the Rokko Mountains on one side and the Inland Sea on the other.

The city overcame its disadvantages by turning them into assets. Beginning in the 1970s, planners began to develop the entire region—some 546 square kilometers—with great attention to detail. "You might say we have a passion for planning," comments Isao Mizohashi, head of the Asian Urban Information Center of Kobe, a think tank attached to the mayor's office.

City planners decided that the best way to tackle the lack of space was to create more of it, by extending the city into the sea. Engineers identified three mountains behind Rokko Peak north of the city that were prone to erosion and landslides and ideally suited to be whittled away for landfill.

Between 1966 and 1981, 80 million cubic meters of earth and stone were relocated 2 kilometers offshore from Kobe's port to create Port Island: 436 hectares of human-built island, constructed at a total cost of about $4.8 billion. Besides adding extensive facilities for container cargo, the island has residential areas, businesses (Kobe's fashion industry), schools, a hospital, hotels, and convention facilities (Kobe City Government 1992).

Before one rock was moved, however, environmental impact assessments helped determine the shape of the island so that it would not alter tidal currents. Even the construction techniques were environment friendly. A vast network of underground conveyor belts carried soil and rock to barges, which then dumped their loads in precise offshore locations. The city was not subjected to dust and exhaust fumes from thousands of trucks grinding slowly through its streets.

A second massive landfill project took two decades to finish (1972–92). The completion of Rokko Island added 580 hectares of cargo terminals, including 8,000 housing units and an industrial zone, to the city's waterfront.

Since coastal population densities are so high, Japanese engineers have to use every inch of coastal space. The leveled mountains were converted into bedroom communities. Today, Suma New Town and Seishin New Town each have populations exceeding 100,000 and are connected to the city of Kobe by an efficient road and rail network.

Unlike the residents of other Japanese cities, nearly all of Kobe's residents are connected to a sewage treatment plant, so coastal waters are not polluted with municipal wastes. All garbage is sent to five city incinerators. Industrial wastes are recycled, treated on site, or incinerated. No wastes—either municipal or industrial—are dumped in landfills or at sea. And the city's air is unpolluted, thanks to rigid pollution-control standards and an ingenious freeway system that instead of sprawling over whole sections of the city, stacks access highways one on top of another.

The key to the city's success at managing its coastline is that in the process of developing, explains Mizohashi, "We have taken care of both land and sea." In the lush, pine-covered Rokko Mountains north of the city, little development has been permitted. Landslide-prone hills are now anchored by evergreen forests. The watershed has been protected, with most of the area reserved for parks, hiking trails, botanical gardens, and managed wilderness.

With over 20 million square meters of public parks, Kobe is one of Japan's greenest cities. "In fact, the amount of green space per person is 14 square meters, the highest of all of Japan's major cities," points out Mizohashi proudly.

Its seaside has an equally impressive record, rare in a country not known for protecting coastal resources. Coastal development outside the city is strictly regulated to avoid urban sprawl and the proliferation of holiday cottages. And it has one of the cleanest beaches on the Seto Inland Sea.

Another reason for Kobe's relatively clean water in a sea known for its pollution and red tides is that it has not allowed large-scale mariculture activities in Osaka Bay. While the near-shore waters of many other cities on the Inland Sea are covered with fish pens and mussel cultures, Kobe's mariculture is done on land.

The city's famous sole are carefully nurtured in tanks at the Kobe Municipal Mariculture Center until they are old enough to survive at sea. Instead of raising the sole to marketable size, then harvesting them, Kobe's mariculturists release about 300,000 young sole every year into near-shore waters in an effort to replenish wild stocks. With most commercially important fish and shellfish facing severe harvesting pressures in Japanese waters, this helps to keep Kobe's struggling fleet afloat.

"All told, we have released around one and a half million sole into offshore waters since the center began operating in 1988," explains Susumu Kakimoto, head of the fisheries section in the Agricultural Bureau of the city government. "We calculate that nearly 10 percent of them have been captured by our fishing fleet, for a total value of roughly $2.7 million."

One of the most intriguing of Kobe's plans calls for a ceiling on the city's population. "Kobe's population will not be allowed to exceed 1.8 million," states Mizohashi. "We now have 1.5 million, so our plans call for capping the city's population when we have another 300,000 residents."

Planning has allowed city officials to keep Kobe livable. It has also given them the tools to keep the city one of Japan's greenest. And if it can actually manage to place limits on its population growth, it will be the only city of its size in the world to do so.

In January 1995, Kobe was struck by a major earthquake, measuring 7.2 on the Richter scale. More than 6,000 people perished, and, in its aftermath, entire sections of the city lay in ruins. Rebuilding the city was estimated to cost up to $120 billion and to take a decade to complete (Kristof 1995). However, by early 1997—two years after the quake—over 80 percent of the damage had been repaired. Moreover, planners have built a more quake-proof city. Population growth has been minimal during reconstruction.

"Every city is unique," observes Mizohashi, "but we have put people first, and that has made all the difference."

Sri Lanka's Coastal Management Program

The island state of Sri Lanka covers some 65,000 square kilometers, off the southern coast of India in the Indian Ocean. Its coastline stretches for 1,340 kilometers, with many bays, lagoons, spits, bars, and islets. The country's coastal wetlands, once generous, have been reduced to mere remnants: Only some 42,000 hectares of estuaries and lagoons remain, along with an estimated 6,000–10,000 hectares of mangrove swamps (Lowry and Wickremeratne 1989).

As in much of the developing world, the country's rush to the sea is a relatively recent phenomenon. When Sri Lanka gained its independence from the British in 1948—after four hundred years of colonial rule—many people left the hilly interior of the country and migrated to coastal cities and towns in search of better economic opportunities. Today, 55 percent of the population of 18 million lives on the coast or within a few kilometers of it. Forty percent of Sri Lankans live on 15 percent of the country's land area, in the southwestern coastal districts, which stretch from just north of Colombo to Galle.

In this crowded section of coast, erosion is a problem. Along one 500-kilometer section of coast from the Jaffna Peninsula in the north to Weligama Bay in the south, up to 285,000 square meters of land are surrendered to the sea each year. Nearly half of the total amount of land lost each year—some 145,000 square meters—is in one 137-kilometer-long coastal strip extending from the mouth of the Kelani River, just north of Colombo, to Talawila on the Kalpitya Peninsula.

Sri Lanka's unstable coastline is susceptible to wave erosion, a geological fact made worse by the wholesale mining of sand and coral reefs for cement production and the loss of mangroves for fuelwood, building materials, and urban expansion. Poorly planned and built coastal structures—everything from shacks to resort hotels—have aggravated the problem.

Faced with soaring costs for "reactive measures" like building land-protecting sea walls and breakwaters and moving structures farther inland, the government decided in the early 1980s that it needed a management plan for its coastal areas, a plan that would allow it to prevent further deterioration while encouraging development. Few third world countries have gone as far as Sri Lanka in developing and actually implementing a fairly comprehensive coastal management strategy.

The Management Process
The government of Sri Lanka enacted the Coast Conservation Act in September 1981, and it went into effect in October 1983. The act charged the Coast Conservation Department (CCD), formed in 1978 and located within the Ministry of Fisheries, with the task of drawing up a coastal management plan.

The coastal zone, as defined by the act, includes an area from 2 kilometers seaward to 300 meters inland from the mean high-water line. Although rather narrow in its definition of the "coastal zone," the act prudently allowed the Coast Conservation Department to focus its efforts on immediate threats to the coast, such as erosion, rather than tie it up with an overly ambitious program that might take years to implement.

The Coast Conservation Department has followed a phased, incremental approach to planning and managing the country's coastal zone. During the first phase of the Coast Conservation Act (1983–87), the Planning and Development Branch of the CCD concentrated on two things: the development of an overall coastal zone management plan and the implementation of an interim permit system.

The act requires that anyone proposing to engage in a "development activity" within the defined coastal zone acquire a permit from the director of Coast Conservation. In applying the permit system, the CCD did not prohibit many coastal development activities. Instead, it amended the permits so that they were brought closer to compliance with the intentions of the act. By the mid-1990s, 2,700 permits had been processed, taking an average of only three weeks to respond to each request.

The National Coastal Zone Management Plan, adopted by the cabinet in 1991, builds on nearly a decade of management experience. In its initial phase, the CCD concentrates on four areas: coastal erosion; the degradation of coastal resources such as estuaries, lagoons, mangroves, seagrasses, and coral reefs; the loss of historic, cultural, and archaeological sites; and the loss of physical and visual access to the sea.

According to a review carried out by Kem Lowry of the University of Hawaii and H. Wickremeratne with Sri Lanka's Coast Conservation Department, the strength and relative success of Sri Lanka's Coastal Zone Management Plan are due to six factors: (1) the strong coastal orientation of the country; (2) general agreement about priorities for action; (3) a law that provides a solid legal basis for management; (4) strong program leadership; (5) adequate political support for planning and management initiatives; and (6) an adaptive, incremental approach to planning and management that builds on experience (Lowry and Wickremeratne 1989).

Two Decades of Coastal Management Experience

In 1994, the Coastal Resources Center of the University of Rhode Island, part of the Graduate School of Oceanography, began a lengthy review of its twenty years of practical experience in coastal zone management. The center was intimately involved in planning Rhode Island's coastal management program, under the U.S. Coastal Zone Management Act, and its staff has extensive field experience in the third world, especially in Sri Lanka, Thailand, and Ecuador (Olsen and Hale 1994).

"During our review of what has worked and what hasn't, a number of common elements present in successful programs began to emerge," comments Lynne Hale, the center's associate director. The Coastal Resources Center has identified the following elements as crucial to the sound management of coastal zones:

- Coastal management is above all a process of *governance,* not just a series of technical problems to be overcome. "In other words," explains Hale, "it is a process that takes account of all major uses of the coastal zone and all major users, bringing together communities, businesses, relevant government agencies, NGOs, research and academic institutions, journalists, and others for a common purpose."

- It is important to highlight the main issues facing coastal communities. "The concerns of local people are often overlooked," says Hale, "and this is a serious inadequacy in many programs."

- Constituency building for proposed management plans is essential if the strategy is to succeed. This involves everyone from the prime minister or president down to local tour-boat operators and subsistence fishermen.

- It is often necessary to work on several levels at the same time in order to build support for coastal management plans. "At CRC, we take the two-track approach," says Hale. "We work at both the national and local level to build political support for the management program."

- There should be some mechanisms in place that allow national, regional, and local government agencies to cooperate on evolving a workable management strategy. Without coordinating mechanisms in place, management plans can fall between the cracks or get caught up in endless jurisdictional disputes between competing ministries.

- It is prudent to start management activities on a small scale in local areas, then build on that success. Since most developing countries, in particular, have little or no experience in coastal management, it makes sense to begin the process by funding small-scale projects, like planting mangroves or making artificial reefs. "This strategy builds up credibility and support for the program," says Hale, "vital for broadening its scope later."

- An incremental, evolutionary management approach is often the best way to proceed.

- Developing problem-oriented solutions to management issues allows coastal managers an opportunity to address pressing concerns at an early stage in the process. "Waiting until you have enough resources in place to take on the big-picture issues with big budgets often means that nothing gets done at all, and the whole exercise loses credibility and support," observes Hale. "In the case of coastal zone management, less is often more."

The Law of the Sea

On November 16, 1994, the United Nations Convention on the Law of the Sea quietly came into force, twelve years after it was signed with much fanfare by 159 nations. At the time of signature, in 1982, Secretary General Perez de Cuellar called the convention "possibly the most significant legal instrument of this century." However, when a number of industrialized countries—led by the United States, Great Britain, and Germany—refused to endorse it, objecting mainly to its seabed mining provisions, the convention seemed destined to sink out of sight.

The Law of the Sea is expected to give impetus to coastal area management, since each state with a coastal zone can now claim jurisdiction over an Exclusive Economic Zone (EEZ) that extends 200 nautical miles from the coastline. This greatly increases the sea area under national jurisdiction.

In reality, the convention merely turned into international law what many countries were already practicing. In 1945, U.S. President Harry Truman, responding to pressures from domestic oil companies, unilaterally asserted economic control over America's entire continental shelf, which extends more than 200 nautical miles from shore. In October 1946, Argentina did the same thing, followed by Peru and Chile in 1947 and Ecuador in 1950.

For centuries, most nations have recognized the "right of innocent passage," that is, the right of a foreign ship to sail through a strategic strait or narrows under their jurisdiction, so long as the interloper's intention is peaceful. What that means in practice is that a Japanese oil tanker picking up a load of crude from the Arabian Gulf, for example, would not have to detour some 5,000 kilometers in order to avoid the territorial waters of Indonesia.

In order to be effective, the convention had to recognize and respond to such traditional uses of the ocean. The challenge was to forge a law of the sea that could be accepted by virtually every coastal state, developed and developing alike.

Highlights of the Convention

In terms of other international marine agreements, the Law of the Sea is unique in several respects (United Nations Division for Ocean Affairs and the Law of the Sea 1993):

- Its provisions include recognition of the right of every coastal state to declare jurisdiction over all resources in an EEZ that may extend up to 200 nautical miles from shore. The coastal state may also exercise jurisdiction over the resources of its continental shelf (assuming there is one) but not the waters above it. In any case, shelf jurisdiction may not exceed a limit of 350 nautical miles from shore, or 100 miles beyond the depth of 2,500 meters.

- EEZs grant each coastal state the right to exploit, develop, manage, and conserve all resources, from oil and gas to fish and gravel, found within these zones. As a result, about 87 percent of all known and estimated hydrocarbon reserves now fall under some state's jurisdiction, along with some 90 percent of all commercial fisheries.

- While maintaining the traditional concept of "innocent passage" through a territorial sea, the convention introduced a new notion called "transit passage" for navigation through narrow straits. This doctrine allows unimpeded passage through straits used for international navigation, as well as overflight privileges. The convention further secures navigation rights by giving all states the privilege of "transit passage" through archipelagic sea-lanes.

- The convention established the International Seabed Authority in an effort to bring future deep-seabed mining activities under the rule of law and spread the benefits around as a "common heritage." Under a compromise, reached in July 1994, seabed mining will be developed by private consortia, under license
- from the Seabed Authority. Two areas must be included in each application:

one to be developed by private interests and one set aside by the authority for its own operations.

- When enforcing pollution standards, the convention gives broad discretionary enforcement powers to the "port state." A port state has the right to impound in its harbor any ship that may have polluted the seas *anywhere*. A port state may take enforcement action against a vessel under its jurisdiction at the request of another state whose waters were polluted by that vessel.

This "Constitution of the Oceans" leaves compliance up to each signatory state. However, unlike its predecessors, this convention contains provisions for enforcing international pollution standards, including a binding dispute settlement procedure that makes the Law of the Sea a unique instrument in international law.

The convention cannot compel coastal states to better manage marine resources. It only gives them the opportunity to evolve management strategies. In the short run, the convention has permitted coastal nations to exert more control over marine resources within their 200-mile EEZs. But what that means in practice is that many third world countries, especially small island states with huge EEZs, now collect annual fees from foreign trawlers, purse seiners, and factory ships operating in their waters. Licenses to Taiwanese, Korean, and Japanese fishing fleets around the Falkland Islands, for instance, bring in over $45 million a year, more than any other single source of revenue.

Similarly, the sixteen independent countries comprising the South Pacific Forum, fourteen small island states plus Australia and New Zealand, have established a formal registry system under the auspices of the South Pacific Forum Fisheries Agency (FFA). All foreign fishing vessels must now apply for permission to operate in any of the EEZs under the jurisdiction of forum member states (United Nations Division for Ocean Affairs and the Law of the Sea 1993).

This has given rise to a number of fisheries treaties, the most notable of which is the Multilateral Fisheries Treaty concluded in 1987 between the Pacific Forum member governments and the United States. Under its conditions, thirty-five U.S. registered purse seiners are allowed to fish for migratory tuna inside South Pacific EEZs. The United States paid an initial five-year license fee of $60 million, which is divided among the sixteen countries according to a formula they worked out themselves. As a result of this and other agreements resource-poor island states such as Kiribati, Solomon Islands, and Tuvalu now earn more from fisheries than from any other source of income.

Although fisheries agreements or treaties do regulate the number of fishing vessels and the type of gear they are allowed to operate, they often contribute little to the actual management of fish stocks. This has been underscored in recent years as fish catches have stagnated, or even dropped, while the size of the world's fishing fleet doubled—from 585,000 large boats in 1970 to 1.2 million in 1990.

CHAPTER 4

The Baltic and North Seas

In Dire Straits

The Baltic and North seas are among the most overexploited and polluted seas on earth. Both are fringed by dense human populations crowded along highly industrialized coastlines. As of 1997, some 240 million people lived in their combined catchment areas. Demographic projections indicate that over the next thirty years northern Europe's coastal zones will continue to harbor most of its population and industry (see table 4.1).

This entire region has been a center for trade and industry since the early Middle Ages, when the Hanseatic League stitched together a loose-knit federation of independent trading cities stretching from the British Isles in the West to Russia's Novgorod in the East. At the height of its influence, in the fourteenth century, the league incorporated two hundred mercantile cities.

Despite the proliferation of heavy industries along the Baltic and North seas in the nineteenth and early twentieth centuries—particularly iron and steel mills, metal smelters, chemical plants and refineries, pulp and paper mills, cement factories, and big engineering firms—the water quality in the seas did not begin to deteriorate dramatically until Europe's postwar economy accelerated into high gear after the formation of the European Common Market in 1958. At the same time, eastern Europe was being fitted with an "iron corset" of heavy, polluting industries stretching in a wide arc from the Baltic port city of Rostock (in former East Germany) through Gdansk, Poland, and the Baltic states to Leningrad (now St. Petersburg) in the former Soviet Union. Rapid industrialization and the continued concentration of economic activities in coastal areas resulted in tremendous pressures on coastal resources (see figure 4.1).

The Baltic

Upwards of 88 million people live in the Baltic's catchment area, which covers more than 1.3 million square kilometers in northern Europe (see figure 4.2). The sea is surrounded by nine highly industrialized countries: Sweden, Finland, Russia, Estonia, Latvia, Lithuania, Poland, Germany, and Denmark (Helsinki Commission 1993).

Over 25 million people live on the coast, around 50 million within 200 kilometers of the coast. Virtually the entire populations of Denmark and the Baltic

*Table 4.1. Length of Coastline and Population Data for Fourteen Countries
Bordering the Baltic and North Seas*

COUNTRY	COASTLINE (KM)	POPULATION 1997 (MILLIONS)	POPULATION GROWTH (%/YR)	% URBAN
Belgium	64	10.2	0.1	97
Denmark	3,379	5.3	0.1	85
Estonia	1,393	1.5	−0.5	70
Finland	1,126	5.1	0.3	65
France	3,427	58.6	0.3	74
Germany	2,389	82.0	−0.1	85
Latvia	531	2.5	−0.7	69
Lithuania	108	3.7	−0.1	68
Netherlands	451	15.6	0.3	61
Norway	5,832	4.4	0.3	74
Poland	491	38.6	0.1	62
Russia	100	147.3	−0.5	73
Sweden	3,218	8.9	0.0	83
United Kingdom	12,429	59.0	0.2	90

Sources: World Population Data Sheet, 1997, Population Reference Bureau, Washington, D.C., 1997; and *World Resources 1996–97*, Oxford University Press, New York, 1996, p. 268.

Notes: Length of coastline for France includes Mediterranean. Population figure for Russia is entire population, not Baltic catchment population.

states—some 13 million people—are coastal, the majority of them living in towns and cities. Nearly half of Denmark's population of 5.2 million lives in its capital, Copenhagen, and the city's sprawling suburbs. Nearly 75 percent of all Swedes live in urban areas along the Baltic and the Kattegat (the body of water that leads out to the North Sea).

The southern coast of the Baltic is the most crowded, with the entire population of Poland—38 million—living in its drainage basin. The wide coastal belt from Kiel, Germany, to St. Petersburg, Russia, holds approximately 40 million people, roughly half the total of the entire sea.

Fortunately, the region is not plagued by high population growth rates. A number of countries—Russia, Germany, Estonia, and Latvia—are actually losing population. There are simply more deaths than births every year. Of those remaining countries with a measurable growth rate, Sweden has the lowest: 0.0 percent a year. At that pace the country will take over 4,000 years to double its current population of 8.9 million. At Poland's rate of growth—a mere 0.1 percent a year—it will take 573 years for its current population to double (Population Reference Bureau 1997).

The entire Baltic is girded by a ring of heavy industries. By the late 1980s, it contained some two hundred major industrial complexes—mostly steel and metal works, chemical and petrochemical industries, and pulp and paper mills. Stringent environmental standards introduced in the 1970s in Sweden, Denmark, Finland, and the former Federal Republic of Germany resulted in significant reductions in both industrial and municipal effluents dumped into the Baltic. Concerted efforts were also made to bring down air pollution levels. However, the gains in the

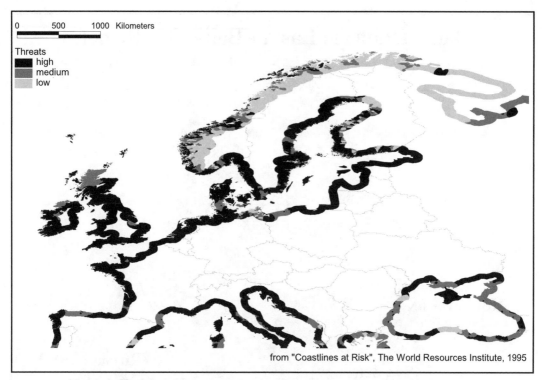

Figure 4.1. Coastal Ecosystems Threatened by Development—North and Baltic Seas.

Source: D. Bryant, E. Rodenburg, T. Cox, and D. Nielsen, "Coastlines at Risk: An Index of Potential Development-Related Threats to Coastal Ecosystems." World Resources Institute, Washington, D.C., 1995.

western half of the Baltic were nearly offset by increased pollution loads generated by the eastern half's smokestack industries.

Geography

The Baltic is a victim of its own geography. It is a small, shallow, tideless, relatively enclosed sea, its 366,000-square-kilometer surface area barely accounting for 0.1 percent of the world's oceans. It is divided into seven distinct parts: Bothnian Bay at its northernmost end, the Bothnian Sea, the Gulf of Finland, the Gulf of Riga, the Baltic Proper, the Belt Sea—including the sound around Denmark—and the Kattegat (Leppakoski and Mihnea 1996).

Despite the fact that the Baltic is rejuvenated every year by 450 cubic kilometers of freshwater brought in by 250 river systems, it still takes nearly thirty years for its brackish waters to renew themselves through the Skagerrak, which connects to the North Sea. With a drainage area four times the size of the sea itself, a mean depth of only 60 meters, a long water-retention time, and thousands of highly polluting industries and municipalities using it as a receptacle for all manner of wastes, it is not hard to imagine how the Baltic became severely degraded in four decades.

Pollution

What has proved hard is halting the Baltic's deterioration, despite more than twenty years of cleanup efforts carried out under the umbrella of the Helsinki

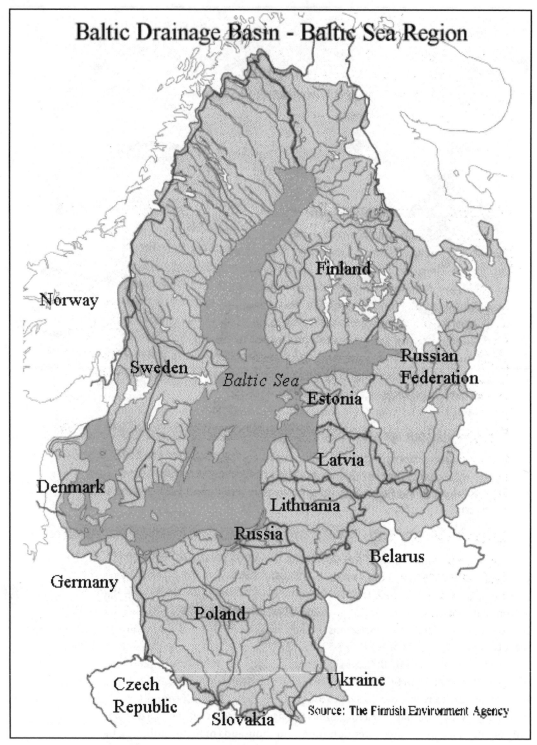

Figure 4.2. *Above:* Geopolitical Map of the Baltic Sea Drainage Basin. *Opposite:* Estimated Population Densities for the Baltic Sea Drainage Basin.

Sources: MAPBSR Project, National Land Survey of Finland, 1997; and Sindre Langaas, *The Global Demography Project,* Baltic Drainage Basin Project, UNEP GRID-Arendal, Beijer Institute, Stockholm University, 1995.

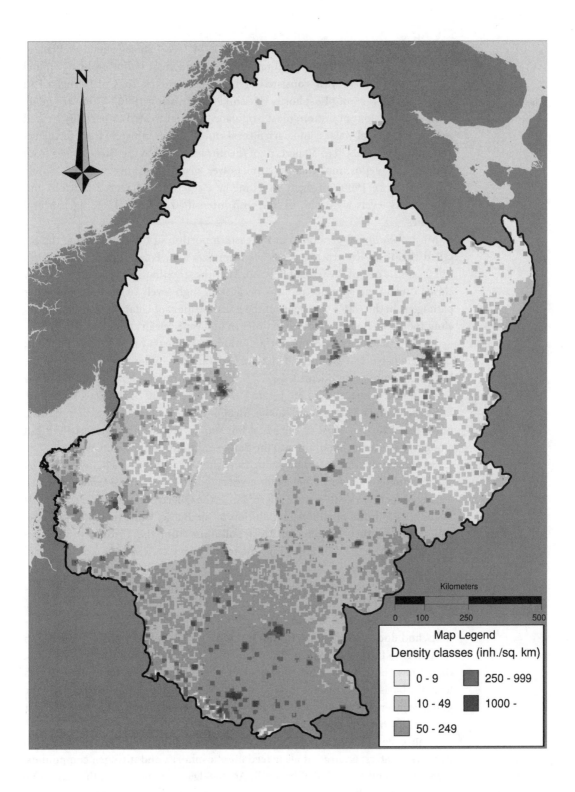

Kilometers

0   100   250   500

Map Legend

Density classes (inh./sq. km)

	0 - 9		250 - 999
	10 - 49		1000 -
	50 - 249		

49

Convention, signed by all Baltic states in 1974 (see "Management: A Plan of Rescue," p. 53). As of 1990, the total amount of nutrient pollution being spilled into the Baltic every year consisted of 1.2 million metric tons of nitrogen and 80,000 metric tons of phosphorus (Helsinki Commission 1994). There are many sources of nutrient pollution; most originates from untreated or partially treated sewage, municipal wastes, and agricultural runoff, predominately from fertilizers and animal wastes. Up to 30 percent of the nitrogen load to the Baltic comes from atmospheric deposition, mainly from power plant emissions and automobile exhausts. Since 1900, nitrogen levels in the Baltic have increased fourfold and phosphorus levels eightfold, in line with intensified industrial and agricultural activities.

Most pollution is generated in the eastern portion of the Baltic, along the so-called Pollution Riviera that stretches from Szczecin, Poland, to St. Petersburg, Russia, at the end of the Gulf of Finland. Many beaches in this region are routinely closed during the summer months because of high levels of bacteria and other pathogens in the water. Residents of the Polish seaport of Gdansk are used to traveling 160 kilometers (100 miles) in order to find water clean enough to swim in.

In 1990, scientists carrying out pollution assessments for the Baltic Sea Joint Comprehensive Environmental Action Programme, under the auspices of the Helsinki Commission, identified 132 "hot spots" of point-source pollution to the Baltic. Of those, nearly 100 are located in the eastern half of the sea—in Poland, Lithuania, Latvia, Estonia, and Russia. Most of the 47 hottest spots were in Eastern Europe and Russia. Most of the worst hot spots were not on the coast itself but along river systems that drain into the Baltic (Helsinki Commission 1993).

So much pollution enters the Baltic from the Vistula River in Poland—some 34 square kilometers a year—that a plume spreads all the way into Swedish territorial waters. In fact, the Vistula, once referred to as the Queen of Polish Rivers, accounts for fully two-thirds of all Polish pollution to the Baltic, while the Oder River accounts for another 25 percent. Most pollution from the Vistula, which runs through the heartland of Poland, is due to the heavy concentration of industries along its route to the sea.

The Vistula meets the Baltic in the Gulf of Gdansk. Here its pollution load is compounded by the untreated wastes generated by a complex of coastal industries, including oil refineries, fertilizer and sulfur-purifying plants, coal-fired power stations, and docks, where ships routinely discharge oily ballast waters. The water in the gulf is so thick with toxic pollution, including hydrocarbons and heavy metals, that it has snuffed out most phytoplankton and rockweed communities, which form the basis for the marine food chain. As a result, fish and shellfish stocks in the gulf have been virtually wiped out. Fish that have been caught often lack scales or fins, while others are covered with tumors.

High levels of nutrient pollution have also turned much of the gulf eutrophic. In its murky waters, an orgy of algae feed on phosphorus and nitrogen compounds, turning the gulf the color of broccoli. As the algae die and decay, they consume oxygen. As a result, "the seabed in the Gulf of Gdansk is said to resemble a graveyard," notes Polish environmental journalist Eugeniusz Pudlis. "Nothing but micro-organisms can survive in its turgid waters" (Pudlis 1993).

Farther east, the condition of coastal waters is just as bad, if not worse. Murray Feshback reports in his book *Ecocide in the USSR* that in 1989, before the disintegration of the Soviet empire, about 3.6 million tons of wastewater, from both municipal and industrial sources, were discharged into the Baltic from the cities of Leningrad, Riga, Klaipeda, and Tallinn. Nearly one-fifth of all organic matter was traced to fourteen pulp-and-paper mills in the region, operating with no pollution controls whatsoever (Feshbach and Friendly 1992).

At Riga, the capital of Latvia, the sewage system has not been overhauled since 1938. As a consequence, the city dumps over 100,000 metric tons of pollutants a year into the Daugava River, which then empties them into the Baltic. The nearby Leilupe River, which empties into the Daugava, may be biologically dead. It is so polluted with municipal and industrial wastes that it has been closed to swimming since 1982 because of coliform counts that reach half a million bacteria per liter of water in the summertime—among the highest levels ever recorded. In 1990, Latvia dumped some 250,000 metric tons of hazardous industrial waste and toxic sludge from wastewater treatment plants and hospitals into its surface waters; nearly all of it eventually entered the Baltic.

In 1990, Lithuania was discharging nearly 400 million cubic meters of highly polluted wastewater, most of it from cities with little or no sewage treatment facilities. The country's two largest cities, Vilnius (the capital) and Kaunas, account for 167 million cubic meters of this waste, all of which is discharged into the Nemunas River, which empties into the Baltic. Vilnius gives its sewage mechanical treatment only, and Kaunas has no sewage treatment plant at all (Stockholm Environment Institute 1994) (see table 4.2).

St. Petersburg, Russia, often referred to as the Venice of the North with its charming canals and eighteenth-century architecture, remains a hot spot of pollution. Before the breakup of the Soviet Union, Peter the Great's "window on the west" had some two thousand significant sources of pollution. Today, three hundred remain—the reduction a consequence of economic restructuring and the closure of unprofitable state-run factories, not cleanup measures.

The rapid buildup of pollution, particularly oxygen-consuming phosphorus and nitrogen, has turned about 100,000 square kilometers of the Baltic's deeper waters into an oxygen-starved dead sea. "In less than fifty years, its deep waters have been transformed from an oxygenated environment with a normal fauna of fish and invertebrates, to an anoxic desert without any life higher than bacteria in the entire waterbody deeper than 50 meters," points out Swedish marine ecotoxicologist Olof Lindén.

The rising tide of toxic pollution, especially elevated levels of mercury, DDT, and PCBs, which bioaccumulate up the food chain, has affected not only fish and shellfish but many marine mammals and seabirds as well. Populations of grey and ringed seals have plummeted as a result of contamination from PCBs, DDT, and heavy metals. Exposure to these dangerous pollutants causes widespread reproductive disorders in seals, triggering spontaneous abortions and birth defects, among other abnormalities. In 1988, over 20,000 seals in the Baltic and North seas perished from a distemper virus, since linked to high levels of organochlorines, including DDT and PCBs, found in the animals' blubber. Until recently, top predators, like the white-tailed sea eagle (*Haliaetus albicilla*), were nearly extinct over much of the Baltic because of environmental contaminants. Continuous exposure to heavy metals such as mercury caused the birds to lay thin-shelled eggs, which never hatched. By the mid-1990s, however, white-tailed eagles and harbor seals

*Table 4.2. Total Nutrient Loads to Baltic from Three East European
Countries, 1990–91 (in metric tons per year)*

COUNTRY	YEAR	BOD	NITROGEN	PHOSPHORUS
Lithuania	1990	165,212	18,639	3,499
Poland	1991	282,000	124,000	15,000
Estonia	1990	67,100	59,180	2,753

Source: Adapted from *Beauty and the East: An Evaluation of Swedish Environmental
Assistance to Eastern Europe,* Stockholm Environment Institute, Stockholm, September 1994.

were both making comebacks, thanks to reductions in organochlorines and other
pollutants.

The Marine Chernobyl

In May and June of 1988 in the waters between Denmark, Sweden, and Norway,
high nutrient pollution levels and warm weather precipitated a population explosion of toxic algae with the scientific name *Chrysochromulina polylepis.* Although
algal blooms are common in the Baltic, especially in shallow coastal bays, this one
was monstrous. The initial bloom, 50 kilometers long, 10 kilometers wide, and 10
meters deep, quickly spread through the Kattegat and Skagerrak and along the
North Sea coast of Norway. It eventually covered 75,000 square kilometers. As the
blooms died and decayed, consuming huge quantities of oxygen, nearly 1,000 kilometers of coastline were blighted by the corpses of millions of marine organisms
that washed up on shore. Many beaches—from the Danish archipelago to north of
Stavanger, Norway—had to be closed as a result (Underdal et al. 1989; MacGarvin
1990).

The death tolls were so high that scientists refer to this incident as the marine
Chernobyl. Hundreds of thousands of invertebrates perished, including cat
worms, whelks, topshells, periwinkles, limpets, sea urchins, and common starfish.
Five species of snails were completely eradicated in affected areas, and mortality
among bivalves was also very high.

Fish fared little better. Although strong-swimming fish such as adult salmon,
cod, and herring were able to flee, young cod and whiting, too small to escape, died
by the thousands. Others, such as wrasse and gobies, perished as they tried to hide
under rocks along the coast. Around 600 metric tons of farmed fish, mostly
salmon, died in their pens as the algae suffocated them. Many more would have
succumbed if Norwegian fish farmers had not taken the costly step of towing the
massive cages out of the fjords and into cleaner waters offshore.

Fisheries

Not long ago, one of Poland's daily newspapers ran a satirical cartoon showing the
Baltic's last codfish being pursued by an enormous collective fishing fleet representing all the Baltic states. Seeing the hopelessness of flight, the fish headed
straight for the Gulf of Gdansk, apparently—according to the caption—"prefer-
ring a merciful quick death from toxic pollution to an agonizingly slower one in
someone's skillet."

Baltic fish stocks have been overexploited and mismanaged for decades. And
rising pollution levels have taken a toll, especially on eggs and young. Moreover,

catch quotas assigned for the major commercial fisheries—cod, herring, and flat-fish—are routinely ignored.

Over the past decade, catches of cod, the major commercial fishery of the Baltic, have fallen off dramatically. The heavily exploited cod fishery dropped from 387,000 metric tons in 1980 to a mere 39,000 metric tons in 1993. By 1994, cod stocks were so depleted that no catch quotas could be assigned.

The Baltic's herring stocks, however, have recovered and are now considered underexploited by the International Baltic Sea Fishery Commission. The catch in 1993 was only 345,000 metric tons out of a total allowable catch of 650,000 metric tons (as recommended by the International Council for the Exploration of the Sea). According to the commission, catches of herring have fallen for purely economic reasons.

With the demise of the cod fishery, Baltic fishermen have switched to taking less valuable sprat. That fishery increased from 62,000 metric tons in 1980 to 182,000 metric tons in 1993. The harvest of salmon also increased marginally, from 2,500 metric tons in 1980 to 4,200 metric tons in 1993; but 90 percent of the catch is farmed fish. The 1994 harvest figures for salmon fell off. A mystery disease, named M74 and possibly pollution linked, decimated hatcheries and fish farms and continues to threaten remaining wild stocks (MacKenzie 1993).

"The state of Baltic fisheries is a scandal," notes Eugeniusz Pudlis. "Polish cod is often loaded with heavy metals such as zinc, cadmium, lead, and mercury. If we were to judge the effects of the twenty years of governmental work in this area in light of the condition of fish and other fauna and flora in this sea, our conclusion wouldn't be favorable."

Management: A Plan of Rescue

By the early 1970s, it had become clear that the Baltic's environment was deteriorating rapidly. Urged by Sweden and Finland, all Baltic states met in Helsinki in May and June of 1973 to hammer out a legal framework for protecting the Baltic. A year later, in April 1974, the Helsinki Convention on the Protection of the Marine Environment was signed by all Baltic states: Sweden, Finland, the Soviet Union, Poland, the German Democratic Republic, the Federal Republic of Germany, and Denmark. It took another six years—until May 3, 1980—before the convention entered into force, when the former Federal Republic of Germany finally ratified it (the last signatory to do so). Despite its shortcomings, the Helsinki Convention remains one of the first comprehensive agreements on the marine environment and is often cited as a model of international environmental cooperation.

The convention's twenty-nine articles and six annexes call on the contracting parties to implement "individually or jointly all appropriate legislative, administrative, or other necessary measures to prevent and/or reduce pollution to the Baltic." Specifically, it covers the following areas of concern: discharges from land-based sources of pollution, including atmospheric deposition; discharges from ships, including oil, chemicals, sewage, and garbage; dumping of hazardous substances; and international cooperation to combat oil spills. It also created the Baltic Marine Environment Protection Commission, known as the Helsinki Commission (HELCOM), to oversee its implementation.

Like most international commissions, HELCOM has no enforcement power. It cannot compel signatory states to abide by the terms of the convention. Instead, unanimous decisions reached by the members of HELCOM are regarded as recommendations to signatory governments. Recommendations are supposed to be translated into national policies and laws as soon as possible.

According to Bertil Hagerhall, a former representative of the Swedish Agriculture Ministry who participated in the negotiations, the Helsinki Commission had five main duties: "to keep the implementation of the convention under continuous review; to recommend measures relating to the purposes of the convention; to recommend amendments to the convention and its annexes; to define pollution control criteria and objectives for the reduction of pollution; and to promote scientific and technological research."

Right from the beginning, however, the convention was given substance in only half the Baltic; implementing the terms of the convention split along east-west lines. The western half of the Baltic, particularly the Nordic countries, pushed ahead with innovative, technology-forcing environmental legislation. The eastern half issued supportive political statements but did little to reduce pollution levels. Investments in environment protection were not given political priority by the relevant ministries, and authorities charged with environment protection were notoriously short of personnel and money.

During the convention's first two decades, some progress was made in reducing the amount of toxic pollution entering the Baltic. Mercury concentrations in fish in the western part of the sea dropped by two-thirds by the mid-1980s. Concentrations of DDT also fell significantly. In the past few years, decreases in PCBs have been noted. All Baltic states banned the use of DDT, and the use of PCBs was severely restricted. Only the former Soviet Union and Poland continued to use them until recently.

Oil pollution remains a problem in a number of areas, but it is being systematically reduced. In the early 1980s, oil spilled or discharged from routine shipping operations, plus oily residues from urban runoff, amounted to around 40,000 metric tons a year. By 1994, that figure had been cut nearly in half due, in large measure, to three initiatives: the building of reception facilities in many major ports to handle ships' wastes, particularly petroleum residues; the establishment of rules and guidelines for cooperation in combating oil spills; and the creation of a reporting system for ships transporting hazardous or toxic cargoes.

With the condition of the Baltic worsening—and the political landscape of Eastern Europe altered radically with the breakup of the Soviet Union—Sweden and Poland lobbied hard for a revised convention. The result was a meeting of Baltic state prime ministers in Ronneby, Sweden, in 1990. That meeting issued the Baltic Sea Declaration, calling for stepped-up efforts at pollution control, particularly from land-based sources. Two years later, in April 1992, a conference of environment ministers, held in Helsinki, adopted the Baltic Sea Environment Declaration and a revised and strengthened Helsinki Convention, which was promptly signed by all states bordering the Baltic and the European Economic Community.

The 1992 conference was heralded as the beginning of a new era of cooperation. Significantly, the Baltic Sea Environmental Declaration calls for renewed efforts to "assure the ecological restoration of the Baltic Sea, ensuring the possibility of self-restoration of the marine environment and preservation of its ecological balance."

As a result, the Baltic Sea Joint Comprehensive Environmental Action Programme was launched in 1993, under the auspices of HELCOM. The Action Programme has a specific agenda, a timetable for accomplishing its objectives, and financial mechanisms. It includes six main components: policy, legal, and regulatory measures; strengthening institutions and human resources development; investment activities; special management programs for coastal lagoons and wetlands; applied research; and public awareness and environmental education (Helsinki Commission 1993).

The first order of business is pollution-abatement measures designed to clean up 132 hot spots of point-source pollution using the "best available technologies." The programme also includes regulatory measures to reduce nonpoint-source pollution, better management of resources (including fisheries), the establishment of protected areas representing various Baltic ecosystems, and related activities.

A unique aspect of the new Action Programme is that it greatly broadens the Baltic constituency of countries (and institutions) involved in the convention. It encompasses all countries in the Baltic's vast catchment area, adding parts of Belarus, the Czech Republic, Norway, the Republic of Slovakia, and Ukraine. The Commission of the European Communities and the International Baltic Sea Fisheries Commission, along with four major international financial institutions—the European Investment Bank, the European Bank for Reconstruction and Development, the Nordic Investment Bank, and the World Bank—are also on board.

The entire Action Programme will be phased in over a twenty-year period. The first phase, from 1993 to 1997, was estimated to cost just over $6 billion (1992 U.S. dollars); the second phase, from 1998 to 2012, is expected to run close to $16 billion. Financing works out to a little more than $1 billion a year, expensive but not astronomical, considering that some Western consultants have estimated a cost of over $600 billion to return the Baltic to its pre-1950 state.

Unfortunately, the first conference arranged to discuss financing for the programme—held in Gdansk, Poland, in March 1993—was a failure. Despite promises by the highly developed countries along the Baltic to shoulder a larger share of the cleanup expenses for their poor eastern cousins, the conference turned up only a fraction of the $6 billion needed to kick-start the programme. At the end of the day, only Sweden and Denmark were able to commit specific amounts to pollution-control initiatives. Both the World Bank and the European Bank for Reconstruction and Development maintained what one Polish journalist described as a "paralyzing calm" in the face of overwhelming need. They were unable to pledge any financial support, insisting that the proposals needed more consideration. This resulted in the reduction of the number of targeted hot spots from forty-seven to sixteen.

"This lack of genuine interest in assistance for our country is disquieting," comments Piotr Krzyzanowski, chief of the Polish Secretariat of the Helsinki Convention. "Frankly speaking, in the long run it's outright dangerous because the recession in Poland, a country of nearly 40 million people, won't last forever. And the increase in industrial and agricultural production in the basins of the Oder and Vistula rivers, which collect most of Poland's sewage and dump it directly into the Baltic, can put us right back where we were in the late 1980s, when pollution levels were at their highest."

Nevertheless, some countries are forging ahead with their own programs. Since the reunification of Germany, authorities have moved quickly to upgrade the eastern half's antiquated infrastructure. Some 6 billion DM ($3 billion) have been earmarked to provide two-stage sewage treatment plants for the highly polluted Oder River basin (shared with Poland) and the underserved part of the Baltic coastline. At the end of 1991, only 12 percent of the population in those regions were connected to biological sewage treatment plants.

Between 1989 and 1995, Sweden allocated around 160 million Swedish kronor (about $20 million) to assist Poland. The bulk of Swedish environmental aid was channeled through BITS (the National Board for Investment and Technical Support). According to Swedish Foreign Ministry counselor Ake Peterson, that kind of international environmental assistance is nothing new. "In addition to assistance motivated by the needs and priorities in recipient countries, Sweden decided some time ago that it can also be economic for the country to support pollution-control strategies in Poland, and other Baltic Sea countries as well, rather than for the Swedish Ministry of Environment and Natural Resources to take all measures inside Sweden."

Despite setbacks, the Environmental Action Programme is advancing. Anticipated investments should result in substantial reductions in pollution loads to the Baltic by 2012, eliminating 550,000 metric tons of BOD (biological oxygen demand), 70,000 metric tons of nitrogen, and 15,000 metric tons of phosphorus (Helsinki Commission 1994).

Shoring Up the North Sea

About 160 million people live in the North Sea's catchment area, which drains 850,000 square kilometers of western Europe, including a quarter of France, most of Switzerland and the United Kingdom, a tiny portion of Austria, all of Belgium and the Netherlands, three-quarters of Germany, half the Danish peninsula of Jutland, and the southern portion of Norway. This vast region contains the largest assemblage of industrial might on the face of the earth. Collectively, the region accounts for 15 percent of the world's industrial output.

Nearly 80 million people live on the North Sea's extensive coastline or within 150 kilometers of it. Most of the population of the Netherlands and Belgium—some 26 million people—live and work in this broad coastal zone, the overwhelming majority in urban areas. The highest coastal population densities are found in the Netherlands, Belgium, Germany, and parts of the United Kingdom; both the Netherlands and Belgium have large sections of coast with more than 1,000 people per square kilometer.

The coastline from Hamburg, Germany, to Brussels, Belgium, is highly industrialized. This region contains gigantic chemical and petrochemical complexes, steel mills, metal smelters, engineering and electrical works, cement and power plants, and ship building, along with industries producing construction materials, automobiles, textiles, and foodstuffs. The region probably has the highest concentration of heavy industries in the world. Germany's powerhouses of production—the Rhine and Ruhr River valleys—drain into the North Sea. The Rhine's watershed contains 42 million people, three-quarters of them living in Germany (MacGarvin 1990).

As the amount of cropland has shrunk, parallel with industrial growth and the helter-skelter expansion of towns and cities, agriculture has become much more intensive. By the late 1980s, both the Netherlands and Belgium were applying over 500 kilograms of petroleum-based fertilizers on each hectare of cropland. Heavy chemical use in agriculture has resulted in tremendous amounts of nutrients, especially phosphorus and nitrogen, leaching into the North Sea.

Over the past 1,000 years, many of the sea's rich estuaries and coastal wetlands have been lost to development. Land reclamation at the edges of the Wash, on the east coast of England, started around A.D. 900. Today, only half of the Wash's rich salt marshes remain intact; 20,000 hectares have been converted to agriculture since 1950. Similarly, nearly 40 percent of the wetlands along the French coast of Brittany have been lost to urban expansion and agriculture over the past twenty years.

Geography

The North Sea is one of the largest coastal seas in the world, covering over 750,000 square kilometers in northern Europe. It is also one of the oldest seas, dating back 350 million years. However, over the past century, it has been fundamentally reshaped by the seven highly industrialized countries that surround it: Norway, Denmark, Germany, the Netherlands, Belgium, France, and the United Kingdom.

Like the Baltic, the North Sea is relatively shallow, averaging only 25–55 meters in the south and falling to 100–200 meters in the north, between Norway and the Shetland Islands. Thanks to generous tidal currents that course through the sea twice daily and vast expanses of shallow coastal waters laced with eelgrass and kelp forests, its nutrient-rich waters support a wealth of marine life.

The Dutch have been battling the North Sea for centuries, reclaiming huge tracts of land from it in the process. Not only did they drain and dike coastal wetlands, but they advanced into the sea itself, building huge polders, or impoundments, which they drained using windmill-powered pumps. Besides Holland's natural dune system, the frontline against the sea is a vast series of high dikes, which guard the country's lowlands (some up to 5 meters below sea level).

In 1932, a drainage scheme to surpass all others was launched in a portion of the Zuider Zee. Some 2,700 square kilometers of this rich, shallow estuary were cut off by a 28-kilometer-long dam. Once partially drained, the new shrunken sea was renamed the IJsselmeer. On the reclaimed land, farmers now raise 70,000 tons of pork, replacing some 12,000 tons of herring that used to be caught in the Zuider Zee. "Although this may be seen as a triumph and justification for the reclamation," explains Dr. Malcolm MacGarvin, a zoologist and consultant for Greenpeace, "in reality it represents a twofold loss for the North Sea: not only has a vast expanse of sea vanished forever, but the pigs are reared on great quantities of fish meal, a product of the industrial fisheries that are currently exploiting the waters of the North Sea" (MacGarvin 1990).

Pollution

Water quality in the North Sea began to deteriorate dramatically about thirty years ago. By 1980, scientists calculated that coastal waters were receiving roughly 1.1 million metric tons of nutrient pollution in the form of nitrogen every year, mostly via the many river systems that carry much of the agricultural, municipal, and industrial wastes of western Europe into the North Sea. Of that amount, between 200,000 and 400,000 metric tons were estimated to have come from atmospheric deposition. Around 137,000 metric tons of phosphorus also entered the sea that year from anthropogenic sources, 10,000 metric tons of it from the atmosphere (North Sea Task Force 1993a, b, c).

In 1990, as many as 1.5 million metric tons of nitrogen pollution were estimated to have entered the sea, along with 100,000 metric tons of phosphorus. In some highly polluted areas, like the German Bight, nitrogen concentrations have tripled over the past thirty years. Despite a decade of efforts aimed at cleaning up the Rhine River, it was still responsible for 33 percent of the phosphorus, and 28 percent of the nitrogen entering the North Sea in 1990 (OECD 1993).

The wastes of industry and agriculture, especially heavy metals and organochlorines (PCBs and DDT), remain a problem because of their long retention time. In 1989, an estimated 23,000 metric tons of zinc, nearly 7,000 metric tons of lead, 4,400 metric tons of copper, 4,200 metric tons of chromium, 1,450 metric tons of nickel, 820 metric tons of arsenic, 150 metric tons of cadmium, and 50 metric tons of mercury were being dumped into the North Sea each year, most brought in by rivers that drain the sea's huge watershed. Worse, most of those toxic metals become bound to near-shore sediments, where they remain for decades. Subsequent studies have found that zinc concentrations in coastal waters were twenty to fifty times higher than natural background levels found in the North Atlantic. Similarly, lead was found to be fifteen times higher, nickel nearly five times higher, copper three to six times higher, cadmium three to ten times higher, and mercury ten times higher (MacGarvin 1990).

"The problem with these numbers," notes Paul Johnston, an aquatic toxicologist at the University of Exeter in England, "is that they are based on rather general, sometimes crude, estimates of pollution inputs to the sea. The real figures could be much higher."

Heavy metals and organochlorines bioaccumulate up the food chain, concentrating in top predators including mackerel, cod, haddock, and whiting, as well as marine mammals and birds. Over the past thirty years or so, cod and mackerel fisheries have been closed in various parts of the sea, due to elevated levels of heavy metals, particularly mercury, in their fish. As in the Baltic, PCBs have been linked to reproductive disorders among marine mammals—seals and otters—and death from viral infections brought on by impaired immune systems. In 1988, some 18,000 harbor seals, out of a total population of 45,000, perished from a distemper virus that devastated entire colonies in the southern part of the North Sea.

Hydrocarbons

The North Sea is dotted with some 150 oil and gas production platforms, which, from the air, look like large carbuncles on the skin of the sea. Most lie in United Kingdom and Dutch waters, not more than 50 meters deep. Oil and gas are extracted from wells, then brought to shore through more than 9,000 kilometers of pipeline. In 1989, at the height of production, around 130 million metric tons of oil and 7.3 billion cubic meters of gas were sucked out of the seabed. Currently, these wells supply about 30 percent of North Sea countries' energy needs.

One platform can drill up to fifty different wells by angling the bits outward like so many spikes on a wheel. This so-called deviated drilling often uses oil-based drilling muds, containing up to 10 percent or more of oil, to ease the passage of the drill and shaft through deep layers of bottom sediment and rock. This has proved to be a significant source of pollution. Once the muds are used up, they are dumped overboard with little or no treatment. Around most production platforms, within a radius of 500 meters, the seafloor is completely smothered by oil-contaminated muds and cuttings that kill off most marine life. Studies reveal that three-quarters of all hydrocarbon pollution in the North Sea is due to drilling muds and cuttings, estimated at nearly 23,000 metric tons in 1988. Oil that spills or seeps from wells is thought to amount to another 500 metric tons a year. But a Dutch estimate in 1987 put the total amount of oil lost through routine production (small spills and seepage) at close to 35,000 metric tons a year (MacGarvin 1990).

According to reports compiled by the North Sea Task Force, established in 1987 as a result of the Ministerial Conference on the Protection of the North Sea, about 2,000 square kilometers of seafloor in the English part of the North Sea are contaminated with oiled drilling muds and cuttings. Biological effects on marine organisms have been noted up to 5 kilometers from oil and gas platforms. Moreover, fish caught in the vicinity of platforms—including cod, saithe, and haddock—have been found to contain elevated levels of hydrocarbons in their flesh (North Sea Task Force 1993b).

Filter feeders, like mussels, make good biological indicators of pollution levels, since they take in a great deal of seawater in order to "filter out" food. Research carried out in the Brent oilfield, belonging to the United Kingdom, showed that mussels had accumulated ten times more oil (from oil-based muds) in their tissues than background levels (normal, unpolluted areas) (North Sea Task Force 1993).

The other source of oil pollution to the North Sea is from routine shipping operations. During normal operations, a ship's engine generates oily wastes. Ships also fill empty fuel and cargo tanks with seawater to act as ballast or to flush out oily residues and chemicals. Often, these wastes are discharged directly into the sea, despite international conventions designed to curb the practice. Every year it has been estimated that nearly 40,000 seabirds perish along the Dutch coast because of such illegal discharges. So far, North Sea states have been unable to agree on how much oil from routine shipping operations is dumped into the sea every year, and estimates vary a great deal—from 1,000 to 60,000 metric tons a year. If the upper level is accurate, normal shipping is responsible for as much of the oil in the North Sea as the entire oil and gas industry.

Dumping

In July 1971, a coastal freighter, the *Stella Maris,* left Rotterdam harbor with 600 metric tons of highly toxic wastes from the Dutch HKZO vinyl chloride factory. The ship intended to sink its cargo near the Halten Bank fishing grounds, 60 nautical miles west of Norway, but the Scandinavians objected strenuously to the disposal of the toxic junk in their backyard. The ship then tried to dock in Iceland, the Faroe Islands, and the Hebrides but was refused permission. After a month of sailing around the North Atlantic and North Sea, it returned, still loaded with its dangerous cargo, to Rotterdam.

That episode precipitated the Oslo Dumping Convention of 1972, which is limited to the North Sea states. It, like the London Dumping Convention of the same year, limits the kinds of substances that can be dumped at sea, banning a wide range of industrial and radioactive wastes. However, a loophole in both conventions allows the disposal of sewage sludge, dredge spoils, and certain kinds of industrial wastes.

Throughout the 1970s and 1980s, all North Sea countries except Norway continued to dump millions of metric tons of sewage sludge and dredge spoils in the North Sea and North Atlantic. All of those types of waste contain substances harmful to marine life, including PCBs and DDT, heavy metals, chlorinated hydrocarbons, and organic and inorganic matter.

Dredge spoils consist of the muck scraped off the bottom of harbors, canals, and ship channels. In 1981 alone, Belgium and the Netherlands collectively dumped close to 100 million metric tons of dredge spoils in the near-shore waters of the North Sea. That same year, the United Kingdom dumped nearly 10 million metric tons of sewage sludge in the North Sea.

A decade later, in 1990, the situation had improved somewhat. That year the United Kingdom, the Netherlands, and Belgium together deposited 43.3 million metric tons of dredge spoils into the North Sea and North Atlantic. The same countries collectively dumped nearly 2 million metric tons of sewage sludge, followed by West Germany with 1.7 million metric tons, Denmark with 1.2 million metric tons, and France with 600,000 metric tons.

At the 1990 ministerial meeting, signatories to the London Dumping Convention (endorsed by sixty-five countries in 1996) agreed to phase out the dumping of *all* industrial wastes by 1995.

Surfers Against Sewage

The United Kingdom's beaches are among the dirtiest in Europe. Seaside resorts, like Brighton on Britain's south coast, get only a fraction of the beach traffic they attracted in the 1950s and early 1960s.

"We calculate that every day, Britain's coastal waters receive some 300 million gallons of sewage water, much of it untreated, or only partially treated," states British surfer Chris Hines. "Britain's beaches meet only 2 of the European Community's 19 criteria for safe bathing waters. Raw sewage going into our coastal waters from outlet pipes can contain coliform counts of up to 10 million per 100 milliliters of water. Once in the sea, we routinely measure coliform counts of over 10,000 per 100 milliliters in bathing waters in resorts like Blackpool and Torbay" (Surfers Against Sewage 1993).

Fed up with dirty water and the lack of response from government authorities, Hines decided to organize a citizen's action group to put pressure on public and private utilities, particularly the country's regional water companies. In the summer of 1990, Surfers Against Sewage was formed. The group now has 17,000 active members and is still growing. The group made its mark when Hines led a group of surfers into Parliament wearing wet suits and flippers in order to hand in a petition urging the government to clean up the country's coastal mess. The group attracted national media attention by taking dozens of journalists and TV crews to see the country's most degraded beaches.

Resources

The Wadden Sea is not so much a sea as a rich mosaic of coastal estuaries and tidal wetlands, stretching 500 kilometers from Esbjerg, Denmark, on the Jutland Peninsula to Den Helder, Holland. It is one of the most important and productive wetland ecosystems in Europe, covering 10,000 square kilometers of shallow nearshore waters. Hundreds of low islands are strewn close to shore, extending like giant stepping stones from Den Helder along the coast to Wilhelmshaven, Germany. Thousands of sand spits throughout the region make navigation, even by small sailboats, a hazardous undertaking.

The Wadden Sea contains the largest expanse of salt marshes in Europe. Its generous tidal flats and marshes provide breeding, feeding, and staging areas for some 12 million waterfowl and shorebirds, including all knots migrating to Africa from northern Siberia, nearly 90 percent of European oystercatchers and shelducks, and more than 50 percent of the East Atlantic flyway populations of grey plover, curlew, dunlin, and bar-tailed godwit. It is a critically important spawning, nursery, and feeding area for commercially important North Sea fish, such as herring, plaice, sole, and flounder, and shellfish, mainly shrimp, cockles, and mussels. In fact, some 80 percent of plaice, 50 percent of sole, and in some years a large part of North Sea herring reach maturity in the Wadden Sea. It is also an area of unsurpassed natural beauty.

Over the past five decades, nearly 50 percent of the estuaries, salt marshes, and unique dune ecosystems of the Wadden Sea have been either lost to agriculture, expanding towns and cities, and tourist resorts or degraded beyond recovery. The sea is plagued with increasing incidences of toxic algal blooms due to excessive nutrient pollution, most of it brought in by three river systems, the Elbe, the Weser, and the Ems, that drain a large portion of Germany and the Netherlands. Thousands of hectares of its eelgrass beds have been destroyed or degraded by cockle fishing and the dumping of dredging spoils and harbor sludge. It is chronically overfished, with shellfish stocks (cockles and mussels) reaching record low numbers. What's more, increasing loads of chemical pollution have reduced species numbers and diversity and decimated marine mammal populations, especially seals (World Wide Fund for Nature 1991).

In 1981, the Netherlands awarded most of its Wadden Sea territory protected status, followed by Denmark in 1982. Germany's three Wadden Sea states—Schleswig-Holstein, Niedersachsen, and Hamburg—created national parks in the area in the period 1985–90. Most important, according to the 1982 Joint Declaration on the Protection of the Wadden Sea, these three countries agreed to cooperate in preserving what was left of it.

However, since 1982, when joint protection efforts started, the Wadden Sea has witnessed the following:

- nearly 7,000 hectares have been embanked and drained;
- an oil drilling platform was erected in the German Wadden Sea, near a national park;
- fishermen removed nearly all mussels in the Danish Wadden Sea;
- the last undisturbed seagrass beds in the Dutch Wadden Sea were virtually wiped out by cockle fishing in 1990; and

- fishing for mussel seeds in the Dutch part of the sea was rendered impossible in 1991 because of insufficient reproduction and overfishing in previous years (World Wide Fund for Nature 1991).

According to a white paper produced by a team of experts working under the auspices of the World Wide Fund for Nature (WWF), the three Wadden Sea countries need to evolve a management strategy that is more ecosystem based, rather than end-use oriented. With backing from more than fifty conservation NGOs, the authors of the report, called *The Common Future of the Wadden Sea,* recommended that, above all, states sharing the sea should develop a Wadden Sea Charter in an effort to harmonize conservation policies. The ultimate goal, according to the WWF white paper, is "to achieve a complete, natural and sustainable ecosystem in which natural processes proceed in an undisturbed way" (World Wide Fund for Nature 1991).

Fisheries

In the plenty years of the 1960s and 1970s, North Sea fishing fleets collectively harvested more than 3 million metric tons of fish a year. By 1990, the total take was down to 2.4 million metric tons. All of the main North Sea commercial fisheries—herring, mackerel, haddock, cod, saithe, plaice, and sole—have registered reduced catches since the 1970s. Despite repeated warnings by scientists that North Sea fleets were consistently overharvesting commercial stocks, heavy fishing pressures continued throughout the 1980s. By the end of the decade, mackerel stocks were so low that some marine biologists were predicting mackerel's virtual extinction in the North Sea, "leaving only visitors from the Atlantic in northern waters." In 1991, both cod and haddock stocks collapsed as spawning biomass reached record low numbers. By 1992, saithe stocks were also considered below renewable biological limits. The amount of sole landed fell by two-thirds between 1960 and 1990: from 150,000 metric tons to under 50,000 metric tons. Of all the major fisheries in the North Sea, only herring has recovered significantly; however, sole is expected to recover in the near future.

Industrial fisheries didn't fare well either. The total catch—consisting mostly of sprat and Norway pout—dropped from 1.8 million metric tons in 1974 to 1 million metric tons in 1985. By the end of the 1980s, both stocks had reached rock bottom. In his book *The North Sea,* Dr. Malcolm MacGarvin reported, "There are so few sprat left that it is no longer possible to carry out an accurate survey of their numbers" (MacGarvin 1990)

Industrial fishers quickly switched to sandeels. Between 1987 and 1991 the sandeel harvest averaged about 1.3 million metric tons a year, accounting for fully two-thirds of total industrial fish catches. Throughout the North Sea food chain, however, many species of seabirds depend on bountiful sandeel populations to raise their young. With far fewer sandeels available, the populations of puffins, arctic terns, and kittiwakes, once found in huge colonies along the rocky shores of Scotland and Norway, have been drastically reduced. Monitored kittiwake colonies, for example, were down to one fledgling for every ten nests in 1988.

One beneficiary of industrial fishing is the mariculture industry in Norway and Scotland, which raises tons of farmed salmon. In 1989, more than seven hundred fish farms, most of them in the sheltered fjords of Norway, produced around 80,000 metric tons of salmon. In the Wadden Sea and the Dutch Delta, mussels and oysters are cultured.

The argument that farming fish will take pressure off wild stocks has not proved true in the North Sea. "The greatest irony is that salmon farms are dependent on industrial fish for food," writes Dr. MacGarvin. "Every metric ton of salmon produced requires 3.3 metric tons of sandeels or other industrial species to be processed into pellets. In other words, the 1988 Norwegian crop of salmon required one quarter of a million metric tons of fish to be taken from the sea" (MacGarvin 1990).

Management: A New "Governance"?

The seven nations that border on the North Sea (plus Switzerland) banded together in the early 1970s in an effort to reverse its deterioration. In addition to the London Dumping Convention of 1972, which covered all the world's seas (at least in principle), the Oslo Convention of the same year was concerned with reducing the dumping of harmful and toxic substances in the North Sea. Two years later, in 1974, the Paris Convention attempted to regulate pollution from land-based sources. Those early attempts, though laudable, were flawed. Both the Oslo and Paris conventions assumed the sea had an assimilative capacity to absorb wastes. Both assumed pollutants were innocent until proven guilty. Hence, regulators concentrated instead on what they considered to be the most "persistent, toxic and bioaccumulative" substances, such as DDT and PCBs. Even with those highly toxic compounds, however, researchers had trouble linking unhealthy and diseased organisms with a particular chemical culprit.

With the health of the sea continuing to deteriorate, the North Sea states launched a series of North Sea Conferences, beginning in 1984. It wasn't until the second conference, held in London in 1987, that some genuine initiatives were taken to tackle the sea's swelling tide of pollution. The eight nations agreed to adopt the "precautionary principle" as their guiding philosophy. This was considered a real about-face from previous marching orders. Now a pollutant was considered potentially harmful unless there was evidence to the contrary. The nations also agreed to reduce discharges of twenty-four polluting substances, including dioxins, heavy metals, and nutrients (nitrogen and phosphorus), from land-based sources to 50 percent of their 1985 levels by 1995; end the dumping of industrial wastes by 1989; phase out ocean incineration by the end of 1994; and establish an interagency scientific task force to harmonize research efforts, focusing on the causes and effects of pollution on North Sea ecosystems and their communities of plants and animals.

The 1990 meeting, held at The Hague (Netherlands), went even further. The North Sea states agreed to cut atmospheric pollution by 50 percent (of 1985 levels) by 1995. At the same time, they decided to expand the list of twenty-four polluting substances to thirty-six, and singled out mercury, cadmium, lead, and the chlorinated dioxins for 70 percent reductions by 1995. The deadline to end ocean incineration was moved up two years, to 1992.

Implementing some of those goals, however, has proved more complicated than expected. Reducing toxic pollution from point sources has proceeded, but coming to grips with nonpoint sources of pollution, particularly the hundreds of thousands of tons of nutrient pollution from agriculture, has proved impossible. The Netherlands has discovered, to its embarrassment, that it simply cannot regulate nitrogen inputs from diverse agricultural sources unless it revamps its entire agricultural base.

Conclusions

- Neither the Helsinki Convention nor the Oslo Convention has been fully implemented despite decades of work. The former socialist states of eastern Europe and the Soviet Union need to commit personnel and money to reversing five decades of environmental abuse and neglect. By the same token, the North Sea's states need to adhere to pollution abatement deadlines and agreements made in the first half of the 1990s to cut discharges of untreated industrial and municipal wastes into coastal waters. They also need to tackle the contentious issue of nonpoint-source pollution from agricultural runoff.

- The commercial fisheries in both seas are in critical need of rational management plans. There are simply too many fishermen chasing too few fish. Whatever solutions are attempted, they must have the backing of the fishing industry and above all be *enforceable*.

- Coastal wetland preservation should be a top priority for both seas. It is time to launch a pan-European initiative to conserve and manage remaining wetlands under the supervision of the EC's environment directorate.

- The process established to oversee the cleanup of the Rhine River could be adopted by other European regions struggling with impaired surface water quality. As two of the major sources of pollution to the Baltic, both the Vistula and the Oder rivers would be prime candidates for such an approach.

- Europe's main environment and resource NGOs might want to consider combining their efforts. Such a pan-European initiative could be launched under the joint leadership of Greenpeace, the World Conservation Union, and the national offices of the World Wide Fund for Nature. Creating long-term political support requires a long-term advocacy strategy that engages parliamentarians and international civil servants in an evolving process. Public information campaigns need to be designed so that national grassroots constituencies can be created to press for change from the bottom up.

CHAPTER 5

The Black Sea

Facing Collapse

The Black Sea has been settled for millennia. Remains of prehistoric settlements dating back at least 60,000 years have been unearthed around its shores. For most of its long history, it was a healthy ecosystem supporting great quantities of fish, including sturgeon, along with a diverse assortment of plant and animal life. Over the past thirty years, however, the sea has been heavily polluted.

It's not difficult to see how this happened. Although the Black Sea is surrounded by only six countries, all with battered economies—Bulgaria, Romania, Ukraine, Russia, Georgia, and Turkey—its vast drainage basin covers some 2.1 million square kilometers of intensively farmed and highly industrialized portions of central and eastern Europe. More than 162 million people live in that region, which embraces virtually all of Austria, Hungary, the former Yugoslavia, Bulgaria, Romania, Moldova, and Ukraine, along with significant chunks of southern Germany, the Slovak Republic, Belarus, Russia, Georgia, and Turkey (with minor inputs from Switzerland and Albania). The watershed of the Danube, the largest river system in Europe, contains 82 million people (compared to 42 million for the Rhine). Another 68 million live in the Ukrainian and Russian parts of the Black Sea, while the Turkish portion contains 12 million inhabitants, about 20 percent of the country's total population in 1996 (Mee 1992) (see table 5.1).

The sea itself is another victim of unfortunate geography. It is relatively small, nearly landlocked, and covers 420,000 square kilometers, slightly more than the Baltic. Its only connection to the world ocean is through the narrow Bosphorous Straits, which links it tenuously to the Mediterranean. With practically no turnover of oxygenated water from the Mediterranean, the Black Sea's great depth—averaging nearly 1,300 meters—works against it. Nearly 90 percent of its 537,000 cubic kilometers of water is anoxic. Virtually all of its water mass below 150–200 meters is devoid of oxygen. Only 73,000 cubic kilometers of near-surface water contain enough oxygen to sustain life higher than microorganisms and bacteria (Mee 1992).

To make matters worse, the Black Sea has an extensive, relatively shallow northwestern continental shelf less than 200 meters deep. Three of Europe's major rivers drain into this continental shelf: the Danube, the Dniester, and the Dnieper.

Table 5.1. Length of Coastline and Population Data for Six Black Sea States

COUNTRY	COASTLINE (KM)	POPULATION 1997 (MILLIONS)	POPULATION GROWTH (%/YR)	% URBAN
Bulgaria	354	8.3	−0.5	68
Georgia	310	5.4	0.3	56
Romania	245	22.5	−0.2	55
Russia	460	147.3	−0.5	73
Turkey	1,600	63.7	1.6	63
Ukraine	2,782	50.7	−0.6	68

Sources: World Resources 1996–97, Oxford University Press, New York, 1996, pp. 268–69; *World Population Data Sheet, 1997*, Population Reference Bureau, Washington, D.C., 1997.

Notes: The figure for Turkey's coastline includes only the length of its Black Sea coast. The figure for Russia is an estimate of its Black Sea coast only. The population figure for Russia is the total population of the entire Russian Federation; roughly 25 million live in Russia's Black Sea region.

With their waters—some 266 cubic kilometers a year—come enormous loads of sediments and pollution. The Danube alone accounts for 203 cubic kilometers, more than the entire freshwater flow into the North Sea on an annual basis.

A number of Europe's major cities are located in the Black Sea's drainage basin, including: Munich, Vienna, Budapest, Belgrade, Sofia, Bucharest, Minsk, Kiev, and Odessa, with Istanbul and Ankara in Turkey. Most population centers on the sea itself are squeezed along its highly developed western and northern coasts, from Burgas and Varna in Bulgaria, through Constanta, Romania, to Odessa, Ukraine, and Rostov, and Krasnodar in Russia (see figure 5.1).

The sea has been overexploited for decades. Its commercial fisheries have collapsed from a combination of poor management and pollution. It is plagued by intensive and largely unregulated shipping activities. On top of those problems, it is the ultimate receptacle for the industrial and municipal wastes of over one-third of Europe, much of it untreated or only partially treated. Although the Black Sea's huge anoxic water layers are mostly a result of natural processes, pollution loads have aggravated the problem in the western part of the sea.

Given the facts that most of the Black Sea is already biologically dead and that the remainder is threatened by excessive nutrient and toxic pollution, scientists are increasingly concerned that without effective management the sea may eventually die completely. "Clearly, the threat to the Black Sea from land-based sources of pollution is potentially greater than in any other marine sea on our planet," says Dr. Laurence Mee, coordinator of the Black Sea Environmental Programme, based in Istanbul (Mee 1992).

Pollution

Over the past thirty years the deep waters of the Black Sea have become oxygen-starved deserts, riddled with pockets of hydrogen sulfide gas. Fishermen often report seeing sheets of flame on open waters, evidence that combustible sulfides have escaped to the surface and been accidentally ignited.

The sea's northwestern continental shelf is critically polluted with nutrients and chemical wastes; much of the shelf has become eutrophic. In the forty-year period

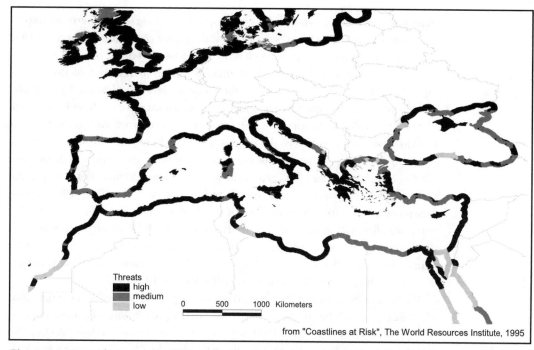

Figure 5.1. Coastal Ecosystems Threatened by Development—Black and Mediterranean Seas.
Source: D. Bryant, E. Rodenburg, T. Cox, and D. Nielsen, "Coastlines at Risk: An Index of Potential Development-Related Threats to Coastal Ecosystems," World Resources Institute, Washington, D.C., 1995.

from 1950 to 1990, the phosphorus load transported by the Danube alone increased from 13,000 metric tons to 60,000 metric tons, while the nitrogen load soared from 140,000 metric tons to over 340,000 metric tons, most of it the result of intensified agricultural development and the widespread use of phosphate detergents (Mee 1992). Over the same period, the amount of organic matter discharged from the Danube into the Black Sea increased five times, to 10 million metric tons. Since the 1950s there has been a threefold increase in nitrates and a sevenfold increase in phosphates in the Dniester River as well (Leppakoski and Mihnea 1996).

Liquid waste discharges to the sea from municipalities and industries have increased tremendously over the past three decades, in line with the growth of towns and cities in the region and an expanded industrial base. In 1992, Romania dumped some 170 million tons of liquid waste into the sea, mostly untreated or partially treated sewage and industrial effluents. That same year, Turkey was responsible for 500 million tons of domestic and industrial waste ending up in the Black Sea, including 200,000 tons of nutrient pollution as measured by biological oxygen demand (BOD).

The Danube, Dniester, and Dnieper rivers bring in massive amounts of pollution in their sediment-laden waters. In 1990, those three river systems dumped approximately 69 million tons of dissolved salts into the Black Sea, along with 130 million tons of suspended solids and 40,000 tons of detergents (Hinrichsen 1991, 1994).

As a result of the glut of nutrients entering the sea, principally from the western and northern continental shelves, algal blooms have reduced the amount of sunlight penetrating the water column. This has resulted in the mass mortality of shallow-water macrophytes such as phyllophora (red algae), formerly an important component of Black Sea ecosystems and a major economic resource. In 1992 only one 500-square-kilometer patch remained, 5 percent of the original area.

Widespread eutrophication has dramatically altered the base of the marine food chain in the Black Sea. In addition to dense blooms of nonoplankton and algae, explosions of a species of zooplankton (*Noctiluca miliaris*) have been noted. By the mid-1980s *Noctiluca* made up 95 percent of the entire zooplankton biomass (Balkas 1990). Unfortunately, the species has a low food value for higher species.

When those organisms die and fall to the shelf floor, they begin to decay, consuming huge quantities of oxygen in the process. Consequently, about 95 percent of the Ukrainian coastline and the entire Sea of Azov (shared by Ukraine and Russia) now suffer from hypoxia (a critical lack of oxygen). That has resulted in the complete elimination of a large proportion of macrobenthic organisms and the death of formerly rich fisheries, particularly clams, oysters, and mussels. One anoxic event in 1991 wiped out 50 percent of the remaining benthic (bottom-dwelling) fish along the Romanian coast (BSEP 1994; FAO 1993).

While eutrophication has severely reduced the populations of many indigenous Black Sea species, it has triggered explosions in the populations of a number of exotic imports. The comb-jelly (*Mnemiopsis leidyi*), for example, a large jellyfish-like ctenophore native to the North Atlantic, arrived in the Black Sea around the mid-1980s, most likely having hitchhiked in the bilge water of a cargo ship. Within a few years, the population of this opportunistic invader swelled to immense proportions. Dr. Yuri Zaitsev, a marine biologist working at the Odessa branch of the Institute of Biology of Southern Seas, reports counting up to five hundred individuals of all sizes in every cubic meter of seawater in Odessa Bay in the late 1980s. In 1990, researchers estimated that no less than 700 million tons of the gelatinous pest were thriving in the sea, accounting for nearly 90 percent of the total biomass in some areas (Zaitsev 1993). Worse, this particular comb-jelly likes to dine on small crustaceans, mollusk larvae, and the eggs and larvae of many species of fish. A dramatic fall in anchovy and sprat stocks is thought to be due largely to predation by *Mnemiopsis* (Zaika 1993).

The Toxic Sea

In 1990, the combined load of heavy metal pollution from the Danube, Dniester, and Dnieper rivers amounted to around 1,800 tons of arsenic, 5,500 tons of lead, 7,500 tons of copper, 65 tons of mercury, 170 tons of cadmium, 17,000 tons of zinc, and 45,000 tons of iron.

Pollution from pesticide residues is also a growing concern in the region. In the early 1980s, measurements carried out in Romania's Danube Delta revealed DDT and lindane concentrations 1,000 times higher than those observed in Mediterranean coastal waters. In the early 1990s, levels of DDT in the Kerch Strait, the narrow passage between the Azov and Black seas, were found to be between eight and twenty nanograms per liter, two orders of magnitude higher than in the Mediterranean. Since water from the Dnieper River is used to irrigate extensive rice paddies, where insecticides and herbicides are known to be used in huge quantities, this river is thought to bring in tremendous amounts of pesticide residues (FAO 1993).

Pollution in the Sea of Azov

The Sea of Azov, squeezed into a pocket between the Crimean peninsula to the south and Russia and Ukraine in the north, is a distinct subunit of the Black Sea. Its only connection to the Black Sea is through the very narrow Kerch Strait. During the 1950s and early 1960s, this was a healthy body of brackish water supporting a wealth of marine life, including highly productive commercial fisheries. In 1960, for example, Azov Sea fishing fleets hauled in 140,000 tons of just three species: pike-perch, bream, and roach. By the mid-1980s, land-based pollution had turned the sea into a large, highly eutrophic lake. As oxygen levels fell, the annual take of pike-perch dropped from a high of 70,000 tons in 1965 to just 3,500 tons in 1985. Over the same period, bream plunged from 46,000 tons to 800 tons and roach from 5,600 tons to 1,000 tons (Griffin 1993).

During the late 1960s and 1970s, agricultural production, particularly of rice, expanded rapidly in the region. So did heavy industry. By 1989, there were over 1,300 point sources of pollution in the Rostov region alone, mostly along the Don River. That same year, Krasnodar pumped around 15 million cubic meters of untreated sewage water into the Kuban River, which quickly dumped it in the Sea of Azov.

The most serious polluter of the sea proved to be agricultural chemicals. Throughout the 1970s and 1980s, massive quantities of fertilizers and pesticides were dumped on rice paddies in a desperate effort to meet yield targets set by central planners in Moscow. In 1989, some 60,000 tons of pesticides were sprayed on rice paddies and other crops in the Rostov region; another 40,000 tons were used on the paddies around Krasnodar. The result is a legacy of heavily contaminated soils and surface waters. Up to 5 billion cubic meters of pesticide-contaminated water is discharged into rivers and streams in the Krasnodar region every year, two-thirds of it from rice paddies. Much of that pollution ends up in the sea's coastal waters. In Russia and Ukraine, the total amount of pesticide residues entering the Sea of Azov annually has been estimated at up to 100,000 tons, around 10,000 tons of which may consist of chloro-organic compounds such as hexa-chlorohexane and DDT.

Hydrocarbon pollution is also a problem. Around 5,000 tons is thought to enter the sea every year from rivers and streams, with another 500 tons dumped directly by coastal towns and cities. But the record for oil pollution is held by the shipping industry, accounting for some 7,000 tons a year. With no port facilities available to reclaim oil wastes, ships routinely empty their oily bilge slops and engine wastes into the sea.

Over the past three decades, heavy metals have entered the sea, mostly from land-based sources, in increasing quantities. From data gathered between 1986 and 1990, researchers have estimated that up to 395,000 tons of heavy metals could reside in the sea, a huge amount for such a small body of water (FAO 1993) (see table 5.2). The effects on marine organisms remain largely unstudied.

In the Danube Delta

The wastes of over 80 million people collect in Romania's Danube Delta, the end of Europe's longest pollution pipeline. Because the vast wilderness of reeds and wetland forest that once dominated the area has been whittled away over the years, the delta's capacity to filter those wastes has been greatly reduced (Hinrichsen 1994).

Table 5.2. Heavy Metal Content in the Water Column and Sediments of the Sea of Azov, 1986–90

HEAVY METALS	WATER COLUMN (TONS)	BOTTOM SEDIMENTS (TONS)
Copper	2,000–6,000	10,000–20,000
Zinc	6,000–15,000	10,000–40,000
Lead	400–1,000	2,500–6,000
Nickel	1,500–3,000	5,000–10,000
Chromium	400–600	24,000–33,000
Cadmium	35–140	—
Mercury	60–200	15–30
Manganese	2,000–10,000	100,000–250,000

Source: "Fisheries and Environment Studies in the Black Sea System," *General Fisheries Council for the Mediterranean,* Food and Agriculture Organization of the UN, Rome, January 1993, p. 23.

"There is no doubt that our biggest problem is pollution," confirms Dr. Grigore Baboianu, director of the Danube Delta Research and Design Institute in Tulcea, Romania. "But it has been made much worse by the draining of huge sections of the delta for agriculture and forestry—nearly 75,000 hectares of polders [reclaimed land] have been created out of prime marshland—and by the construction of dams and dikes along the Romanian floodplain leading into the delta."

Indeed, nearly a third of the original ecosystem has been converted to agriculture, fish ponds, managed forests, orchards, and settlements. Agriculture's gain has been a loss for water quality. As more pollution builds up in the delta's sediments, more of it reaches the coast. Every year the Danube discharges around 600,000 metric tons of nutrient pollution into the Black Sea. In all, the Danube is responsible for nearly 50 percent of the nutrient load to the Black Sea.

Though a direct causal link is difficult to prove, this pollution is doubtless one of the reasons for the devastation of commercial fisheries in the delta. "Most of our men now have to eke out a thin existence catching anchovies or sprat in small nets," complains an elderly women in Sfintu Gheorghe, a small fishing village that clings to a sandy spike of land where the Danube meets the sea. "We used to catch sturgeon and were able to save some money, but no longer. Now we live from hand to mouth," she says in heavily accented Russian. Official statistics bear her out. Catches of pike-perch, sturgeon, even carp have dropped off since 1970. Landings of sturgeon fell from 191 metric tons in 1971 to 19 metric tons in 1989, scuttling an industry that once employed thousands.

Resources

The combination of pollution and introduced species continues to take a tremendous toll on Black Sea marine life. In the coastal waters and estuaries along the western and northern coasts—where algal blooms often turn waters the color and consistency of pea soup—species diversity and numbers have been drastically reduced. Entire groups of organisms have been eliminated. Copepods, tiny crustaceans important as food for larger fish, have all but disappeared, along with a number of gastropods (snails), such as *Patella tarentina* and *Melarphe neritoides*. Clams, oysters, and mussels have been reduced to remnant populations. Hermit crab populations have decreased by an order of magnitude. And brown algae (*Cystoseira barbata*), a species known to be sensitive to detergents, has disappeared entirely.

Populations of cetaceans, particularly dolphins and porpoises, have plummeted over the past forty years. In 1950, around 1 million common and bottlenosed dolphins thrived in the Black Sea; by 1995, not enough could be found to do a proper population survey.

"These examples of local changes in the number and morphology of coastal marine communities can be considered direct evidence of the effects of toxic substances on Black Sea inhabitants," observes Yuri Zaitsev. "Regrettably, it is impossible to assign changes in Black Sea biomass to any particular polluting substance" (Zaitsev 1993).

Fisheries

The UN Food and Agriculture Organization reports that between 1982 and 1992, the Black Sea's total commercial fish catch plummeted from close to 1 million metric tons to 100,000 metric tons, a tenfold drop in a decade. Over the thirty-year period 1960–90, the number of commercially valuable species plunged from twenty-six to just five (FAO 1993).

In the 1950s and 1960s, most of the Black Sea catch was composed of larger species such as Atlantic bonito, mackerel, bluefish, three species of grey mullets, turbot, red mullet, pike-perch, and bream. Only around 35 percent of the catch consisted of smaller forage fish such as anchovy and sprat. By the mid-1980s, nearly all of the larger fish had disappeared, victims of pollution, overfishing, and the population explosion of exotic species such as *Mnemiopsis* that feed on eggs and fry. Nearly 80 percent of the total take of fish consisted of anchovies, sprats, whitings, and the Black Sea horse mackerel.

By 1992 even those stocks were severely depleted. The Turkish share of the anchovy fishery fell to 15 percent of its 1985 level, from 300,000 metric tons to 66,000 metric tons. On the northwest shelf the anchovy catch declined tenfold, and after 1989 it ceased altogether in the Sea of Azov. By 1991, Romanian catches of sprat declined by nine-tenths compared to the early 1980s (Mee 1992).

All bottom-dwelling species living in near-shore waters had declined precipitously by 1994, including rays, sole, turbot, flounder, red mullet, weever, stargazer, and sturgeon fry. On the northwest shelf, landings of sturgeon fell from 300 tons in 1970 to 65 tons by 1980. Now, virtually no sturgeon are taken.

At the beginning of 1995, commercial fishing operations in the Black Sea and the Sea of Azov more or less ceased, idling around 150,000 fishermen and inflicting direct economic losses of more than $200 million (UNDP, UNEP, and the World Bank 1993).

"The Black Sea is the first sea to have suffered a collapse of almost its entire commercially exploitable fish fauna and creates an unprecedented situation that must be tackled cooperatively by all coastal states," concludes an FAO document published in 1993 (FAO 1993).

Management: Can the Sea Be Saved?

The Black Sea states finally banded together in 1985 in an attempt to negotiate a legal convention to protect the sea, modeled along the lines of the Barcelona Convention, forerunner of the Mediterranean Action Plan. All six Black Sea states signed the Convention on the Protection of the Black Sea Against Pollution, known as the Bucharest Convention, in April 1992. The convention took seven years to negotiate, under the auspices of the United Nations Environment Programme's (UNEP's) Oceans and Coastal Areas Programme Activity Center (OCA/PAC), based in Nairobi. The six states immediately requested help from UNEP in designing an action plan to implement the convention. At the same time, they also made a formal request to the Global Environment Facility (GEF)—the green fund jointly managed by the World Bank, the United Nations Development Programme, and UNEP—seeking funding for an initial three-year program, Environmental Management and Protection of the Black Sea.

The World Bank responded in 1992 with an initial GEF grant of $9.3 million, intended as seed money to mobilize funding from other sources in an effort to strengthen support for the convention. This initial grant did earmark money for data collection and analysis on sources and biological effects of pollution. Information was gathered on nutrient loads, waste stream discharges, inflows of freshwater from feeder rivers, coastal marine pollution from industries and municipalities, offshore dumping of wastes, and the extent of overfishing.

The action part of the GEF grant began in 1993. It allocated funds for institution building on both the national and the regional levels. The World Bank also identified thirteen priority investment projects in the region involving industrial pollution control, treatment of municipal wastes, the construction of port reception facilities for handling wastes, and a survey of biodiversity and conservation needs.

The World Bank managed to broaden the number of donors contributing to the rescue program, adding the European Community, the European Bank for Reconstruction and Development, and several bilateral funders. The funds committed so far, however, some $30 million, are a fraction of what is needed to save the sea (Sorensen et al. 1997).

Managing the Danube River Basin

Until recently, a whole watershed management effort was absent from the package of activities aimed at resuscitating the Black Sea. However, "1992 signaled a breakthrough," says David Rodda, coordinator of the Environmental Programme for the Danube River Basin. With its Secretariat based in Vienna, Austria, the programme embraces all eleven countries of the Danube's vast watershed (excluding only Serbia, under censure from the United Nations) and is supported by a multitude of international funders, including the World Bank (through the GEF), the European Community, USAID, and the governments of Austria and the Netherlands. Collectively, they have committed $35 million for an initial four years. Most of the funds will go to produce baseline studies, strengthen institutional capacities for environmental management, and prepare an action plan that focuses on pollution hot spots (Hinrichsen 1994).

With the exception of Germany and Austria, however, most of the states in the region lack the institutional capacity, personnel, and funds for a long-term cleanup program. Some development experts foresee a system of "pollution assessments" levied proportionate to each nation's contribution of contaminants, but success may hinge on whether the Danube basin countries can reverse the recent deterioration of their economies.

On the plus side, less industrial activity means less pollution being dumped into the Danube and its three hundred tributaries. In July 1994 the watershed states endorsed the Convention on Cooperation and Sustainable Use of the River Danube. Despite its impressive name, the convention requires merely that the signatories share their knowledge of water management and conservation. By contrast, the 120-page draft action plan calls for measures to protect water resources and major ecosystems, identification of short- and long-term problems, selection of priorities for action, and the setting up of financial mechanisms to pay for the program.

Despite its obvious limitations, "the convention is a start," insists Rodda, "a very important platform for broadening the terms of cooperation later on." He foresees a cleanup effort that could take two decades before it produces any tangible improvements in water quality (Hinrichsen 1994).

Conclusions

- The regional seas program, under UNEP's supervision, needs to be given top priority by the region's governments. Only a regional approach to the sea's many ailments is likely to yield lasting results.

- As discussed above, the Danube River Basin Management Program needs to be strengthened considerably and all watershed states involved in the cleanup. So far, only Germany and Austria, along with an impressive list of bilateral and multilateral funding agencies, have committed funds to salvaging the Danube.

- A workable fisheries management regime for the entire Black Sea needs to be worked out and implemented without delay. The FAO has been involved in previous attempts and might constructively contribute to such a process if all Black Sea states could agree on some basic ground rules.

- Major pollution sources to the Black Sea and its primary feeder rivers must be brought under control by any regional program. Unfortunately, this is one crucial area where there is little agreement on priorities. If the Black Sea states cannot make progress on this issue, it is unlikely that they can deal effectively with any of the other contentious issues facing the sea.

CHAPTER 6

The Mediterranean Sea

Caught Between Two Worlds

The Mediterranean Sea has been celebrated for centuries as the cradle of Western civilization. Along its sun-drenched shores flourished some of the world's greatest empires: Egyptian, Persian, Phoenician, Assyrian, Minoan, Macedonian, Greek, and Roman. The sea was a great highway that facilitated the spread of modern mercantile and political systems, new ideas and religions, to far corners of the earth. Homer's "wine dark sea" seemed limitless, unconquerable.

By the early 1970s it was clear that the Mediterranean Sea was in serious trouble. Its shores were crowded with largely unregulated tourist development, its cities and towns bursting with people and pollution. Increasingly, its beaches were littered with debris and tar balls, its near-shore waters contaminated with untreated industrial and municipal wastes.

Geography

The Mediterranean is Europe's largest sea, covering just under 3 million square kilometers and holding close to 4 million cubic kilometers of saltwater. It is divided almost in half, east from west, by a submarine ridge that runs between the south coast of Sicily and Tunisia. These two major basins are subdivided into smaller, regional seas, extending from the Alboran Sea near Gibraltar to the Levantine Sea off the coast of Lebanon and Israel. The Mediterranean is also deep, averaging 1,500 meters. In the Ionian Sea, between Greece and the heel of Italy, it is over 5,000 meters deep.

Like a huge, clogged bathtub that drains very slowly, the Mediterranean is bottled up behind the narrow Strait of Gibraltar, its only outlet to the world ocean. Since the strait is only 14 kilometers wide and 320 meters deep, the Med's hydrological pump works so slowly that most of the sea's waters are never renewed. Since it is a largely enclosed sea, it lacks tides. Hence, there are no tidal currents to flush away pollution. It also lacks extensive continental shelves—shallow coastal waters can suddenly drop to depths of 1,000 meters or more—so it is not blessed with vast submarine meadows capable of sustaining a wealth of marine life. It is too far north to support biologically diverse mangrove swamps or coral reefs. And there are no upwellings of deep, nutrient-rich waters. The sea is nutrient starved and relatively fish poor.

The lands around the Mediterranean tend to be as barren as most of the sea. Low rainfall in many areas, particularly on the south rim, means that thirsty crops must be constantly irrigated, placing emormous strains on limited supplies of freshwater. Except for oil and gas reserves in Egypt, Libya, and Algeria (with lesser amounts in Tunisia and Syria), the region is poor in mineral and energy resources. In fact, in the bone-dry eastern and southern parts of the Mediterranean Basin, water is often more precious than oil.

Population

The twenty-one countries that now surround the Mediterranean have a collective population of over 400 million people (see table 6.1). Nearly 146 million of them, along with 150 million sun-seeking tourists, are crammed along the sea's 45,000 kilometers of narrow coastal plain, much of which is hemmed in by mountains and hills. Demographic projections provided by the Blue Plan (see page 15) estimate that the Mediterranean Basin's resident population could reach as high as 555 million in 2025 (depending on when population growth rates begin to level off along the southern and eastern rims) (Batisse 1990, 1994).

The Med's urban coastal population, which hit 100 million in 1995, is growing more rapidly than the rural one. According to Blue Plan projections, the urban population of coastal Mediterranean administrative regions is expected to be at least 128 million by 2025, though it could go as high as 176 million, with up to 350 million seasonal tourists crowding its shores. The northern rim's urban population is expected to increase significantly—from 52 million in 1985 to as many as 76 million by 2025. Over the same period, the southern rim's increase will be enormous—from 30 million to upwards of 101 million.

In 1990, nearly 60 percent of the Mediterranean's population lived in urban areas, a figure that is expected to increase to 76 percent in 2025. In 1995 there were 540 towns and cities with populations of 10,000 or more squeezed around the sea's serpentine shore.

The urbanization of the Mediterranean has left entire coastlines purged of their natural habitats. As figure 6.1 illustrates, close to two-thirds of the Med's coasts are under severe stress from development. For nearly one-third of the year the population of the region doubles. The result has been widespread devastation of coastal resources, particularly wetlands, forests, and the Mediterranean maquis (a shrub-dominated ecosystem between a grassland and a forest). Croplands have also been lost to urban expansion and the unplanned development of resort hotels and other tourist accommodations. In Spain, France, Italy, and parts of Greece there are practically no undeveloped coastlines. Only a small fraction of the 580-kilometer-long Catalonian coast of Spain has been spared from urban, port, industrial, or tourist development. Over 75 percent of Italy's Romagna coast is developed. Twenty years ago the south coast of Attica between Athens and Cape Sounion was covered with lush olive groves. Today it is dominated by interconnected tourist complexes and summer villas.

Coastal population densities are already high, averaging 250 to 300 people per square kilometer along much of the coast from Spain to Israel. Three-quarters of Syria's population is squeezed along its Mediterranean coast and around the

Table 6.1. Demographic Data for Twenty-One Mediterranean Countries

COUNTRY	POPULATION 1997 (MILLIONS)	POPULATION GROWTH (%/YR)	POPULATION 2025 (MILLIONS)	% URBAN
Spain	39.3	0.1	39.0	64
France	58.6	0.3	62.7	74
Monaco	0.027	—	0.036	100
Italy	57.4	−0.0	54.8	67
Yugoslavia	10.6	0.2	11.4	51
Slovenia	2.0	0.0	2.0	50
Bosnia-Herz.	3.6	0.6	3.9	34
Croatia	4.8	−0.0	4.2	54
Albania	3.4	1.7	4.6	37
Greece	10.5	0.0	10.2	72
Turkey	63.7	1.6	89.9	63
Syria	15.0	2.8	26.3	51
Cyprus	0.7	0.8	1.0	53
Lebanon	3.9	2.2	6.1	87
Israel	5.8	1.5	8.0	90
Egypt	64.8	2.1	97.6	44
Libya	5.6	3.6	14.4	85
Tunisia	9.3	1.9	13.5	58
Algeria	29.8	2.4	47.7	50
Morocco	28.2	2.0	39.9	51
Malta	0.4	0.5	0.4	89

Source: Based on data from *World Population Data Sheet, 1997,* Population Reference Bureau, Washington, D.C., 1997.

Note: Serbia and Montenegro formed a new state in 1992, the Federal Republic of Yugoslavia. Population projections for 2025 based on UN medium-term growth rates, as recalculated in 1996.

The
Mediterranean
Sea

capital, Damascus. In Egypt's Nile Delta there are over 1,000 people per square kilometer. Projections indicate that the south rim's cities will continue to draw people out of the countryside, becoming even more crowded than they are now.

"This littoralization [of the population] must be considered in any assessment of the future," says Michel Batisse, president of the Mediterranean Blue Plan. "In the Blue Plan scenarios, for instance, calculations have been attempted to determine the implications of this population increase on domestic water demand, waste disposal into the sea, and coastal land use. In many ways, littoralization is likely to dominate the future of the whole basin and the different economic sectors."

Take Istanbul as an example. Turkey's largest city has doubled its size about every fifteen years since the late 1950s, swollen by migrants from the countryside. Today, nearly two-thirds of Istanbul's 10 million residents come from the poverty-stricken regions of eastern Anatolia and the Black Sea coast.

When the modern Turkish Republic was founded in 1923, 15 percent of its 13 million people lived in cities. Today, over 60 percent of the country's 62 million

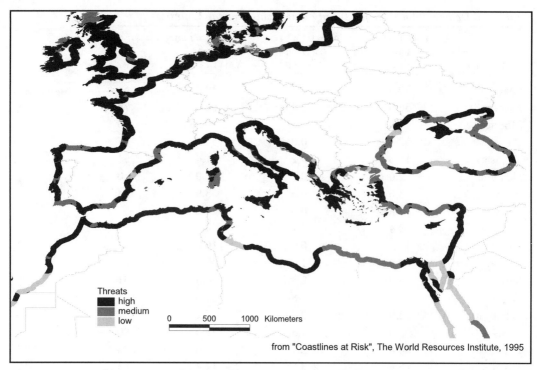

Figure 6.1. Coastal Ecosystems Threatened by Development—Mediterranean and Black Seas.
Source: D. Bryant, E. Rodenburg, T. Cox, and D. Nielsen, "Coastlines at Risk: An Index of Potential Development-Related Threats to Coastal Ecosystems," World Resources Institute, Washington, D.C., 1995.

inhabitants are urban. Ankara, the capital, grew by 1 million residents between 1980 and 1994: from 2.5 million to 3.5 million. It is a trend that shows no signs of slackening.

The next century will see the aging of the north rim's population; in some states, such as Italy, the population will actually decline as there will be more deaths than births every year. The developing south rim, however, will be dominated by a youthful population, most under the age of thirty. The huge increase in the southern rim's population—from about 224 million in 1996 to 351 million by 2025—will create a population divide in addition to the religious divide that already separates Muslim south from Christian north (see table 6.1).

Pollution

By 1970 the Mediterranean had acquired the reputation of one of the most polluted seas on the planet. The famous marine explorer Jacques Cousteau described it as a "garbage dump." Despite nearly twenty years of cleanup efforts, it is still polluted with millions of tons of municipal and industrial effluents, along with agricultural chemicals and solid wastes, most of it brought in by the region's seventy major river systems. In addition, only around 30 percent of the sewage from over seven hundred coastal towns and cities is treated before being discharged directly into coastal waters. Since many sewage outfalls don't extend far enough from land, these wastes can linger in near-shore waters creating local health hazards (Jeftic 1990).

In the late 1970s, researchers estimated the total amount of wastewater (from industries and municipalities) discharged into the Mediterranean from its watershed, via rivers and streams, as some 420 billion cubic meters a year. By 1990, those loads had been reduced but not significantly. Between 30 and 50 million metric tons of untreated or partially treated sewage is still pumped into the Med's coastal waters every year. This load, along with diverse agricultural inputs, translates into 1.5 million metric tons of nutrient pollution as measured by BOD, including slightly over 1 million metric tons of nitrogen and 355,000 metric tons of phosphorus pollution. (Pearce 1995; Jeftic 1990).

In 1983, marine scientists in the Gulf of Trieste (northern Adriatic) watched as over 90 percent of bottom-dwelling fish, shellfish, and invertebrates perished in just one week from a lack of oxygen. Similar events have been reported in the Sea of Marmara, south of Istanbul, and in the Saronikos Gulf, off Athens.

Heavy nutrient pollution in the northern Adriatic has generated algal blooms nearly every year during the decade of the 1980s and early 1990s. Some algal blooms in the Gulf of Trieste have reached densities of 100 million cells per liter of water. In the Venice Lagoon, algal growth has reached upwards of 1 million metric tons a year, coating the entire surface with a thick soup of algae. Nearby bathing beaches have been stained brown with algal scum, which often resembles the suds left in a tub by dirty bath water (Pastor 1991).

Toxic pollution continues, despite attempts to curb it. Heavy metal inputs, in particular, have not decreased significantly since the 1970s. In 1990, nearly 130,000 metric tons of heavy metals, including mercury, lead, chromium, zinc, copper, nickel, and cadmium, were finding their way into the sea via rivers and streams, though nearly a third came from atmospheric deposition. In addition, up to 570 metric tons of organochlorine compounds, including DDT and PCBs, continue to enter coastal waters.

Organochlorines and heavy metals are persistent poisons that enter food chains and accumulate in bottom sediments. Once in coastal ecosystems, they can create long-term hazards for marine life, along with severe health problems for people who dine regularly on seafood. Filter-feeders like clams, oysters, and mussels often have high levels of heavy metals in their tissues. High concentrations of mercury—up to 47 ppm (parts per million)—have turned up in bottom sediments along the Gulf of Trieste. Not surprisingly, elevated mercury levels have been noted in bluefin tuna, striped mullet, and Norway lobsters. In the waters off Sousse, Tunisia, tuna and pilchard regularly contain such high levels of mercury that, if the law were to be enforced, they could not be sold. Mercury is so toxic that the World Health Organization considers levels over 1 ppm in seafood to be unfit for human consumption.

Nearly every one of the Mediterranean's seventy major cities (those having more than 100,000 people) creates local hot spots of pollution from the discharge of untreated sewage or industrial wastes. Chronically high pollution levels have been recorded for the entire northern Adriatic; Izmir Bay in Turkey; Elevsis Bay, Greece; the Lagoon of Tunis; Tunisia; and the near-shore waters of Alexandria, Egypt.

The waters around Alexandria, Egypt, used to be polluted with more than 1 million cubic meters of untreated (or partially treated) municipal and industrial effluents every day, along with garbage, algal scum, household slops, construction debris, and toxic industrial sludge. By the mid-1980s, bathers along the city's beaches were complaining of skin rashes, sores, and intestinal disorders. A joint U.S.–Egyptian project, called the Updated Alexandria Wastewater Master Plan, renovated two antiquated sewage treatment plants and built four new pumping stations in an effort to provide at least primary treatment for most of the city's sewage. The entire system began operating in the autumn of 1993. But one problem has proved intractable: Like Cairo, Egypt's second largest city is growing too fast for planners to keep up with demand for new infrastructure and services. A second phase has been approved by the U.S. Agency for International Development in an effort to carry out environmental assessments and increase the capacity of local institutions to make further improvements (USAID 1994).

Nearly one-fifth of all the Med's beaches remain periodically unsafe for bathing. Athenians are accustomed to driving 70 kilometers to find water clean enough to swim in. Many of the region's shellfish-growing areas do not produce seafood fit for the table. Typhoid and periodic outbreaks of cholera have been reported over the years. In the summer of 1973, for instance, a cholera epidemic struck Naples and southern Italy, with 325 reported cases and 25 deaths. The cause was eventually traced to mussels contaminated with raw sewage.

De-oiling the Mediterranean

The Mediterranean is one of the major shipping routes for Mideast oil and gas destined for Europe. Some 600 million metric tons of petroleum products—some 22 percent of the world's total—are shipped into and through the Med every year. Roughly half ends up at eighteen major port cities. In the mid-1970s, the sea was covered with so much oil that researchers trying to study marine life often pulled in nets filled with tar balls instead of fish. In 1977, during a research cruise in the eastern Mediterranean, scientists from the International Atomic Energy Agency's Marine Laboratory at Monaco made the following observation: "Between Crete and Libya a 30-minute neuston (surface) tow completely filled a one-liter collecting jar with tar balls."

In 1979, scientists estimated the amount of oil discharged into the sea every year at somewhere between 500,000 and 1 million metric tons, most of it from shipping operations. By 1990, the amount spilled into the sea was thought to be at least 635,000 metric tons. Of that amount, some 330,000 metric tons were accounted for by routine shipping operations discharging dirty ballast waters, bilge slops, and oily wastes. Another 270,000 metric tons came from land-based sources, mostly via rivers and streams, and 35,000 metric tons from atmospheric deposition. In all, the Med, which accounts for only 1 percent of the world's sea surface, receives nearly a fifth of all oil spilled or discharged into the world's seas (Pastor 1991; Pearce 1995; Buckley 1992).

Although a Regional Oil Combating Centre was set up on Malta to coordinate efforts at fighting oil pollution from accidents, that strategy left unsettled the matter of routine oil pollution from normal shipping activities. Action on that front got underway in the early 1980s, led by Greece, whose maritime tradition goes back three thousand years.

The Greek government decided to promote the construction of floating facilities for the reclamation of waste oils and other petroleum residues. Today, the country has ten such facilities moored at major ports like Piraeus, Patras, Siros, and Thessaloniki. Both the International Maritime Organization (IMO) and the European Community support the construction of floating facilities. Given the success of the Greek operations—in the late 1980s Greece was able to collect nearly 3 million metric tons of oily wastes a year, from which they extracted about 30,000 metric tons of recovered oil—more of them are expected to be built in ports around the Mediterranean.

The first oil reclamation facility in Greece was built by a young entrepreneur named Denis Yatras, using, appropriately, a retired oil tanker. His ship, the *Delta,* is usually anchored in Elevsis Bay between huge ocean-going tankers. Yatras explains how his system works: "In the late 1980s we were processing over 400,000 metric tons of tanker slops and bilge water and were able to get nearly 20,000 metric tons of recovered oil. We then deliver our oil, which still contains 3 percent water, to a local refinery, where it is reprocessed. In return for the oil, we receive 75 percent of the value of the oil in the form of fuel oil, which we market ourselves." This swap system seems to work well. So far, the *Delta* more than pays for itself.

Yatras is convinced that floating oil recovery systems are the wave of the future. "The point is not just recovering oil, but pollution prevention," he explains. According to the IMO, all major ports are required to have oil-waste handling facilities, but high start-up costs have prevented many from fulfilling this requirement.

Resources

Like marine mammals everywhere, those in the Mediterranean are under siege. Dolphin populations have fallen sharply in recent years, despite conservation efforts. So too have the numbers of seals and sea turtles. Marine mammals (and amphibians) are losing on two fronts: They are running out of breeding areas as habitats are destroyed or disturbed to make room for more people, and increasing exposure to toxic pollutants is weakening their immune systems.

The populations of all three species of dolphins found in the Mediterranean—common, bottlenosed, and striped—have dropped off dramatically in the past twenty years. Striped dolphins, in particular, are being killed off by a mysterious viral disease linked to elevated levels of organochlorines such as DDT and PCBs and to heavy metals. Since 1990, scientists estimate that as many as 6,000 striped dolphins have died. Autopsies performed on some of the victims have revealed extremely high levels of PCBs and mercury in their fat deposits and vital organs. PCB levels of 400–500 ppm have been found in blubber, along with mercury concentrations of more than 800 ppm in livers.

Perhaps the most critically endangered marine mammal is the Mediterranean monk seal (*Monachus monachus*). Its population is hovering around three hundred to four hundred individuals, and its range is now restricted to a few isolated Greek islands and the Aegean coast of Turkey, except for tiny remnant populations in Morocco and Algeria. The monk seal faces the dual threats of dwindling habitat and intense commercial fishing pressures. Fishermen view the seal as a competitor for limited stocks of fish and kill them whenever they can. Others are strangled and drowned when they become entangled in fishing nets.

Fisheries

The Mediterranean has always been a relatively fish-poor sea. The 800,000 metric tons captured, on average, every year, represent a mere 1 percent of the world's total take of fish and shellfish. As a result, some 90 percent of all fishing activity is labor intensive and small scale. Most fishing boats are no longer than 15–20 meters, with crews of two to four people. Net fishing is concentrated in shallower, near-shore waters and around the sea's numberless islands.

Still, over the past forty years a trawling industry has evolved. Concentrated around Sicily, these vessels can stay at sea for weeks harvesting great quantities of pélagic fish. The trouble is, over half of what they catch is tossed overboard as trash fish, since only certain species, like tuna, fetch the higher prices needed to pay over-head on large trawlers. As more fishers go after fewer fish, trawling pressures have mounted.

Trawling in shallow, coastal waters, along with construction of boat marinas and docking facilities, and sand excavation, has also precipitated the destruction of the Mediterranean's distinctive *Posidonia* (seagrass) meadows, on which many species of fish, shellfish, and mollusks depend for breeding and nursery areas.

Pollution has taken a toll on marine life, as well. Massive breeding failures have been reported for bonito tuna and mackerel in the Sea of Marmara, near Istanbul. Similarly, populations of grey mullet have plummeted in Abu Qir Bay, east of Alexandria. Most fish in the Gulf of Naples, Cagliari, and the Venice Lagoon are unfit for human consumption. And the Bay of Muggia at Trieste in northern Italy has been turned into a desert from overfishing and chronic pollution.

Because the Med's meager fish catches cannot keep up with demand, nearly two-third of all fish consumed in the region are imported, most commonly from the Atlantic.

Management: The Mediterranean Action Plan

In 1975, sixteen of the Mediterranean's eighteen states gathered in Barcelona, Spain, for a historic occasion: the launching of the Action Plan for the Protection of the Mediterranean Environment. This proved to be only the first in a series of "regional seas" action plans negotiated under the auspices of the United Nations Environment Programme.

The plan called for a broad, three-pronged approach to controlling pollution and managing the sea's ravaged resources. First, it initiated a series of legally binding treaties to be drawn up and signed by Mediterranean governments. Second, the plan created a regionwide pollution-monitoring network and provided for the coordination of major scientific research efforts in the sea. Third, it created a socioeconomic program that would attempt to reconcile vital development priorities with a healthy Mediterranean environment.

The following year, the same countries, plus the European Community and Syria, returned to Barcelona in February to sign the Barcelona Convention for the Protection of the Mediterranean Sea Against Pollution. The convention committed the Mediterranean states to "take all appropriate measures . . . to prevent, abate and combat pollution . . . and to protect the marine environment." It provided the legal muscle for the action plan that had been adopted the year before. Two significant protocols were also signed on February 18, 1976. The Protocol for the Prevention of Pollution of the Mediterranean Sea by Dumping from Ships and Aircraft black-listed dangerous substances such as mercury, cadmium, crude oil, chlorinated hydrocarbons, pesticides, and radioactive waste, prohibiting their disposal in the sea. A gray list of substances that could be dumped under special circumstances (by permit only) was attached to that protocol. The Protocol concerning Cooperation in Combating Pollution of the Mediterranean Sea by Oil and Other Harmful Substances in Cases of Emergency committed governments to cooperate in combating oil and chemical spills from ships and land-based facilities. Later that year, a regional oil-combating center was established in Malta.

In a number of countries, such as Greece and Italy, the protocol on oil pollution buttressed national legislation already in place; in others, particularly on the south rim, it initiated contingency plans. Greece, for example, already had an extensive set of laws in place to protect the country's marine environment from oil pollution. "Every port authority in Greece—some ninety in all—has plans and equipment for combating oil spills," says Captain D. Doumanis, former head of the Marine Environment Protection Division in the Ministry of Mercantile Marine. "This has helped enormously in lessening the coastal impacts of oil spills."

The first steps in regional cooperation involved the setting up of the Mediterranean Pollution Monitoring and Research Programme (MEDPOL). The first phase ran from 1976 to 1980 and involved scientists from eighty-three research centers and laboratories from sixteen Mediterranean countries. Initially, research concentrated on finding out just how polluted the sea really was by conducting numerous baseline studies. The action plan countries established standardized analytical procedures so that data gathered by one institution, in, say, Egypt or Tunisia, could be used and understood by colleagues in France or Italy.

Standardizing data collection methods facilitated the establishment of a scientific network and helped avoid the costly and unnecessary duplication of research efforts. In the mid-1980s, 102 research projects were being carried out by sixty-two research centers in sixteen countries. By 1995, virtually all of the action plan's original signatories had implemented national monitoring programs in cooperation with UNEP.

In 1979 a Blue Plan for the long-term management of the region's coastal areas was launched as part of the socioeconomic component of the action plan. It was intended to "take the long view," according to Michel Batisse, its chief architect, "integrating future development plans with environment protection measures. But above all, its main purpose is to promote sustainable development in the entire Mediterranean Basin" (Batisse 1994). (See chapter 1 for more details.)

A year later, in 1980, following lengthy negotiations, the seventeen members of the action plan signed the Protocol for the Protection of the Mediterranean Sea Against Pollution from Land-Based Sources. This landmark agreement identified measures to control coastal pollution from municipal sewage, industrial wastes, and agricultural chemicals. Copying the dumping protocol, it also included a list of black and gray substances. Each government was responsible for implementing the protocol in its own way and on its own time frame. With the costs of cleanup estimated to be on the order of $15 billion over a fifteen-year period, little wonder that this protocol barely got off the ground.

Undaunted by their inability to fund major pollution abatement initiatives, all Mediterranean governments met in Geneva in 1982 and approved another protocol, this one providing special protection for endangered species of fauna and flora as well as critical habitats. Officially called the Protocol Concerning Mediterranean Specially Protected Areas, it entered into force in the spring of 1987, after nine countries and the European Community ratified it. The agreement bound signatories to set up areas of biological interest—fisheries, breeding grounds for protected species, and monk seal and sea turtle sanctuaries—as soon as possible. A number of marine parks and protected areas were also created. A regional center was set up in Tunis to assist governments in applying it.

Despite the importance of the action plan to the economies of the region, financial setbacks have plagued it from the beginning. A Mediterranean Action Plan (MAP) coordinating unit was eventually established in Athens and is currently operating with an annual budget of $5 million. Although the action plan functions as part of UNEP, program decisions are made by the governments of the region. The eighteen countries of MAP (now twenty-one) set up a Mediterranean Trust Fund to underwrite costs for the secretariat in Athens and the five regional centers and to provide core funding for the plan's far-flung research, monitoring, and conservation programs.

In 1985, all MAP countries met in Genoa and agreed to sweeping measures to make the Mediterranean Sea cleaner and safer. Ten priority targets were set for the second Mediterranean decade (1985–95). Included were commitments to establish more reception facilities for dirty ballast waters and oily wastes from shipping; to construct sewage treatment plants in all cities with over 100,000 inhabitants, as well as to install outfalls and/or treatment plants for all towns with more than 10,000 residents; to apply environmental impact assessments in coastal development; to cooperate to improve the safety of maritime navigation, particularly for those ships carrying dangerous or highly polluting cargoes; and to take measures to reduce industrial pollution of coastal waters and to see to the proper disposal of solid wastes.

By 1995, none of those ambitious goals had been reached. Only nine of nineteen major oil-loading terminals had installed facilities to handle waste oil. Fisheries were still largely unregulated. Although a few of the larger Mediterranean cities had built sewage treatment plants, it will likely take decades before all seventy cities with over 100,000 people have two-stage treatment plants in operation.

Still, not all reneged on their commitments. Istanbul invested $2 billion in cleaning up the Golden Horn, horribly polluted with raw sewage, slaughter house offal, and industrial wastes. Marseilles inaugurated a giant $180 million underground sewage treatment plant to cover its entire population. Athens built a huge sewage treatment plant on the island of Psitalia to handle all of the domestic wastes of the city's 4 million inhabitants. The European Investment Bank provided Italy with $5 billion to clean up the Po River, a major source of pollution to the northern Adriatic. And between 1975 and 1985 Spain spent over $120 million on water supply, sanitation, and sewage networks for 181 communities along its Mediterranean coast (Ress 1986, 1988).

In order to revive a largely moribund Genoa Declaration, the European Union initiated a meeting of all Mediterranean countries in Cyprus in 1990. Collectively, they issued another missive: the Nicosia Charter. This new declaration of good intent went several steps further than all previous pronouncements. It not only reiterates the countries' commitment to the aims of the Genoa Declaration, but also provides for taxation and other financial incentives to pay for it. More realistically, the cleanup programs are being stretched out over thirty-five years.

Meanwhile, another initiative, known as the Mediterranean Environmental Assistance Program (METAP), jointly funded by the European Investment Bank, the World Bank, and the Commission of the European Communities, is entering its second phase. According to the World Bank, a total of $30 million has been made available to finance over one hundred separate activities in the region, including projects in Albania, Algeria, Egypt, Tunisia, Turkey, Lebanon, and Syria. Initial investments are concentrating on municipal and solid-waste management and treatment, the conservation of water resources, and the strengthening of institutional capacities for environmental management (World Bank 1993–94).

The region's most promising breakthrough came during a high-level summit meeting on the future of the Mediterranean, attended by foreign ministers from fifteen member countries of the European Union, eleven Mediterranean states, and Palestine in November 1995. They agreed, in principle, to turn the entire Mediterranean into a giant free-trade zone by 2010. The European Union quickly agreed to provide $6 billion in aid over the next five years (*Economist* 1995).

Conclusions

- Existing fisheries regulations are inadequate to meet new challenges posed by drift netters and others using illegal gear or ignoring quotas and violating seasonal bans. Stiff penalties need to be introduced for offenders.

- The Blue Plan needs to be reinforced by deeds. It is an extremely useful planning tool and should be put to good use while designing the future, especially one so dependent on international tourism.

- Coastal resource management needs to be institutionalized within the context of each nation's structure of governance. Taking management plans and operationalizing them is a prerequisite for sustainable managment, but the Med's governments must begin to develop a process of coastal management that actively involves local communities and other major resource users in a constructive and ongoing process.

CHAPTER 7

The Pacific and Atlantic Coasts of North America

When the first waves of colonists arrived in North America from the Old World, they found a continent rich in natural wealth beyond their dreams. Its broadleafed forests were full of wildlife, its rivers and lakes teemed with fish, its estuaries and coastal wetlands were marvels of biological diversity. The Hudson River was so full of fish in the mid-1600s that Indians told of scooping them out by the basket full. In the early 1700s, the Chesapeake Bay, the greatest and most productive estuary in North America, was so clogged with oyster reefs, some of them 20 feet thick, that navigation was imperiled. Indians referred to the Chesapeake as the Great Shellfish Bay. In fact, the entire east coast of the continent was laced with generous, pristine salt marshes and coastal salt ponds, stretching from Labrador to Florida.

The west coast of North America remained largely unexplored and unsettled by Europeans until the late 1700s. Spanish missionaries had founded a few missions in the mid-1700s, from San Diego through Monterey and ending at San Francisco. But Spain failed to colonize California in any real sense. The Danish explorer Vitus Bering, in the employ of the Russian Czar, discovered the strait between Siberia and Alaska that bears his name in 1741. Russian explorers then pushed down the coast of Alaska and Canada to what are now Washington, Oregon, and northern California, bent mostly on exploiting the sea's wealth of marine life.

In 1779, the famous English captain James Cook sighted the coast of Oregon. Cook's arrival was followed in the early 1790s by that of another Englishman, Captain George Vancouver, who explored the "savage coast" of western Canada and northwest America. Both Cook and Vancouver marveled at the abundance of marine life they found along the Pacific coast of North America.

Those cold, nutrient-rich waters were filled with probably the greatest concentration and variety of marine mammals the world has ever seen. From Alaska south along the Canadian coast to the Pacific Northwest and California, the waters teemed with life—walruses, fur seals, sea lions, harbor seals, elephant seals, Steller's sea cows, sea otters, dolphins, and whales. Within 150 years (1750–1900), however, the populations of most fur-bearing mammals had been harvested to the brink of extinction. Steller's sea cow—named after Georg Steller, the German naturalist attached to Bering's expeditions—was nearly extinct a mere twenty years after its discovery; a few individuals are thought to have survived until the mid-1800s. Sea otters, prized for their lustrous fur coats, were nearly wiped out by hunters within 150 years of their discovery by Bering (McDougall 1993).

Two centuries of rapacious and uninterrupted exploitation left its mark on America's coasts. By the early 1970s, unbridled development, including rampant port and city expansion, coupled with gross pollution, had ruined many coastal environments, including coastal wetlands and estuaries.

These diverse ecosystems serve multiple functions as wildlife housing, fish hatcheries, biomass farms, bird motels and diners, flood- and erosion-control buffers, air purifiers (by absorbing carbon dioxide), and water treatment plants. Dr. Robert Costanza of the University of Maryland and his colleagues attempted to calculate the economic value of U.S. wetlands in the late 1980s. Using Louisana wetlands as the basis for their calculations, they figured that each acre provided up to $846 in commercial fishery benefits (including shrimp), $401 from trapping, $181 from recreation, and just over $7,500 in storm protection benefits. The total—just under $9,000—is considerably above the $200–400 each acre fetches on the real estate market.

Despite their value, in the two hundred years between 1785 and 1985, the United States lost more than half of its coastal wetlands, and many remaining sites were degraded beyond repair. According to the Fish and Wildlife Service's National Wetlands Inventory, in the twenty-year period from 1970 to 1990, an average of nearly 300,000 acres (circa 125,000 hectares) were lost *every year.* Both California and Connecticut lost more than half of their wetlands during that period. In the mid-1970s, the state of New Jersey was losing nearly 4,000 acres of coastal wetlands a year, Virginia 2,300 acres a year, North Carolina 40,000 acres a year, and the Chesapeake Bay around 500 acres a year. The record continues to be held by Louisana, which is still losing coastal wetlands, especially in the Mississippi River Delta, at the rate of 60 square miles a year.

By the early 1970s, some keynote legislation began to address the dreadful state of the country's coastal environments. The Clean Water Act of 1970 (and subsequent amendments) was followed by the Coastal Zone Management Act of 1972, the Marine Protection, Research and Sanctuaries Act of 1972, and the Wetlands Act of 1973. These laws helped forge a new consensus in favor of managing coastal areas more effectively and protecting them from some of the excesses of short-sighted development. These initiatives came just in time, for a new demographic trend was beginning to make itself felt; a renewed rush to the coast was on. Many Americans were moving to coastal areas from the interior of the country, or out of big cities and into expanding suburbs and bedroom communities located along or near coasts.

Population

The collective population of North America—Canada and the United States, excluding Mexico, which is covered elsewhere, under The Wider Caribbean and Latin America—amounts to about 295 million people, with 30 million in Canada and 265 million in the United States. As of 1997, somewhere between 55 and 60 percent of all Americans—roughly 157 million people—lived on or close to the Atlantic and Pacific oceans, the Gulf of Mexico, and the Great Lakes. Of Canada's popula-

*Table 7.1. Length of Coastline, Loss of Coastal Wetlands, and
Population Data for the United States and Canada*

	U.S.	CANADA
Coastline (km)	152,000	244,000
Population 1997 (millions)	265.7	30.1
Coastal Population (millions)	157	20
Coast Wetlands Lost %	50	—
% Urban	75	77

Sources: World Population Data Sheet, 1997, Population Reference Bureau, Washington, D.C., 1997; Brian Needham, ed., *Case Studies of Coastal Management: Experience from the United States,* Coastal Resources Center, University of Rhode Island, Narragansett, 1991, p. 1; and C. Fraser, *Integrated Coastal Area Management—A Canadian Retrospective and Update,* paper presented at the UN Commission on Sustainable Development, New York, 1996, p. 2.

Notes: Length of coastline for both Canada and the United States includes both the Atlantic and Pacific coasts, the Great Lakes, and the Arctic. The U.S. figure also includes the Gulf of Mexico. Islands are counted. The lower 48 states of the U.S. have an estimated 31,000 km of coastline. Canada has lost most of its prairie wetlands in the grain belt. Coastal wetlands are harder to quantify. In populated areas, close to 40 percent have been lost, according to some estimates; but the figures are very tentative, based on limited surveys.

tion, only one-quarter (about 7.5 million) live along the coast of the country's maritime provinces. However, if the population living on the Canadian side of the Great Lakes is counted as coastal, then close to 70 percent of all Canadians—roughly 20 million—are coastal dwellers (Culliton et al. 1990) (see table 7.1).

According to the 1990 U.S. census, populations rose dramatically in twenty-five out of thirty coastal states during the 1980s. The largest increases were noted for Alaska, which grew by 36 percent; Florida, with a 31 percent increase; and California, with a 24 percent rise. By 1988, population density in coastal counties had reached 341 people per square mile, more than four times the national average. Overall, coastal areas have been growing about three times faster than the nation as a whole (Clark 1996).

Projected U.S. trends over the next three decades show steadily rising coastal populations, particularly in southern California, the Pacific Northwest, and Alaska and from North Carolina to Florida. By the year 2025, nearly 75 percent of all Americans are expected to be living and working in coastal areas.

Eight coastal counties in California and Florida will be in the top ten in absolute population growth between 1988 and 2010. Los Angeles, Orange, and San Diego counties in California will grow by 2.6 million between 1988 and 2010, while the Miami area (Dade, Broward, and Palm Beach) will add 1.2 million people over the same period. In fact, only 5 percent of all coastal counties, mostly in the Northeast and the Great Lakes, will lose population over this period (Culliton 1990).

Pollution

With North America's coastal population growth have come increased pollution and degradation of coastal areas. At the same time that towns and cities have been rapidly expanding along coasts, agricultural production has intensified. Tons of point-source pollution from municipal and industrial wastes and nonpoint-source pollution—agricultural runoff from fertilizers, animal manure, and pesticides, as well as runoff from city streets—end up in near-shore waters.

Every year around 2.3 trillion gallons of partially treated sewage water is discharged into U.S. coastal waters, along with more than 2.2 million metric tons of chemical wastes. Most large-scale municipal sewage treatment plants in the United States use a two-stage treatment, which removes nearly all solid materials and most of the BOD in the effluents. However, few states have the funds to install three-stage, or tertiary, treatment, which removes secondary nutrients like nitrogen and phosphorus and chemical pollutants. Despite secondary treatment, older big-city plants, which were constructed to handle both storm water and sewage, often have to contend with so much effluent that they are overwhelmed; accidents are common, and during heavy rains storm-water runoff can surge through treatment plants, washing tons of raw sewage into rivers and coastal waters. In New England, municipalities and industrial plants discharge some 575 billion gallons of contaminated wastewater into coastal waters every year; with that goes another 700 billion gallons of polluted storm water from the region's highways and streets, much of it contaminated with heavy metals, oils, and salts.

In the early 1990s, experts contended that pollution from untreated (or partially treated) sewage and industrial wastes, in combination with agricultural runoff, was largely responsible for the closure of over 50 percent of America's productive shellfish beds on the Pacific and Atlantic coasts and nearly 60 percent along the gulf coast.

A Natural Resources Defense Council survey of twenty-nine coastal states and territories revealed 3,522 beach closings and pollution advisories for 1995, a 50 percent increase over 1994. Most of the closures were traced to high coliform counts in the water, linked mainly to untreated or partially treated sewage, storm runoff, and other municipal wastes (NRDC 1996).

In 1995, the EPA reported that, despite two decades of cleanup efforts, some 40 percent of the country's surface waters (both inland and coastal) were unfit for bathing or fishing. The report, based on extensive surveys conducted in 1994, indicated some improvement in the condition of the twenty-eight estuaries that are now part of the National Estuary Program. Out of 34,388 square miles of estuaries, the EPA surveyed 26,847 miles, or 78 percent of the total. Two-thirds of those highly productive ecosystems had good water quality; one-third were polluted to varying degrees. However, samples taken along 5,224 miles of U.S. Great Lakes coastline out of 5,559 miles—94 percent—revealed that 97 percent of the water was polluted, 63 percent in the poor category. Only 3 percent met U.S. national water-quality standards (EPA 1995).

North of the border, Canadian communities along the Atlantic and the Pacific are pumping raw sewage into shallow coastal waters. In cities like St. Johns on Newfoundland; Halifax, Nova Scotia; and Victoria, British Columbia, most sewage goes in untreated. The main reason is lack of money for proper treatment plants. Despite the fact that pollution is now a transboundary issue between Canada and the United States—the Strait of Juan de Fuca, which separates British Columbia from

Washington state, is only 20 miles wide—Victoria residents have consistently refused to vote funding for a sewage treatment plant (Farnsworth 1995). City officials claim that municipal wastes are dispersed by currents, but currents take the wastes along the coast not out to sea. It is a problem that plagues many Canadian coastal towns and cities. Only those cities along the Great Lakes have two- or three-stage sewage treatment plants.

Oil Pollution

Although most oil in coastal waters comes from land-based sources, particularly runoff from city streets, oil spills have been increasing. According to the U.S. Coast Guard, between 1981 and 1986, the number of oil and hazardous-waste spills into U.S. waters averaged over 10,000 a year. The average quantity spilled amounted to over 56,000 tons, with oil accounting for 70 percent of the total. Worse, most of the accidents occured in river channels, ports, and harbors not the open sea.

The worst oil spill in U.S. history struck Alaska's Prince William Sound in March 1989, when the ill-fated *Exxon Valdez*, in an attempt to change course to avoid icebergs, slammed into a reef, dumping 257,000 barrels of crude into the sound. Currents eventually deposited the oil along 1,500 miles of pristine shoreline. At one point, an oil slick the size of Vermont covered 10,000 square miles of the sound (Lewis 1989). In the aftermath of the spill, nearly 5,000 sea otters perished, along with 300,000 seabirds (mostly common murres), and as many as 300 bald eagles. Some estimates put the death toll of seabirds at over half a million.

Oil is known to linger longer in cold northern waters than in tropical seas, where it breaks up and evaporates faster. Commercial fishing in the sound, which brought in $131 million worth of salmon, herring, halibut, and shellfish in 1988, crashed completely. The herring season was cancelled in 1989, costing fishermen $12 million in lost income. Moreover, millions of herring larvae that hatched in oil-fouled water developed defects like crooked spines; most did not survive. Salmon fry were also found to be genetically defective.

Although long-term damage is hard to quantify, ecologists claim that some shoreline communities of plants and animals have not recovered completely, six years after the accident. Rick Steiner, a marine biologist with the University of Alaska, insists his post-spill studies of the sound show that the ecosystem is still unbalanced (Steiner 1995).

Resources

Along with pollution, sedimentation, habitat degradation, and loss of biodiversity, another problem has arisen: loss of coastal land to the advancing sea. Scientists estimate that nearly a quarter of the 95,000 miles of U.S. coastline is subject to serious erosion rates. Along the eastern seaboard, some 90 percent of the three hundred barrier islands, which stretch like a string of pearls from Maine to Florida, are suffering from erosion. Some shorefront property along the U.S. Gulf Coast is losing out to the sea at the rate of 100 feet a year, though the average for the worst hit state, Louisiana, is 30–50 feet (Dean 1989; *Economist* 1993).

Over the past century, the U.S. Atlantic coast has, on average, lost 2–3 feet a year to the sea. The south shore of Martha's Vineyard, a popular summer resort island off Massachusetts, is losing nearly 10 feet a year, and in Cape Hatteras, North Carolina, the sea is gaining 12–15 feet a year, much to the alarm of local residents, who depend on the generous beaches to draw in tourists.

As the sea continues to devour shoreline, many resort towns have resorted to engineering solutions—building barriers, like seawalls and jetties, or constructing groins—in an attempt to hold the sand in place. Unfortunately, many of those solutions fail outright or backfire in unexpected ways. When Ocean City, Maryland, built groins to hold its beaches in place, shifting currents began eating away at the undeveloped beaches in nearby Assateague Island, a national wildlife refuge, increasing erosion from 2 feet to nearly 40 feet a year (*Economist* 1993).

The dynamics of wave action on beaches are not understood completely, but geologists have recently gained a better grasp of how the process works. Jetties are particularly destructive, because they break up the profile of the sand, actually accelerating the rate of beach erosion in front of them. A study carried out by Orrin Pilkey, Jr., a geologist with Duke University, notes that beaches that had been "stabilized" by engineering works such as seawalls and other barriers had no beach left, while those beaches left to nature's will were wider, with more sand (Dean 1989).

Other communities—such as Hilton Head, South Carolina; Padre Island, Texas; and Sea Bright, New Jersey—have responded to the crisis by launching expensive beach nourishment programs, trying to replace the sand the sea has reclaimed. Sea Bright is spending $100 million to replenish 12 miles of beach. But beach nourishment is an ongoing process. Beaches depleted of sand by such artificial means as coastal construction or channel dredging, for instance, need repeated infusions of sand if they are to survive. Miami, Florida, renewed its beaches a decade ago at a cost of $65 million. The city must now sink millions more into its continuously eroding beaches, to keep the tourist dollars rolling in along with the surf.

Fisheries

In 1610 the English captain John Smith reported that the Grand Banks, off Newfoundland, were "so overlaide with fishers as the fishing decayeth and many are constrained to return with a small fraught."

The root cause of the decline in North America fisheries is twofold: chronically poor management of stocks by national, state, and provincial authorities; and the overcapitalization of American and Canadian fishing fleets. With the declaration of 200-mile EEZs, Canadian and American fishers were encouraged to increase their efforts, investing in bigger boats and better equipment. Subsequently, with the resounding collapse of fisheries on both coasts, the same fishermen are being told to scale back their operations. But now it is too late for many of them. In debt to banks and loan associations, they have to keep fishing to pay their bills or lose everything.

In Canada's maritime provinces, 50,000 fishermen and their families are now off the sea, subsisting on government handouts. American fishermen have no such cushion. According to a report from the Massachusetts Offshore Groundfish Task Force, the region's collapsing fisheries cost New England's economy $350 million in lost revenue and 14,000 lost jobs each year since the late 1980s.

The severity of the crisis is noticeable in any fishing community on the East Coast of North America. The demise of New England's groundfish resources—cod, sole, flounder, halibut, haddock—on Georges Bank, off Massachusetts, has increased competition in other areas. At Point Judith, Rhode Island, the tail end of a picturesque peninsula near the border with Connecticut, commercial trawlermen complain bitterly about fleets from Maine and Massachusetts plying their traditional fishing grounds. Says one: "They are refugees from years of 'catch as catch can'; it's not our fault that they fished out their own waters. Now they come into our fishing areas to repeat the same mistakes, but at our expense!"

The "trespassers" from other states claim they have no choice. "We've got to go somewhere," says Howard Nickerson, head of the Offshore Mariner's Association of New Bedford, Massachusetts. "We could be closed off from our own grounds for years."

Rhode Island fishermen asked the Atlantic States Marine Fisheries Commission to ban any new boats from going after squid and monkfish. The Georges Bank fishermen, on the other hand, want anyone who has caught as little as five pounds of monkfish or squid over the past few years to be allowed to exploit southern New England's waters off Rhode Island and Connecticut. "I think we may see a real war over remaining fish stocks in our waters," says one disgruntled fisherman from Point Judith.

The West Coast has fared little better. Fisheries in the Northwest developed more recently than those in New England, but pollution has already ruined some traditional fishing grounds in Puget Sound. Washington State fishermen routinely haul in fish with tumors or missing fins. And like nearly everywhere else, catches are on the decline.

Off the coast of Southern California, warming seas (perhaps triggered by the El Niño phenomenon) have severed a crucial link in the marine food chain: zooplantkon populations have been reduced by 80 percent. Zooplankton communities are composed of a wide range of tiny animals on which hundreds of species of fish and birds dine. Low numbers of zooplankton mean less life in the sea ("Marine Life . . ." 1995).

Meanwhile, fishermen in the Gulf of Alaska and the Bering Sea are rushing to cash in on North America's last great seafood bonanza—a fisheries resource estimated at 16.3 million metric tons, consisting mostly of bottom-dwelling cod, flounder, sole, perch, and pollock. In 1980, fishing fleets hauled in 100,000 metric tons of pollock and other groundfish; by 1990, they were taking 2 million metric tons worth $2 billion (Griffin 1992a).

"The National Marine Fisheries Service views Alaska as their showcase," points out Ben Deeble, a fisheries expert working for Greenpeace in Seattle. "But we have a different view. The fisheries look better, but that's because Alaska had more fish to start with." Deeble and others think the industry is overcapitalized, with too many boats and too much waste.

Anxious to exploit the fishery while there are still big profits to be made, some seventy factory ships and thousands of smaller vessels are competing fiercely for remaining stocks. The North Pacific Fishery Management Council has responded to the intense fishing pressure by shortening the season; the previously year-round pollock fishery is now down to five months.

Other fisheries have been depleted so throughly that the legal season is measured in days. The entire year's quota for Pacific halibut, once harvested over a six-month season, is now taken in two frantic twenty-four-hour periods. Some 6,000 boats compete to haul in as much as they can get in the time allotted. As a result, carcasses are badly handled, many are not cleaned properly, and nearly all of the landed fish have to be frozen because shore-based processing plants cannot handle the enormous volume that hits them all at once. The halibut season is not even the shortest one. So many boats go after herring that the season in some areas is just twenty minutes.

Short seasons and intense competition among boats have also created enormous waste. In 1990, for instance, trawlermen fishing for cod and pollock jettisoned over 20 million pounds of halibut, worth about $24 million, because they were an unwanted by-catch. The same year, Alaskan fishing fleets threw away 550 million pounds of other groundfish because they were either too small or the wrong species. In an attempt to salvage what's left of Alaska's rich fisheries, the North Pacific Fishery Management Council in 1992 voted to institute a three-year moratorium on new boats.

According to the National Marine Fisheries Service, close to 50 percent of U.S. commercial stocks are overexploited, with another 26 percent listed as fully utilized. Only 12 percent are considered underexploited. Nearly half of all commercial finfish species found in coastal waters are overexploited. Fisheries scientists claim that fourteen of the most valuable species—including New England groundfish, red snapper, swordfish, striped bass, and Atlantic bluefin tuna—are threatened with commercial extinction, meaning that there will soon be too few to justify the expense of trying to catch them.

Conservation: Reviving Wetlands and Estuaries

Although North America's coastal areas are among the world's most battered, some large-scale ecosystem restoration efforts are perhaps showing the way ahead. Rebuilding damaged resources is much costlier and time consuming than protecting them in the first place, but, in many cases, it may be the only option left.

Putting the Ever Back in Everglades

Florida Bay, a huge, shallow, brackish water estuary, squats at the southern tip of Florida before sinking ever so slowly into the Gulf of Mexico. "This is the tail end of the Everglades watershed," explains Amy Knowles, an avid fly fisher and environmental activist, "which extends just south of Orlando to Lake Okeechobee, then south through the Everglades Conservation and Agricultural Areas and Everglades National Park, ending here."

What also ends in the bay is South Florida's "pollution pipeline," which brings in increasing loads of nutrients flushed off agricultural lands and dairy farms, along with chemical pollutants from industries and urban areas. Even though the last two years have been unseasonably wet, the two previous decades of lower than average rainfall and periodic drought drained water levels in South Florida. With less freshwater entering the bay, salinity levels tripled, snuffing out many of the lush seagrass meadows that carpeted the bay's bottom. Nutrient pollution combined with masses of dying seagrass triggered gigantic algal blooms, which covered sections of the bay with green slime, reducing oxygen levels and eventually killing some 100 square miles of seagrasses.

Seagrass beds not only provide vital nursery and feeding areas for pink shrimp, spiny lobsters, stone crabs, and commercially valuable fish, they also absorb nutrients from the land and stabilize sediment, keeping the water clear. With many seagrass beds dead, sponges of all species perished and spiny lobster populations fell by 60 percent. The pink shrimp fishery in the nearby Dry Tortugas collapsed in 1993; catches dropped from 10 million pounds to 2 million (Reiss 1992).

Since the health of the bay is intimately linked to the health of the Everglades, it makes "dollars and sense" to fight for its preservation, claims Knowles, a member of the National Wildlife Federation and coordinator of the Water Quality Joint Action Group, a grassroots umbrella organization whose members represent about 90 percent of the population of the Florida Keys. "When the state first circulated its plans to restore the Everglades, the bay wasn't even on the agenda," she says pointedly. "Our economy, which is increasingly based on tourism, including recreational fishing and diving, depends to a great extent on a healthy Florida Bay, and a healthy bay is dependent upon the health of the entire Everglades. You can't have one without the other," she continues.

That viewpoint, looking at South Florida as one interlinked ecological system, represents a revolution in thinking. "You have to remember, that for the past one hundred years or so everyone wanted to destroy the Everglades and convert it into something useful, be it shopping malls, housing subdivisions, or agricultural land," points out Maggy Hurchalla, a long-time Florida resident and environmental activist.

Floridians are reversing course. Today, federal, state, and local government agencies are busy trying to restore and rebuild South Florida's wetlands. The state has allocated over $30 million to buy up wetlands not yet ruined. Rivers, like the Kissimmee north of Lake Okeechobee, that were artificially straightened by the Army Corps of Engineers, are being made crooked again. Levees are being breached and canals filled in.

Although rescue plans for the Everglades have surfaced periodically over the past two decades, none of them survived Florida's gridlocked political process, which pitted environmentalists against growers, sport fishers against commercial, conservationists against developers, federal agencies against state. Ironically, it was a lawsuit that changed South Florida's political course. In 1988 the federal government sued the Florida Department of Environmental Protection and the South Florida Water Management District for allowing polluted water to flow from agricultural land into Loxahatchee National Wildlife Refuge and Everglades National Park. In 1991, Governor Lawton Chiles walked into the courtroom and "surrendered" to an astonished team of federal lawyers. He agreed to tackle the problems by launching a multifaceted rescue effort for South Florida's remaining wetlands.

In 1994, Governor Chiles established the Governor's Commission for a Sustainable South Florida, in order to build broad public support for a comprehensive Everglades restoration program and make recommendations to the state on specific measures to be taken. Once the political bandwagon to restore the Everglades was up and rolling, "everyone wanted in on it," says Hurchalla. "It became a bellweather issue that no political figure could afford to oppose."

The broad-based coalition of environmental organizations, consumer groups, businesses, and politicians that was stitched together to save the Everglades may still unravel. Many contentious issues remain to be resolved. Foremost among these is which restoration plan the state and federal government will eventually try to implement. The Army Corps of Engineers is currently reviewing six different options, varying from the band-aid approach of opening a few more locks or punching holes in some levees, to a complete ecosystem restoration that will mimic the Everglades' original hydrology.

The Everglades Coalition, a loose confederation of some forty environmental organizations, favors plan six—"the cadillac version," as Lewis Hornung, the Everglades project manager at the Corps of Engineers, calls it. Among other things, plan six would reconnect Lake Okeechobee to the Everglades by restoring seasonal water flow from the lake southwards through the canal system that now bisects the Everglades Agricultural Area (presumably, less water would be available for agriculture); store in impoundments water that is now simply dumped in the lake, for use during the dry season or in periods of drought; restore sheetflow (water with no definite channel) from the Water Conservation Areas south into Everglades National Park; improve the quality of the water flowing into and through the Everglades by using degraded wetlands as pollution filters; and build a flow-way along the eastern fringe of the Everglades to increase freshwater flow into the bay (Hinrichsen 1995).

Meanwhile, the state is busy carrying out restoration efforts mandated by the Everglades Forever act, passed overwhelmingly by the state legislature in 1994. At the core of the act is the Everglades Construction Project, which includes eleven major projects. The liquid heart of these projects is the conversion of more than 40,000 acres of degraded wetlands into "constructed wetlands" called stormwater treatment areas (STAs). Between now and the year 2010, six STAs will be built, ranging in size from 800 acres to over 16,000 acres, in an effort to decrease the amount of phosphorus and other pollutants percolating into the region's marshes and sloughs from the Everglades Agricultural Area south of Lake Okeechobee. Phosphorus is thought to be altering Everglades ecology, encouraging the growth of cattails and other nutrient-loving plants and shouldering aside native sawgrass, which prefers a low-nutrient environment. (Hinrichsen 1995; Everglades Coalition 1993).

According to preliminary plans, existing canals will channel polluted irrigation water and runoff from agricultural fields into constructed marshes—huge ponds with earthen walls, filled with cattails and other nutrient-absorbing plants. These plants will then filter out pollutants, such as phosphorus, making the water clean enough to be discharged into an outflow canal that will send it back into the Everglades system.

Restoring the Everglades has brought state and federal agencies together. Some eleven federal agencies and six departments are working with forty-two different stakeholders at the state level, including representatives from government agencies, Indian tribes, businesses, environmental NGOs, pensioners, and citizen action groups.

Such a long-term, multi-agency endeavor is fraught with risks. "We have got to be able to maintain political commitment at both the state and federal level over a period of decades," explains Carolyn Waldron, director of the National Wildlife Federation's Southeastern Natural Resources Center in Atlanta. "Considering that most politicians have a time frame no longer than their terms of office, this is going to prove to be one of the most difficult aspects of the restoration efforts—maintaining continuity of political support."

Restoring the Chesapeake Bay

There are plenty of reasons to save the Chesapeake Bay. Its fecundity alone is legendary. "More than 100 million pounds of seafood landed annually with a dockside value of tens of millions of dollars. . . ; more than half the nation's blue crab catch; 90 percent of its soft crabs; 15 percent of its oysters; spawning grounds, until recently, for 90 percent of the East Coast's striped bass," write Tom Horton and William Eichbaum in *Turning the Tide: Saving the Chesapeake Bay* (1991), "all this from a mere pinprick on the North American coastline that contains a few billionths of the planet's water."

By the mid-1970s, it was beginning to become clear that the Chesapeake Bay was in bad need of rescue. Its underlying problems, however, were not clear until you looked at a map. The bay, which covers only 2,200 square miles and averages only 21 feet deep, is dwarfed by its watershed, which sprawls over 64,000 square miles through six states. Moreover, 15 million people live and work in the watershed, one of the most heavily populated areas on the East Coast. The bay was being overwhelmed on all fronts: by nutrient pollution leached into its shallow waters from sewage treatment plants, agricultural land, and livestock farms—especially increasing levels of nitrogen and phosphorus from fertilizers, untreated sewage, and chicken and cow manure; by a noxious mix of chemical pollutants belched into the air from expanding industries and increasing traffic; and by heavy metals, salts, and oils washed off city streets by storm water.

The results of decades of abuse include the following (Horton and Eichbaum 1991; Horton 1993):

- By 1987, the bay was on the receiving end of up to 184,000 metric tons of nitrogen and 74,000 metric tons of phosphorus a year from human and animal wastes and commercial fertilizers. This glut of nutrients triggered algal blooms, which smothered the bay's seagrass meadows, vital habitat for a wealth of marine life.

- Since colonial times, the bay has lost more than half of its tidal and nontidal wetlands and 40 percent of its watershed forests. According to the U.S. Environmental Protection Agency (EPA), between 1950 and 1980 the bay lost 60,000 acres of wetlands, most of them drained for agricultural land and livestock farms or bulldozed under for urban expansion and housing estates.

- Seagrass beds, which once covered over half the bay's bottom, amounting to several hundred thousand acres, had been reduced to no more than 34,000 acres by 1984 (roughly 10–20 percent of their historic acreage). The demise of this rich habitat not only imperiled shellfish beds and fish nurseries, but also contributed to further deterioration of water quality, since seagrasses help filter nutrients and anchor sediment, stabilizing the bottom and making the water cleaner and clearer.

- Commercial catches of rockfish (striped bass) dropped from 2,608 metric tons in 1970 to just 272 metric tons in 1983. American shad were fished so intensely in Maryland that by 1980, when the fishery was closed, landings were less than 1 percent of their historic levels. By 1984, landings of river herring had dropped 93 percent throughout the bay, and catches of yellow perch fell by 90 percent in Virginia and by 71 percent in Maryland. Sturgeon, which used to grow to 12 feet or more and supported a thriving caviar industry at the turn of the century, are no longer caught at all (Horton and Eichbaum 1991; Horton 1993).

• From an all-time peak of 20 million bushels hauled out in 1884, the oyster harvest declined progressively over the years until by the mid-1980s, oyster populations were only 1 percent of historic levels. In 1992, only 168,000 bushels were taken. The oyster's decline left thousands of the bay's watermen without work or income, draining tens of millions of dollars from regional economies. The oyster is an efficient natural filter for cleansing the water. In the 1700s, oysters were so plentiful they could filter the entire volume of the bay's water in five to seven days; today that would take remaining oyster beds over two years.

Foremost among the nonprofit groups working to save the bay is the Chesapeake Bay Foundation, incorporated in 1966, with headquarters in Annapolis, Maryland. Struggling along with only 10,000 members during its first decade, the foundation now has 90,000 members in virtually all lower forty-eight states. It not only monitors the health of the bay, it also works closely with state and federal agencies charged with cleaning it up and has an extensive outreach educational program designed to give people firsthand knowledge of the bay and its complex ecology. Every year, over 34,000 people receive on-the-water instruction in estuarine ecology (Chesapeake Bay Foundation 1991).

The Chesapeake Bay Foundation was one of the key organizations involved in urging the state governments to take a comprehensive watershed approach in combating the bay's problems. After a slow start in the early 1980s, the three key state governments—of Maryland, Virginia, and Pennsylvania—with help from the EPA, agreed to launch an ambitious, integrated cleanup program.

In 1987, when the EPA and the governors of the three main watershed states met at a major strategy conference, nothing on that scale had ever been attempted in the United States. The result was the Chesapeake Bay Agreement between the states and the federal government. "We now have a solid political commitment from all watershed states for a 40 percent reduction in nutrient loadings to the bay by the year 2000, using 1985 as the base year," explains Bill Matuszeski, director of EPA's Chesapeake Bay Program Office in Annapolis. "The year 2000 goal is a permanent cap on emissions of nitrogen and phosphorus," he continues, "in essence a nondegradation policy for the entire watershed, in which each major tributary of the Chesapeake will reduce nutrient pollution by 40 percent."

In order to meet such a far-reaching, ambitious goal, each state crafted its own pollution reduction strategy within the terms of the overall agreement. The Pennsylvania legislature even passed the first nutrient management law in U.S. history. Although neither Maryland nor Virginia took binding legislative steps, both states have encouraged farmers in the Chesapeake watershed to adopt "best management practices." These involve low-till or no-till agriculture along with the judicious use of pesticides and a marked reduction in the application of chemical fertilizers, one of the main sources of nutrient pollution to the bay.

The watershed states have made solid progress in reducing point sources of pollution—effluent from sewage treatment plants and industrial complexes. Maryland, Virginia, Pennsylvania, and the District of Columbia also banned the use of phosphate detergents. These actions cut phosphorus pollution to the bay by 40 percent between 1985 and 1994. But nonpoint-source pollution from diverse agricultural activities (and storm-water runoff from city streets) has proved difficult to regulate. For example, along the Delmarva Peninsula, which encompasses parts of Maryland, Delaware, and Virginia, nearly half a billion broiler chickens are raised every year. Disposing of their daily wastes, the equivalent of that of a city with four million inhabitants, is a daunting task. Previously, chicken farmers got rid of their manure by spreading it liberally over nearby agricultural fields. Although natural manure is better than petroleum-based fertilizers, the farmers spread too much on too little land. Massive amounts of nutrients were released into rivers and streams linked to the Chesapeake or soaked into groundwater aquifers.

With best management practices in place, chicken farmers are still producing manure, but they spread it over much larger areas of cropland, reducing the amount of unused runoff. Actions like this helped cut nitrogen loadings by 16 percent between 1985 and 1994.

"But the real improvement is that we have halted the increase in nitrogen pollution to the Bay in spite of a growing population," claims Matuszeski. "This has come about as a direct result of building three-stage sewage treatment plants around the bay and decreasing the amount of animal manure and fertilizers seeping into the region's surface waters."

Now the EPA and the states are concentrating on trying to coax seagrasses back into areas where they formerly flourished. "We have had some success already," says Matuszeski. "In 1984 only 34,000 acres remained, today we have over 75,000 acres of seagrasses, and they continue to expand." In addition to soaking up nutrients, thriving seagrass communities stabilize the bottom and aid in the recovery of fisheries by providing spawning, nursery, and feeding areas.

Progress has been slower than expected, but at least most trends are in the right direction. According to the EPA, many farmers in Maryland, Pennsylvania, and Virginia have cut fertilizer use dramatically; some use no commercial fertilizers at all, relying instead on animal manure and crop residues. And the use of pesticides has dropped 20 percent since 1985. "Once we can get these natural ecosystems to start working for us, like wetlands and seagrasses," says Matuszeski, "progress in meeting our goals will speed up, too."

Management: The U.S. Experience

The Coastal Zone Management Act, enacted in 1972, was founded on a partnership between the states and the federal government and attempts to balance economic development with proper resource management and conservation. Coastal states can get federal assistance in order to develop coastal management plans, which must conform to federal standards. By 1997, thirty of the thirty-five U.S. coastal states and territories had drawn up coastal zone management plans.

Congress required that, at a minimum, state programs should do the following:

- Protect natural resources, including wetlands, floodplains, estuaries, beaches, dunes, barrier islands, coral reefs, and fish and wildlife habitat.
- Manage coastal development in an effort to minimize the loss of life and property.
- Institute a process for siting major facilities related to national defense, energy, fisheries, recreation, ports, and related development.
- Grant public access to coasts for recreational purposes.
- Assist in the development of urban waterfronts and ports and preserve historic, cultural, and scenic coastal areas.
- Coordinate and simplify government decision making in managing coastal resources.
- Consult and coordinate with federal agencies.
- Ensure public and local government participation in the coastal management process.
- Encourage comprehensive planning, conservation, and management of living marine resources, including proposals to site pollution control and aquaculture facilities in the coastal zone.

According to David Godschalk of the Department of City and Regional Planning at the University of North Carolina, close to 40 percent of the federal implementation grants awarded between 1982 and 1987 were earmarked for improved "decision making through facilitated permit reviews, technical assistance to local governments, plan implementation, building management data systems, and public participation in coastal management." Nearly 30 percent was allocated for better natural resource management with special attention to wetlands and estuarine waters, controlling pollution, and acquiring land for parks and protected areas. "Remaining grant funds," says Godschalk, "have been spent on improving coastal access, natural resource development, hazards mitigation, urban waterfront development, and port projects" (Godschalk 1992).

In practice, implementing coastal management plans has resulted in many states becoming bogged down in interagency turf battles. Most have been unable or unwilling to vest authority over coastal areas in a clear lead agency; instead, they parcel out authority and end up with a fuzzy decision-making process and fragmented programs shared by too many agencies.

On the other hand, California, Alaska, and South Carolina have taken the time and made the effort to enact comprehensive legislation governing coastal areas. Each state established one coastal management agency with broad authority to review and regulate major activities along the coast. California even went so far as to set up the California Coastal Conservancy, charged with acquiring, preserving, and restoring coastal land. It has managed to set aside thousands of hectares of land that would otherwise have been developed, including a strategic wetland in Monterey Bay.

Rhode Island's Coastal Management Plan

When it comes to comprehensive coastal zone management, few states can match tiny Rhode Island's accomplishments. In 1971, a year before the federal legislation was passed, the Rhode Island General Assembly enacted ambitious legislation that created a seventeen-member Coastal Resources Management Council and granted it broad decision-making powers to regulate coastal activities.

In 1980, after studies revealed that the state's unique salt pond environments along the coast were being overwhelmed with housing developments and pollution from municipalities and farms, the state, with help from the Coastal Resources Center of the University of Rhode Island, launched a comprehensive study to determine what could be done to salvage them. Eventually, CRC and the state settled for a Special Area Management Plan for the Salt Pond Region, which includes six major coastal lagoons and smaller ponds, encompassing over 82 square kilometers of watershed in the southern part of the state (Archer 1988; Needham 1991).

A truly comprehensive management plan was finally approved in 1984 after going through a tough public review process. Among other things, it established an action committee within the Coastal Resources Management Council, which included municipal and state agencies; persuaded three municipalities to amend their land-use zoning plans to increase the size of residential lots in some critical areas; encouraged the construction of on-site treatment plants, such as septic tanks, for individual houses; limited extensions of public water and sewage lines in areas close to the ponds, which would encourage residential building or commercial development; launched strong public information and education campaigns, including the formation of a volunteer citizen action group to monitor the health of salt ponds; and established wastewater treatment districts for nonsewered areas.

One of the reasons Rhode Island's coastal management plan works so well is that it is highly dependent on citizen participation. Many states overlook the importance of local government and citizen input into their plans, which condemns them to a kind of enforcement limbo. The plans look fine on the books, but they aren't implemented because local authorities either don't understand the regulations or don't want to enforce them.

Management: The Canadian Experience

Canada's coastline, more than 244,000 kilometers, is the longest one in the world, and throughout the 1970s and 1980s, wide jurisdictional chasms opened up between the federal government in Ottawa and the country's ten provinces. The hostility between federal and provincial government authorities and overlapping jurisdictions involving as many as fifteen different federal agencies and half a dozen provincial and local governing institutions rendered cooperation on coastal management issues virtually impossible.

However, by the mid-1980s, with many of the country's coastal areas in crisis and the advent of a new era of cooperation between Ottawa and the provinces, progress on coastal management issues became possible. The first joint federal-provincial coastal management program was initiated in 1985 for the Fraser River Estuary on Canada's jagged Pacific coast. Thanks to the establishment of a broad coalition of support for sustainable coastal governance composed of groups from industry, business, local communities, and the federal government, the Fraser River Estuary management program proved that complicated coastal management issues could be addressed under a common banner.

By 1995, the process had been extended to other critical coastal areas of the country. The St. Lawrence River in Quebec is now being managed jointly by the federal and provincial governments with a minimum of bureaucratic fuss. And community-based management programs have been set up for thirteen priority hot spots—mostly harbors and estuaries—in Canada's Atlantic provinces.

"Administratively, we had to demonstrate, by small-scale and nonthreatening case studies, that various agencies and stakeholders could work together and that broad participation was not only possible but essential to success," explains Lawrence Hildebrand from Environment Canada. "Only by complementing this bottom-up strategy with top-level federal and provincial policy and program support could national integrated coastal zone management get a foothold."

By the early 1990s, Canada was successfully taking local and regional efforts and "scaling them up" in order to create wider areas of coastal governance. No nationwide coastal management system has emerged, but the federal government has created an Interdepartmental Committee on Oceans, which is now mandated to coordinate and help implement marine programs and policies at the federal level. To ease things along, the Canadian Parliament passed the Canada Oceans Act in 1997. Among other things, this comprehensive piece of legislation designates the Ministry of Fisheries and Oceans as the lead agency to oversee and coordinate provincial coastal area management initiatives. By the turn of the century, Canada is expected to have a functioning coastal management program in each of its maritime provinces (Fraser 1996).

Conclusions

- In both the United States and Canada, states and provinces need to recognize the importance of their coastal zones and craft effective initiatives, with broad public support, to deal with the complex issues affecting them.

- One of the keys to successful coastal management in both Canada and the United States is to find ways to actively incorporate the needs and aspirations of local communities into the process. This helps ensure public participation and broadens the agenda. Too often, special interests dictate coastal zone use.

- The "democratization" of coastal governance is important for the following reasons: (1) to ensure the stability of the programs; (2) to generate government interest and provide avenues for long-term funding; (3) to build up political commitment from the grass roots; and (4) to enhance the capacities of local, provincial, and state governments to plan and implement coastal management strategies.

CHAPTER 8

The Wider Caribbean

Abundance in the Midst of Scarcity

The wider Caribbean, which includes the Gulf of Mexico, encompasses twenty-four island states and territories and twelve mainland nations. It includes the east coasts of Mexico and Central America. The region has long been divided between the Hispanic group of countries and the separate English-, French-, and Dutch-speaking states. Not surprisingly, communication, not to mention cooperation, has been a rare commodity.

Behind the Caribbean's carnival of abundance is a world of scarcity—of jobs, skills, resources, and investments. As in other parts of the world, resource abundance is yielding to degradation and overuse. Despite development initiatives, like President Reagan's proposed "mini-Marshall Plan" for the region, poverty is more entrenched than ever.

Sitting on top of the Caribbean is the United States, the world's most powerful economy, worth some $6 trillion a year, while the southern rim is dominated by the rapidly developing economies of Colombia and Venezuela. In between lies the penury of most of Central America and the islands (with a few notable exceptions, such as Costa Rica, Puerto Rico, and the Cayman Islands).

Despite continuing attempts at economic modernization, many Caribbean states are handicapped by a lack of infrastructure, expertise, and investments. Crushing poverty often means that development schemes contribute to the problems of the environment rather than solutions. And Caribbean ecosystems are plagued by problems: chemical pollution produced by industry and agriculture; silt from dredge-and-fill operations and poor land management; nonsustainable exploitation of coastal resources; and untreated wastes from coastal cities, towns, and tourist centers. Most of the gulf coast of Mexico, as well as the U.S. gulf coast, is facing severe development pressures (see figure 1.3 on page 10).

The wider Caribbean is not nearly as landlocked as the Mediterranean or the Baltic. It consists of two huge basins: the Caribbean Sea proper and the Gulf of Mexico. Its surface area covers more than 4 million square kilometers. The so-called American Mediterranean is also deep, averaging 2,200 meters, with the deepest part—known as the Cayman Trench—plunging to 7,100 meters.

The drainage basin of the sea is equally impressive, encompassing some 7.5 million square kilometers and eight major river systems, from the Mississippi and its

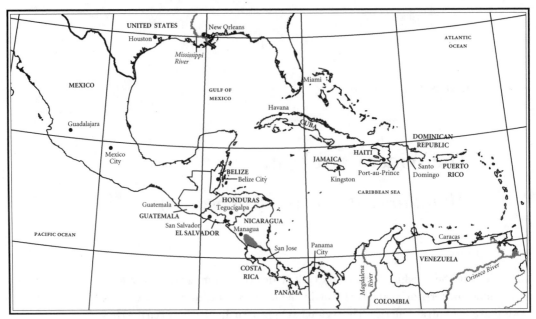

Above: Geopolitical Map of Caribbean and Central America. *Opposite:* Estimated Population Densities for Caribbean and Central America, 1995.

Source: Waldo Tobler, Uwe Deichmann, Jon Gottsegen, and Kelly Maloy, *The Global Demography Project,* National Center for Geographic Information and Analysis, Department of Geography, University of California, Santa Barbara, 1995.

tributaries which drains over one-third of the United States into the Gulf of Mexico, to the Orinoco in Venezuela, which pumps 538 cubic kilometers of freshwater into the Caribbean every year. These rivers, and others, also bring in millions of tons of pollutants scoured out of the hinterlands.

The Caribbean is the origin of the Gulf Stream, the tremendous ocean current that carries more water with it than all the rivers on earth combined. The Caribbean Current and the Antilles Current join at the tip of Florida to form a huge wedge of warm water, which then sweeps across the North Atlantic to northern Europe and finally to Scandinavia and into arctic Russia, modifying the climate and making life in northern Europe more bearable.

Population

The majority of the Caribbean Basin's 200 million permanent residents (including over 20 million people living in ninety-nine coastal counties along the U.S. gulf coast) live on or near the seashore (see figure 8.1). The resident population is swelled every year by the influx of 100 million tourists, nearly all of whom end up on the region's beaches. The Caribbean is now one of the world's premier tourist destinations, second only to the Mediterranean. Most of the smaller island states in the Greater and Lesser Antilles consist of nothing but coastal areas. But even along mainland Central America and the Caribbean coast of Colombia and Venezuela, human numbers continue to increase, often outstripping available resources.

Despite several decades of family-planning efforts, population growth rates for Central America still average 2.3 percent a year, enough to double the region's

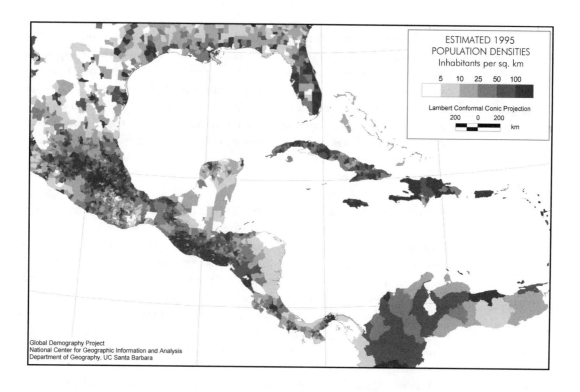

Global Demography Project
National Center for Geographic Information and Analysis
Department of Geography, UC Santa Barbara

population in three decades. Although growth rates for the Caribbean's twenty-four island states are beginning to drop—averaging just 1.5 percent a year—most of the island states and territories already have more people than the land can support. Collectively, they contain only 5 percent of the region's total population on 15 percent of its land area, yet, with the exception of Cuba and the Dominican Republic, population densities on the Antilles islands exceed 100 people per square kilometer. Many islands, in fact, have population densities exceeding 200 per square kilometer: Antigua has nearly 300 people per square kilometer; New Providence (Bahamas) nearly 600; Barbados 543; Martinique (France) 282; Grenada 285; Jamaica 209; Aruba (Netherlands) 367; Saint Vincent 323; and Puerto Rico 379 (WRI 1994).

As a consequence of too many people on too little land, many Caribbean states, especially the islands, must somehow subsist on just one or two primary export commodities. Jamaica's main export is bauxite, Cuba's sugar, Costa Rica's bananas, Trinidad and Tobago's oil. Most must also import food, including fish. On Dominica and St. Kitts, local demand for fish exceeds supply by more than 250 percent.

Constantly expanding populations throughout Central America have given rise to two troublesome trends: the rapid growth of towns and cities and the shrinking of per capita landholdings (Foer and Olsen 1992). Coastal cities, the main economic centers, now act like magnets, drawing people inexorably out of the countryside. Most urban growth has occurred in medium-sized and large cities, those with populations of 100,000 or more. On average, 65 percent of Central America's population and 60 percent of the Caribbean's live in towns and cities. This uneven distribution of population works against sustainable resource management. Since most of the new migrants to urban areas are poor peasants forced off exhausted land no longer capable of supporting them, they join the numberless ranks of the unemployed and underemployed in bulging slums and shantytowns. Here they contribute to the problems municipal governments face in attempting to provide basic services such as potable water, sanitation, health care, family planning, and education. Many beleaguered municipal governments in the region cannot even begin to match services with need. Consequently, the conditions in most slums and squatter settlements continue to deteriorate. So too does the state of coastal resources, overwhelmed by numbers and needs.

The urban revolution has taken place at breakneck speed in Latin America and the Caribbean. The Dominican Republic's urban population, for instance, doubled every decade between 1950 and 1980 (Gilbert 1990) (see table 8.1).

Grinding rural poverty and lack of opportunities continue to fuel the basin's urban explosion. But perhaps nothing underscores it better than the diminishing size of family farm holdings, a direct result of population growth and unequal land distribution. In Guatemala, the average amount of farmland per person dropped from 1.11 hectares in 1964 to 0.79 hectares by 1979. In the early 1990s, the average was down to around 0.22 hectares. Similarly, the size of upland farms in the Dominican Republic shrank from 3.3 hectares in 1985 to 2.8 hectares by 1992, a 20 percent decrease in just seven years.

A more detailed look at Guatemala's agricultural sector reveals a pattern common in the basin. Although 20,000 new smallholder farms, 0.7 to 7 hectares, were created between 1964 and 1979, they totaled just 6,231 square kilometers in 1979. During the same period, farms of more than 45 hectares increased by 55 percent, accounting for 26,531 square kilometers by 1979. In other words, a mere 13,600 farms (3 percent of the total number) controlled 65 percent of the country's productive agricultural land, while over 300,000 small farms, amounting to nearly 60 percent of the total number, subsisted on 15 percent of the country's farmland, much of it consisting of upland jungle plots with poor soils, hacked out of remaining rainforests (Southgate and Basterrechea 1992).

In Nicaragua, two-thirds of the population—about 3 million people—live in the Pacific region on just 15 percent of the total land area. The population is growing rapidly, by close to 3 percent a year, enough to double the country's population in twenty-three years. Moreover, Managua, the capital, contains close to half of Nicaragua's urban population, roughly 1.5 million.

It's a similar picture for the region's other capital cities, nearly all of which are coastal; only Bogota (Colombia), Mexico City (Mexico), and Tegucigalpa (Honduras) are located more than 50 kilometers from the nearest coast.

Pollution

Rapidly growing coastal populations, coupled with municipal and industrial expansion, have contributed to the region's pollution. At present, not more than 10 percent of the wastes generated by the basin's 200 million people receive any treatment. Although more tourist resorts are now putting in septic tanks or small-scale treatment plants before pumping the wastes into coastal waters, much remains to be done.

Table 8.1. Length of Coastline and Population Data for the Greater Caribbean Region

COUNTRY	CARIBBEAN COASTLINE (KM)	POPULATION 1997 (MILLIONS)	POPULATION GROWTH (%/YR)	% URBAN
Antigua & Barbuda	153	0.1	1.2	36
Bahamas	3,542	0.3	1.7	86
Barbados	97	0.3	0.3	38
Belize	386	0.2	3.3	48
Colombia	1,114	37.4	2.1	70
Costa Rica	212	3.5	2.0	44
Cuba	3,735	11.1	0.7	74
Dominica	148	0.1	1.4	61
Dominican Republic	1,288	8.2	2.1	61
Grenada	121	0.1	2.4	32
Guadeloupe	306	0.4	1.2	99
Guatemala	148	11.2	3.0	39
Guyana	459	0.8	1.7	31
Haiti	1,771	6.6	1.8	32
Honduras	820	5.8	3.2	47
Jamaica	1,022	2.6	1.7	50
Martinique	290	0.4	1.0	81
Mexico	9,330	95.7	2.2	71
Netherlands Antilles	377	0.2	1.2	90
Nicaragua	450	4.4	3.1	63
Panama	800	2.7	1.9	55
Puerto Rico	585	3.8	1.0	73
St. Kitts-Nevis	135	0.04	1.2	43
Saint Lucia	158	0.1	1.9	48
St. Vincent & the Grenadines	91	0.1	1.6	25
Suriname	386	0.4	2.0	49
Trinidad & Tobago	362	1.3	0.9	65
Venezuela	2,800	22.6	2.1	85

Sources: World Population Data Sheet, 1997, Population Reference Bureau, Washington, D.C., 1997; *World Resources 1996–97,* Oxford University Press, New York, 1996, pp. 268–269; Gordon Foer and Stephen Olsen, eds., *Central America's Coasts: Profiles and an Agenda for Action,* USAID, Coastal Resources Center, Washington, D.C., 1992, pp. 64, 135, 201, 236.

Notes: Not all 24 island states and territories are included in this table, as data for Grand Caymans, Virgin Islands, and others are not calculated by PRB. The United States is not included. For Mexico and Honduras, length of coastline refers to the Caribbean and Pacific coastlines combined; for Honduras, only about 50 kilometers lies on the Pacific coast. For St. Vincent & the Grenadines, length of coastline refers to St. Vincent only.

Alfred Taylor, a native Barbadian and former president of the Caribbean Hotel Association, points out the profits and perils of mass tourism:

Tourism is the future of the Caribbean. But at the same time, we have to be very careful about our environment. Solid waste disposal is now a serious problem. Our waters are getting more polluted. Our reefs are dying. On a lot of islands the hotels are too close to the beaches. The sewage pollution is killing the reefs, which then causes beach erosion. If we are not careful, we will end up with loads of hotels, but no beaches and tourists.

The environmental effects of coastal crowding are obvious to anyone who has visited the region. Mangroves are giving way to squatter settlements and shanty-towns or are cut down for timber or to make room for tourist resorts. Coral reefs are overfished and overexploited for building materials and coffee table trophies. Without proper regulations and zoning restrictions, coastal development spreads unchecked.

The waters along nearly every urbanized coastline are clogged with raw sewage and municipal garbage. The near-shore waters of Port-au-Prince, Havana, Kingston, and San Juan are so choked by untreated sewage and other municipal wastes that they are becoming devoid of oxygen. In La Paraguera, Puerto Rico, blue-green algae has snuffed out seagrass beds, causing anoxia. Eutrophication of near-shore waters has also resulted in the widespread decline of coral reefs.

In some cases—as on Haiti—health alerts have been issued and bathing beaches closed. There have been periodic outbreaks of cholera and typhoid, and pollution-induced diarrheal diseases are endemic throughout much of the region. In the late 1970s, studies revealed that diarrheal diseases in Guatemala and Nicaragua accounted for 26 percent and 34 percent, respectively, of all deaths registered.

Not only municipal wastes contribute to the pollution problems of the Caribbean. The high organic loads from sugar cane mills and food-processing plants also rob shallow coastal waters of oxygen. In Kingston Harbor, Jamaica, and off the coast of Trinidad, as well as the Mississippi Delta south of New Orleans, toxic industrial wastes kill fish and foul marine habitats. Untreated municipal and industrial effluents flow into Kingston Harbor at the rate of 35–45 million liters per day. In March 1988, a raft of dead fish 1.5 kilometers long and 300 meters wide was found in the Gulf of Paria between Venezuela and the islands of Trinidad and To-bago—victims of industrial wastes and hydrocarbon pollution (Lindén 1990; Lundin and Lindén 1993).

Heavy metals from mining operations and metal smelting pose serious threats to coastal environments in some areas. In the Coatzacoalcos Estuary in the Gulf of Mexico, bottom sediments contain high levels of lead, cadmium, mercury, and copper leached from its watershed. Dangerously high levels of mercury pollution have been documented for the bays of Cartagena, Colombia, Guayanilla in Puerto Rico, and Puerto Moron in Venezuela. Havanna Bay is also full of heavy metals, including mercury, lead, and cadmium.

The widespread use of agro chemicals and pesticides contributes another set of contaminants to the marine environment. Like heavy metals, these persistent poisons are taken up by sediments and bioaccumulate up the food chain. The waters around Puerto Rico and the islands of the eastern Caribbean contain measurable levels of pesticides. Pesticide runoff into coastal waters has killed fish around Jamaica and off the coast of Colombia. Years after both were banned, DDT and DDE have shown up in the tissues of reef-dwelling fish like groupers taken in the Gulf of Mexico and the Grand Bahamas. Shrimp and plankton from the northern Caribbean were found to contain measurable levels of DDT, as well, but not in high amounts.

The Oil Highway

Like all seas from which oil or gas is extracted and through which petroleum products are transported, the Caribbean suffers from chronic oil pollution. Every day around 5 million barrels of oil are transported through the Caribbean, and every year, on average, about 7 million barrels of oil are dumped into the sea. Some 50 percent of that pollution is thought to be accounted for by tankers and other ships discharging oily wastes, dirty bilge waters, and tanker slops in direct violation of IMO treaties. A significant amount of oil also finds its way into Caribbean waters from offshore oil rigs and exploratory drilling. In 1978, one of the few years for which data are available, the equivalent of nearly 77 million barrels of oil leaked into the sea from oil rigs.

The Gulf of Mexico has the distinction of being the scene of the world's worst oil spill. In the early morning hours of June 4, 1979, the Ixtoc 1 exploratory well in the Bay of Campeche blew out. It was not capped until March 23, 1980, 290 days later. During that time 475,000 metric tons of oil were spilled into the gulf's waters. Hundreds of thousands of oil-soaked crustaceans were washed up on gulf beaches for months afterwards, but the full extent of the damage was never assessed (Hinrichsen 1981).

The oil-producing countries of the region (not including the United States)—Colombia, Venezuela, Mexico, Trinidad and Tobago, and Barbados—were extracting oil at the rate of 3.5 million barrels a day during the late 1980s and early 1990s. The big three—Colombia, Venezuela, and Mexico—have petroleum reserves totaling some 12 billion metric tons. Not only does the Caribbean produce oil and gas, exporting most of it, but much of it is also refined in the region. Across the Caribbean, including the Gulf of Mexico, around seventy-three refineries are capable of handling over 12 million barrels of crude per day.

To a large extent, oil has fueled the development plans of Mexico, Colombia, Venezuela, and Trinidad and Tobago, but another result of the oil boom has been devastated coastal ecosystems in major oil-producing areas and along tanker routes. Mangrove swamps, seagrass meadows, and coral reefs have been wiped out by oil spills in many parts of the Caribbean. The windward exposed beaches from Barbados to Florida are heavily contaminated with tar balls and oily resides; in some places, as much as 100 grams of tar have been found per meter of beachfront. Oil not only tars beaches and reduces tourist dollars, it also kills marine life. Sea turtles and marine mammals are especially susceptible, since they sometimes ingest floating tar and die. And once oil hits coastlines, it can devastate marine communities, killing off entire populations of shellfish and crustaceans and polluting habitats for years to come.

One of the longest and most comprehensive studies on the effects of oil on tropical marine ecosystems was carried out in Panama, following a spill at a refinery at

Bahia las Minas on the Caribbean coast in 1986, which leaked between 60,000 and 100,000 barrels of crude oil into shallow coastal waters (Guzman and Jarvis 1996). After five years of study carried out by the Smithsonian's Tropical Research Institute in Panama City, scientists concluded that the long-term damage of oil spills in near-shore tropical waters is much more severe and long lasting than previously thought. Researchers discovered that the oil had ruined much of the infrastructure of the areas's ecosystems, resulting in the loss of mangroves and permanent damage to nearby coral reefs. Not only did coral polyps die as a result of the spill, but, even more ominous, those reefs that survived are not reproducing themselves.

This is leading to some fundamental alterations in the balance of species. Traditional reef species have suffered a serious decline in numbers. Sea urchins and shrimp have disappeared. So too have grazers such as parrot fish, which help keep the reef clear of opportunistic algae. As sections of the reef die, the entire ecosystem becomes impoverished. Moreover, without the shore protection provided by coral reefs, erosion has become a serious threat to the coastline of Bahia las Minas.

"All in all, the prospects for recovery are very long term, and are made worse by the lingering presence of oil trapped in sediments," says Dr. Brian Keller of the Smithsonian. "The time course of the recovery processes goes beyond the life span of most investigators" (Pain 1994).

Resources

Mangrove swamps, seagrass beds, and coral reefs are all coming under increasing stress throughout the Caribbean, victims of pollution, sedimentation, and the direct effects of dredging and coastal land reclamation. If left unchecked, the destruction of these vitally important habitats could sterilize coastal areas, greatly reducing their productive capacities and biodiversity. The loss of these ecosystems also impoverishes coastal communities.

With a few exceptions, mangrove swamps are being degraded and destroyed at unprecedented rates. Puerto Rico's mangrove forests, for example, have been reduced by 75 percent since the first Europeans began colonizing the island. Nearly all of Haiti's mangroves have been felled by poor peasants and sold for timber, fuelwood, and charcoal. Costa Rica has lost perhaps half of its mangrove resources over the past forty years, and 65 percent of Colombia's Caribbean mangrove forests have been indiscriminately exploited; in the Salamanca region, some 70,000 hectares have been killed off by alterations in tidal flows caused when poorly engineered roads were built over landfill without proper drainage. In 1984, 24,000 cubic meters of mangrove wood were removed for domestic consumption in Honduras (Bossi and Cintron 1990).

Increasingly, deforestation of uplands and coastal areas brings in its wake erosion of agricultural lands, coupled to landslides during the rainy season and droughts during the dry season. Rivers draining such denuded areas run swollen with sediments pulled off the land. As more soil is washed into coastal areas, mangroves, seagrasses, and coral reefs are smothered and killed.

Seagrass beds are affected by some of the same kinds of pressures that destroy mangroves. The greatest threats facing seagrass meadows in the Caribbean include dredge-and-fill operations; eroded sediment from coastal deforestation and poor agricultural practices; fishing with bottom trawls; and water pollution caused by industrial and municipal wastes, thermal discharges from power plants, and oil spills.

However, the magnitude of the destruction of seagrass beds is largely unknown. Seagrass loss has been documented in only a few countries. One study carried out in Boca Ciega Bay, Florida, following a dredge-and-fill operation to enlarge a boat harbor, revealed that 20 percent of the seagrass community in the bay had been wiped out, causing an 80 percent reduction in fish species and a fisheries loss estimated at $1.4 million (Thorhaug 1981).

Often the trouble starts when mangrove forests are clear-cut. But probably nothing is more harmful to seagrasses than dredge-and-fill operations. Whether for the construction of harbors, residential estates, coastal industries, or ship channels, dredging churns up enormous quantities of bottom sediment. Water quality is severely impaired and visibility reduced as suspended particles of sand and mud clog the water column. The resulting turbidity interferes with photosynthesis and reproduction, and when the sediment finally settles, it often buries remaining seagrasses. To make matters worse, dredge-spoils are often dumped indiscriminately over seagrass beds.

A number of countries (mainly France, Australia, Britain, the United States, and the Philippines) have launched programs to rehabilitate seagrass beds. So far, restoration efforts have been successful with four varieties, including one of the most common species, known as turtlegrass (*Thalassia testudinum*). However, the costs of reintroduction are high: ranging from $3,000 to $25,000 per hectare.

Dr. Anitra Thorhaug, professor of biological sciences at Florida International University in Miami, has successfully replanted seagrasses in Biscayne Bay, Florida, using volunteer labor. But it is time consuming and hard work since the new shoots have to be planted by hand. According to Thorhaug:

> Seagrasses and the important fish nurseries associated with them have been badly neglected and damaged throughout many areas of the world, particularly in the Caribbean. But efforts are underway to conserve remaining seagrass meadows and to restore others which have been lost. Unfortunately, the combination of seagrass vulnerability to pollutants and their tendency to grow close to shore, where dumping occurs most frequently, has left large parts of the Caribbean denuded of seagrass (Thorhaug 1981).

The Caribbean's coral resources are also under assault, with large areas degraded beyond their capacity to recover. About 14 percent of the earth's coral reefs are found in the Caribbean, and up to three-quarters of them—32,000 square kilometers—are in serious decline or under threat. Nearly all of the region's forty-nine marine parks and protected areas with coral resources have noted significant damage in recent years (Jameson, McManus, and Spalding 1995).

One of the most extensive studies of coral decline was carried out on Jamaica by Terence Hughes, a marine ecologist with James Cook University in Townsville, Australia. Hughes's report, which is based on reef surveys carried out over the course of seventeen years, claims that "the scale of damage to Jamaican reefs is enormous." According to Hughes, from the early 1970s to the early 1990s, coral cover declined from a mean of 52 percent to barely 3 percent across the island. He attributes the rapid decline of Jamaica's coral communities to rampant overfishing, hurricane damage, and excessive nutrient pollution—from municipal wastes (untreated sewage) and agricultural runoff (pesticides and fertilizers)—among other factors (Broad 1994; Hughes 1994).

Meanwhile, the Caribbean's tropical forests are disappearing as well. Every year some 1.2 million hectares of forestland are destroyed, while only around 30,000 hectares are replanted. According to Norman Myers, author and consultant on environmental issues to various UN agencies, the forested area of Central America and Panama decreased by 112,800 square kilometers between 1961 and 1978. Nearly 65 percent of Guatemala's forests have been lost over the past thirty years. Much of this forested land was turned into cattle ranches and cash crops. But slash-and-burn subsistence cultivators and loggers are mostly responsible for the wholesale destruction of the region's upland watersheds. In the rural areas of some energy-poor islands, such as Haiti, Martinique, and Guadeloupe, fuelwood gatherers also contribute to the process of deforestation.

In *Bordering on Trouble,* environmental journalist Larry Mosher describes what is happening to the forests of St. Lucia, a beautiful volcanic island in the Lesser Antilles:

> Much of its forest loss stems from the monoculture export of bananas. In addition to degrading the island's fragile topsoil, such plantation practices have forced many farmers off the best land and into the hills, where they slash and burn to raise food as well as grow more bananas. This exacerbates soil loss through erosion, which in turn forces the farmers to move on to other areas after several growing seasons when their crop yields diminish. The vicious cycle can ultimately change the microclimate, transforming farmland into semi-desert (Mosher 1986).

Once the forest cover has been stripped away, hilly areas quickly shed their topsoil. Panama has nearly 1 million hectares of eroded soils; Venezuela has ten times that amount. Measured rates of soil loss in the Caribbean are as high as 35 metric tons per hectare. Central American soils are in even worse shape. In Guatemala, soils are eroding at rates between 20 and 300 metric tons per hectare per year. Soil stabilization and reclamation measures could rehabilitate some of the degraded land, but the costs are often prohibitive.

The alternative to rehabilitation, however, is often far worse. "Once you've lost the hills, you lose the sea," points out Beverly Miller, chief of external relations at UNEP in Nairobi and formerly a senior program officer in the Regional Co-ordinating Unit of the UNEP-sponsored Caribbean Environmental Programme, based in Kingston, Jamaica. "We are trying to make the critical links between what happens on land and how this affects what goes on in coastal areas," she continues. Most coastal management plans fail to link reforestation of the uplands with efforts to guard against further degradation of near-shore resources.

It has been estimated by the World Conservation Union that roughly 40 percent of global vertebrate extinctions have occurred in the Caribbean Basin. Many of those extinctions are attributed to the indiscriminate destruction of upland and coastal forests (including mangroves), coastal erosion and dredging, and the extensive use (and misuse) of agricultural chemicals.

The Caribbean monk seal (*Monachus tropicalis*) is thought to be almost or entirely extinct. The number of manatees is constantly declining due to habitat loss, entanglement in fishing gear, and collisions with motor boats. Belize may harbor two hundred or more of the West Indian manatee, the largest single population of that subspecies remaining in the Caribbean. Crocodiles and alligators continue to be poached throughout the Caribbean, despite bans on hunting them. Dolphin species are dwindling, victims of fishing nets and pollution.

Virtually all of the region's marine turtles are endangered or threatened: loggerhead, green sea, hawksbill, Kemp's ridley, Central American river, and leatherback. They are running out of nesting areas as their beaches are taken over for tourist resorts, housing, and industries. Adult sea turtles are killed for their meat and shells. And thousands are drowned in fishing nets, particularly shrimp trawls.

In 1974 some 40,000 sea turtles came ashore to lay eggs in the Gulf of Mexico. By 1976 only 700 were counted, and a year later the total was down to 450. Since then, numbers of some turtles—like the green sea and Kemp's ridley—have rebounded, but their overall status remains critical (Williams 1995).

Still, sea turtle conservation programs have had some moderate successes. In 1994, for instance, researchers counted over 1,500 ridley nests on the beaches of Rancho Nuevo, on the gulf coast of northern Mexico. By July 1995, over 1,800 had been spotted. Marine biologists think that the Kemp's ridley can be saved if shrimpers use turtle-excluding devices (known as TEDs) on their nets, as mandated by the National Marine Fisheries Service. Although the devices were developed in the early 1980s, shrimpers were not required to use them until 1989, when the Fisheries Service began enforcing the law (Williams 1995). By 1995 many shrimp boats had either "lost" their turtle excluders or sowed the escape holes shut in defiance of the law.

Fisheries

The Caribbean does not have extensive continental shelves stretching out from land masses and the islands. Because of a pronounced lack of upwellings of nutrient-rich subsurface water, the sea is generally poor in nutrients. Consequently, mangrove forests, seagrass meadows, and coral reefs provide the major breeding, feeding, and nursery areas for many commercially valuable species of fish and shellfish. Seagrass beds alone may account for 80 percent of the breeding grounds for many species of fish and shellfish.

Caribbean fisheries are confined largely to smaller-scale commercial operations and artisanal activities. Still, FAO reported that they hauled in some 2.6 million metric tons in 1984. Since then, landings of commercial varieties have plunged. In 1992, FAO reported that 1.7 million metric tons of fish and shellfish were caught, though that figure does not include the catches made by small-scale, artisanal fisheries. Still, the overall trend for Caribbean fisheries, as elsewhere, is downward. According to FAO, most of the decline can be attributed to the chronic overexploitation of gulf menhaden, American oysters, and calico shrimp.

Years of unsustainable harvesting have precipitated the collapse of key shellfish fisheries in many areas. In Belize, the spiny lobster, conch, and grouper fisheries have been severely overexploited. The conch fishery, in fact, plunged from 1.25 million pounds in 1972 to a mere 365,000 pounds in 1990.

The richest fishing grounds in the Caribbean are found on the Campeche Bank in the Gulf of Mexico, on the Mosquito Bank off the coasts of Honduras and Nicaragua, in the Gulf of Paria between Venezuela and Trinidad and Tobago, and in the coastal waters of Guyana and Suriname. Two of these important fishing areas are also chronically polluted. The Campeche Bank is filled with offshore oil-drilling platforms and is the site of the Ixtoc 1 oil spill. The Gulf of Paria is contaminated with industrial and municipal effluents.

In the past few years, longline fishing has increased significantly in the deeper waters of the Caribbean. Commercial fleets from Japan, Taiwan, and South Korea dominate longline fishing operations. There are fears that this technique might deplete certain stocks of fish, especially tuna and billfish (like marlin).

As wild fish stocks continue to dwindle, aquaculture and mariculture activities have soared. Pond-raised shrimp, prawns, and spiny lobsters are being marketed in Panama, Colombia, Venezuela, Honduras, Mexico, and the U.S. gulf states. The Dominican Republic has introduced the cultivation of Caribbean spider crabs (*Mithrax spinosissimus*), which can grow up to 2 feet long and weigh 2 kilograms.

However, due to environmental problems with a number of mariculture operations, including outbreaks of disease and fluctuating market prices, pond-raised fish and shellfish have not had the tremendous economic impact that was predicted. Moreover, most cultured fish and shellfish are destined for the more lucrative export markets. In general, they contribute little to the diets, or income, of local Caribbean populations.

One exception is Honduras, where shrimp farms covering around 10,000 hectares had generated more than 25,000 new jobs, directly or indirectly, by 1994. The country plans to have up to 30,000 hectares exporting shrimp worth around $110 million a year by the turn of the century. The trouble is that most of this expansion would have to come at the expense of mangrove forests. Although the government has enacted laws to protect the country's remaining mangroves, "the government has neither the tools nor the will for enforcement," claims University of Honduras biology professor Sherry Thorn. "More than half of the mangroves that grew along Honduras' gulf coast in 1950 are already gone" (Wille 1993).

Management: The Caribbean Environment Program (CEP)

In 1976, UNEP, in cooperation with the Economic Commission for Latin America, established a joint project aimed at developing an action plan for the sustainable management of the Caribbean environment. After lengthy consultations with Caribbean governments, the CEP's action plan was finally adopted at a high-level meeting in Montego Bay, Jamaica, in April 1981.

Priority projects were selected for the first phase of the action plan, and a trust fund was established to finance the program. Unfortunately, "for almost three years, the Caribbean Action Plan remained merely a diplomatic vision," writes Larry Mosher. "Its nine member monitoring committee met twice, first in New York City and then in Cartagena, Colombia, mostly to agree that nothing could start until its trust fund grew larger. By mid-1982, a year after CAP's inauguration at Montego Bay, its trust fund had received only $25,355 out of initial pledges that totaled $1.7 million." It was not until the autumn of 1983 that the monitoring committee had enough money to begin work (Mosher 1986).

The committee's first effort suffered from financial difficulties but was nevertheless impressive. An intergovernmental meeting in Cartagena, Colombia, in 1983 adopted two important legal instruments for dealing with common environmental concerns: the Convention for the Protection and Development of the Marine Environment in the Wider Caribbean Region and a Protocol Concerning Co-operation in Combating Oil Spills in the Wider Caribbean Region. The Regional Co-ordinating Unit, set up in Kingston, Jamaica, opened in September 1986, five years after the action plan had been adopted. Both the Cartagena Convention and the protocol on oil spills were signed or acceded to by sixteen countries and the European Community and entered into force in October 1986 (UNEP 1987).

Over the past decade, the Caribbean Environment Programme has received around $8 million, mostly from UNEP's Environment Fund and from contributions by the CEP countries. The United States, by far the dominant economy of the region, elected to give no money to the Caribbean Trust Fund. Instead, the U.S. donates experts to the co-ordinating unit in Jamaica and provides services in kind. Congressional and State Department politics, revolving around Cuba's participation, have prevented the United States from fully participating in the Caribbean Environment Programme.

Despite the lack of U.S. involvement in the CEP, the governments of the region seem to be taking resource management more seriously. "Environmental concerns are now a part of each election in the region," notes Beverly Miller of UNEP. "In fact, the environment is one of the most important political issues in the Caribbean." One of the reasons environmental and resource issues are beginning to make their way onto government agendas is the success of information campaigns conceived and carried out by the largest conservation NGO in the region, the Caribbean Conservation Association, based in Bridgetown, Barbados. The awareness fostered by the association has led to the demand by grassroots environmental organizations and citizen groups for action to solve the pressing resource and environmental problems confronting the Caribbean.

Other Management Efforts

A few countries that depend on tourism for the bulk of their foreign exchange, like Belize, are in the process of launching comprehensive coastal management programs on their own terms. Tourism, especially ecotourism, is now the number one industry in Belize, accounting in 1994 for over $150 million Belize dollars. Currently, over 200,000 foreign tourists visit Belize every year, effectively doubling the population of the country during the winter season. Seventy-eight percent of some two hundred major hotels are located along the coast.

The government of Belize established the Coastal Zone Management Unit (CZMU) within its Fisheries Department in 1990 in order to coordinate coastal management initiatives. It is now the primary agency responsible for implementing coastal management plans. Currently, the capacity of the CZMU to regulate and manage coastal area activities is being strengthened through support from the UNDP/GEF Coastal Zone Management Project. "The overall objective of this five-year project," explains Janet Gibson, a coral reef expert and adviser to the Belize government on coastal management issues, "is to preserve the high biological diversity of the coastal zone of Belize by ensuring the sustainable management of its resources and to assist in the development of an integrated coastal zone management program for the entire country."

A Coastal Zone Management Technical Committee was set up to facilitate a solid working relationship between the many government agencies, NGOs, private-sector interests, and others with a stake in coastal governance. One of the main challenges facing the committee is to sort through some forty major pieces of legislation governing the country's coastal zone in an effort to streamline the process of governance and avoid conflicting laws and regulations.

Realizing that its future economic health depends to a great extent on healthy coastal ecosystems, the country is working feverishly towards this goal.

The 288-square-kilometer, crescent-shaped island of Bonaire, in the Dutch Antilles, some 100 kilometers off the coast of Venezuela, is an example of how a sound coral reef management system can pay for itself while safeguarding coral resources. The waters surrounding Bonaire, from the shoreline to a depth of 60 meters, are officially protected as the Bonaire Marine Park. Divers and snorkelers are required to pay user fees, averaging about $10 each, and commercial dive operators pay for permits that give them access to prime dive sites. By 1992, the park had installed thirty-eight permanent moorings for dive boats, in order to reduce anchor damage to the reef. A snorkel trail was also created, taking pressure off some of the more popular dive sites.

As of 1992, user fees from over 17,000 divers and permits were generating close to $185,000 a year, more than enough to cover the costs of managing the park, which average about $150,000 per annum. In addition, tourists pump thousands of dollars into the local economy. Divers and tour operators alike approve of the fee system, claiming that the island's pristine reefs are now better managed and popular dive sites are not as congested as before. The World Bank, which supported the initial management plan, thinks that Bonaire's management system could be replicated by other small, cash-short island states in the Caribbean (Dixon, Scura, and van't Hof 1993).

The Bonaire model holds much promise, but islands and mainland states alike have been slow to develop special-area management plans. Barbados has struggled with the practical realities of coping with out-of-control coastal development for a decade. Progress has been slow but steady. In 1991, a three-year feasibility study, funded by the Interamerican Development Bank, was launched. One of its main features is the strengthening of local governing institutions to manage coastal areas by turning the Coast Conservation Unit (CCU) into a permanent national agency, with broad authority to regulate all major activities in the island's coastal zone.

Recognizing the bureaucratic pitfalls involved in implementing comprehensive coastal-area management plans, Barbados has elected to take a gradual, incremental approach to coastal governance. Interagency committees have been set up incorporating representatives from the relevant ministries: planning, ports, fisheries, and public works. It is the government's intention to make the Coast Conservation Unit the point agency for the country's ongoing coastal management efforts.

Other countries that have introduced countrywide coastal-zone management plans, like Costa Rica, have not been able to muster the political will power and institutional support to implement a truly comprehensive management system.

Costa Rica's coastal-zone management program is limited in scope, regulating only the first 200 meters of coastline. Consequently, its coastal management plan does not address the most critical threats to the country's coastal and marine resources: deforestation, pollution, and environmentally unsound uses of lands in watersheds. Although legislation is in place that regulates deforestation and pollution in watershed areas, it is not comprehensive or unified, and administration is spread out over a great number of competing institutions. Hence, enforcement is weak or nonexistent (Foer and Olsen 1992).

Conclusions

- Since the region's economy depends increasingly on international tourism, it is imperative that the coastal resources that attract 100 million visitors a year are managed and conserved.

- Old, traditional management methods, based on local community control, could prove to be useful models for managers as they try to figure out a way to manage remaining resources, including near-shore fish stocks.

- Cross-fertilization of experience is an important avenue that is often overlooked because there are no mechanisms in place to allow countries to benefit from the experience of others. UNEP's Caribbean Environment Program can play a catalytic role by disseminating the results of successful coastal and special area management plans.

CHAPTER 9

Latin America

Littoralization

As elsewhere, the overwhelmingly predominate trend in Latin America—for the purposes of this book, both coasts of South America and the Pacific coast of Central America—is the movement of people from the countryside to the cities. This vast region already contains the highest number of urban dwellers in the world: An average of 70 percent of the population overall is urban. As rural economies continue to deteriorate, it is a trend that shows no signs of slackening. Of the region's seventy-seven major cities, nearly sixty are coastal. Consequently, as the population becomes more and more urban, it also becomes more and more littoral.

The serpentine west coast of Latin America, extending from the U.S.–Mexico border south to the tip of Chile, covers more than 20,000 kilometers. The east coast, from Brazil to the tip of Argentina, snakes along for more than 13,000 kilometers (see table 9.1). This region has many of the same problems that beset the Caribbean. It is plagued by rapidly growing coastal populations. Upland forests have been destroyed to make way for subsistence agriculture and grazing lands for cattle and sheep. Mining operations on the west coast have reduced entire mountains to rubble, denuded watersheds, and caused massive erosion and siltation of rivers. Industries pump millions of tons of effluents into coastal waters, while rivers bring in toxic mine tailings and raw sewage. Municipal wastes from coastal towns and cities are dumped untreated into the ocean.

Besides degraded coastlines, two other factors are shared by all Latin American countries: per capita reduction of farmland and a legacy of pervasive poverty. Latin America's coastal states have a collective population of nearly 440 million people; 70 million of them survive on less than $1.00 a day. Such widespread poverty undermines attempts at sustainable resource management. But even in rapidly developing Chile, where unemployment is down to 4 percent, a booming economy has resulted in more unregulated coastal development and more pollution of coastal waters not less.

Population

Close to 75 percent of South Americans are already urban, a figure that is expected to keep climbing.

Table 9.1. Length of Coastline and Population Data for the Pacific Coast of Latin America and Atlantic Coast of South America

COUNTRY	COASTLINE— PACIFIC (KM)	POPULATION 1997 (MILLIONS)	POPULATION GROWTH (%/YR)	% URBAN
Mexico	5,000*	95.7	2.2	71
Guatemala	252	11.2	3.0	39
Honduras	50*	5.8	3.2	47
El Salvador	307	5.9	2.6	45
Nicaragua	460	4.4	3.1	63
Costa Rica	1,164	3.5	2.0	44
Panama	1,690	2.7	1.9	55
Colombia	1,300	37.4	2.1	70
Ecuador	2,859	12.0	2.3	59
Peru	2,414	24.4	2.2	70
Chile	6,435	14.6	1.5	86

COUNTRY	COASTLINE— ATLANTIC (KM)	POPULATION 1997 (MILLIONS)	POPULATION GROWTH (%/YR)	% URBAN
Brazil	7,491	160.3	1.4	76
Uruguay	660	3.2	0.8	90
Argentina	4,989	35.6	1.2	87

Sources: World Population Data Sheet, 1997, Population Reference Bureau, Washington, D.C., 1997; *World Resources 1996–97,* Oxford University Press, New York, 1996, pp. 268–69; Gordon Foer and Stephen Olsen, eds., *Central America's Coasts: Profiles and an Agenda for Action,* USAID, Coastal Resources Center, Washington, D.C., 1992, pp. 63–259.

Note: *Figures are estimates.

The West Coast

With the sole exception of Colombia, the majority of the west coast region's people live along the Pacific coast or within 200 kilometers of it. Roughly 40 million people inhabit this coastal zone, most of them in rapidly expanding urban areas.

As in much of the rest of the region, Ecuador's four mainland coastal provinces—Esmeraldas, Manabi, Guayas, and El Oro—have been growing faster than the country as a whole: their populations have quadrupled since 1950, while the national population increased threefold. There are now twenty-two coastal urban centers, four of them with populations in excess of 100,000: Machala, Guayaquil, Manta, and Esmeraldas. In 1990, half of all Ecuadorans—over 5 million—were coastal dwellers. The percentage is even higher today; close to 60 percent live along or within 200 kilometers of the coast (Foer and Olsen 1992; Population Action International 1990).

Although Ecuador's population as a whole is growing by 2.2 percent a year (as of 1996), its largest city, Guayaquil, has been expanding by 4.5 percent a year since the 1980s. If this rate were to continue indefinitely, the city would double its population every fifteen years. Most of this growth is attributed to migration from the impoverished countryside. Guayaquil now has over 2 million inhabitants in an area that was designed to accommodate around 1.5 million. Many of the new arrivals live in slums and squalid shantytowns built over the blackened remains of mangrove swamps or next to sewage outlets and garbage dumps.

Already, 86 percent of Chile's population is urban, as are 70 percent of Peru's and 70 percent of Colombia's. In Chile, three-quarters of the population lives and works along a 500-kilometer stretch of coastline between Valparaiso and Concepción, on 15 percent of the total land area. The Lima-Callao metropolis contains over 8 million of Peru's people, 30 percent of the country's total. Most Panamanians live on the Pacific coast, 800,000 of them in Panama City. And little El Salvador's 6 million people are all coastal.

Attempts to encourage the growth of secondary towns away from coastal areas have not had much success. Peru launched a decentralization drive in 1959 in an effort to encourage the development of industrial centers in the largely neglected interior. Despite the creation of industrial parks in cities such as Cuzco, Huancayo, and Pasco, the Lima-Callao area still contains 65 percent of the country's industries.

Colombia, with two coasts to choose from, developed its more accessible Caribbean side. As a result, its population is concentrated in a fertile crescent sprawling over the north-central part of the country and along the Caribbean. Only within the last few years has the government come up with plans to develop its sparsely populated Pacific coast, which at last count contained only 615,000 people spread out along 1,300 kilometers of mangrove swamps. Planners in Bogota joke that one of the reasons the Pacific coast has never been developed is because the area's most prominent residents are mosquitoes, many of which carry malaria and yellow fever. But actually, the country's Pacific coast has been neglected because of political indifference, the absence of an infrastructure, and its remoteness from traditional markets. The only city on the Pacific coast is Buenaventura, which is linked by a road network to the rest of the country. Surprisingly, it is in small, out-of-the-way towns like Tumaco, hard against the Ecuadoran border, that new patterns of development are taking shape (see "A Success Story in Tumaco," p. 127).

Given the intense level of development along the rest of the Pacific coast, the governments of Ecuador, Peru, and Chile may wish they had Colombia's options. As more people press upon the coast, pollution mounts and resources are destroyed. "We have heaped this abuse upon ourselves," claims a government planner in Peru. "We have not managed to distribute our growing populations away from the coast. We have not been able to encourage industries to go elsewhere, and so there are fewer jobs in the interior. Our uplands are starving for development, while our coasts are drowning in it."

The East Coast

The combined populations of Brazil, Uruguay, and Argentina amount to a little more than 199 million people, 160 million of them in Brazil. Urbanization rates in those countries are the highest in Latin America: 87 percent of Argentina's citizens are urban, 90 percent of Uruguay's, and 76 percent of Brazil's.

The Atlantic coastline of South America is even more heavily developed and crowded than the Pacific side. Beginning with the Buenos Aires–La Plata conurbation in Argentina, across the La Plata River estuary to Montevideo, the capital of

Uruguay, and up the coast of Brazil to Recife, cities and towns are stacked against each other like cords of wood. Some 15 million people live in the Buenos Aires–La Plata–Montevideo region. But the largest and most crowded coastal area by far is the highly urbanized region stretching from São Paulo to Rio de Janeiro, Brazil. That area already bulges with some 30 million people; if trends continue, it is expected to hold 40 million or more inhabitants by 2010 (Sadik 1993).

By the turn of the century, Brazil fully expects to have around 80 million people living and working in its broad coastal zone. The overwhelming majority of them, close to 90 percent, will be living in large urban agglomerations, compounding the problems already faced by urban mayors struggling to provide their citizens with essential services.

Pollution

There are virtually no working sewage treatment plants on the continent of South America, but most large cities now have municipal sanitation projects, funded by the World Bank and the InterAmerican Development Bank (IDB), underway. Meanwhile, both the Pacific and the Atlantic coasts are on the receiving end of millions of tons of pollutants. Most major river systems bring in poisons from mining operations in the Andes, pesticide residues from agricultural land, and tons of untreated sewage and municipal wastes. Nearly every urban area along both coasts flushes its untreated wastes into coastal waters (UNEP 1988b, 1992).

The Pacific Side

"Basically, there are no functional sewage treatment plants on the entire west coast," according to Jaìro Escobar Ramìrez, former coordinator of the South-East Pacific Action Plan. "Industries don't treat wastes either: everything goes into coastal waters—organic wastes from fish and food processing plants and slaughter houses; mine tailings; and toxins from tanneries, metal smelters and chemical plants."

El Salvador's coastal waters are highly polluted with untreated sewage. Fecal coliform counts in the Gulf of Fonseca routinely exceed 1,000–10,000 per 100 milliliters of water (the standard recommended by WHO is no more than 10 per 100 milliliters). Intestinal disorders and diarrhea from poor environmental conditions account for over 56 percent of all illness (Foer and Olsen 1992).

In the Gulf of Nicoya, on the Pacific coast of Costa Rica, some 51,000 kilograms of untreated industrial effluents per day flow into near-shore waters, along with at least 40,000 kilograms of municipal wastes and 190,000 kilograms of coffee-processing wastes (Whelan 1989).

Panama Bay receives some 34 million metric tons of untreated sewage a year from three rivers and twenty outfalls in Panama City and the port of Balboa. The pollution has created anoxic areas in the bay, where oxygen-starved waters support no fish. Pollution and sedimentation have caused a drastic reduction in the diversity of marine life in the bay (D'Croz 1988).

Pollution is contributing to a decline in the catches of shrimp and anchovies. The Panama shrimp catch alone is worth on average around $70 million a year. Trawlers now have to go farther out into the bay to find economical quantities of shrimp and fish. As catches decline, overfishing becomes rampant. In 1988 the government imposed a two-month closed season on shrimping, as well as regulations on the type of nets used and the number of boats operating in efforts to bring over-harvesting under control. However, slack enforcement has resulted in a continuing deterioration of shrimp and anchovy stocks.

Also in Panama, near-shore mussel beds, too contaminated to be harvested, had to be moved. Poor artisanal fishermen can no longer make a living from the bay's polluted waters. Instead, some have turned to smuggling.

The only heavily polluted area on Colombia's mangrove coast is Buenaventura Bay, contaminated with organic wastes from fish-processing plants, untreated sewage, assorted municipal wastes, and oil. Parts of the bay are now rich in nitrogen and ammonia and low in oxygen, due to the massive discharge of raw sewage from the city and three river systems. Municipal wastes flow into the bay at the rate of 54,000 cubic meters per day. Bacteria counts are exceptionally high, as algae, fecal matter, and garbage collect around boats in the harbor. The major cause of death for children under one year of age in Buenaventura is diarrhea, caused by drinking contaminated water and eating tainted fish and shellfish (Hinrichsen 1990).

Ecuador, with its highly developed coastline, suffers from acute deterioration of its coastal waters. Its coastal cities are responsible for pumping roughly 90 million cubic meters of wastewater a year into the Pacific; 60 percent of that is accounted for by Ecuador's largest city, Guayaquil (Robadue and Arriaga 1993a).

Elsewhere in Ecuador, urban populations are outstripping city infrastructures. The coastal city of Esmeraldas has a sewer system designed to accommodate no more than 30,000 people, while its current population (1997) is just under 200,000. The percentage of coastal residents with access to potable water and adequate sanitation services actually declined over the decade of the 1980s: from 20 percent in 1982 to 17 percent in 1990 (Robadue and Arriaga 1993a; Robadue 1995).

In addition, Guayaquil Bay receives the wastes from the Guayas River Basin, which covers 34,000 square kilometers, the second largest on the west coast of South America. Its waters dump sewage, pesticide residues, and heavy metals from mining operations into the bay. Copper concentrations in the Guayas River and its tributary, the Daule River, exceed safe levels. Some of the toxicity of the copper, however, is thought to be offset by the huge quantities of domestic sewage found in Guayaquil Bay. Health effects from heavy metals have not yet been detected, perhaps because many of the bay's poor people suffer from chronic stomach and intestinal disorders brought on by eating contaminated seafood and drinking dirty water. Viral hepatitis is epidemic, as is malaria.

In Peru, mining operations in the Andes routinely flush millions of tons of polluted mine tailings into rivers and streams, which carry them to the coast. In the mountainous province of Tacna, on the Chilean border, two huge copper mines pump more than 73 million metric tons of mine debris and tailings into local rivers every year. So much of it ends up in the Pacific that near-shore waters contain twenty-one parts-per-billion of copper, enough to endanger local fisheries.

Untreated sewage is a serious problem in the Bays of Valparaiso and Concepción, Chile. But industrial pollution is even worse. Every year, both bays receive a combined total of some 244 million cubic meters of untreated effluents, mostly from copper mines, pulp and paper mills, fish-processing plants, and oil refineries. The Bio Bio River, which flows into the Gulf of Arauco, just south of Concepción Bay, is laden with sewage and heavy metals, particularly mercury and lead, from towns and industries in its watershed. Nationwide, Chile dumps over 700 million cubic meters of domestic waste into its coastal waters every year (UNEP 1988b; Boraiko 1995).

As in Peru, mining is also a major source of coastal pollution in Chile. In the northern part of the country, copper and gold mines flush nearly 12 million metric tons of polluted tailings and process water directly into the sea every year. Since mercury is used in refining gold ore, the mercury level in those waters is extremely high; its effects on marine life have not been studied, however.

Pesticides are yet another source of pollution. The Pacific side of South America uses some 60,000 metric tons of pesticides every year, most of it on croplands. Pesticides banned in North America such as DDT, are still being used. Panama Bay, the estuary of the Guayas River in Ecuador, and Concepción Bay and San Vicente Gulf in Chile are contaminated with pesticide residues. Little has been done to reduce the use of deadly pesticides or to study their effects on coastal ecosystems.

The Atlantic Side

The east coast of South America suffers even more damage from sprawling coastal urbanization and industrialization. The near-shore waters of virtually every coastal city are heavily contaminated with untreated municipal and industrial wastes.

The Rio de la Plata estuary, between Argentina and Uruguay, is so saturated with toxic industrial wastes that tens of thousands of fish perish every year. Even worse, the water supply of Buenos Aires is seriously threatened from industrial poisons, which seep into groundwater aquifers. Bathing beaches near Buenos Aires are routinely closed during the summer months because tons of toxic industrial wastes and untreated sewage linger in shallow coastal waters, posing serious health risks.

Farther north, the heavily populated coastline from São Paulo to Rio de Janeiro discharges millions of tons of untreated industrial and municipal wastes into near-shore waters. Untreated industrial and municipal effluents pour into the Santos Estuary, near São Paulo, at the rate of 90 cubic meters per second (*Economist* 1996). Along the northeast shoulder of Brazil, rivers like the São Francisco bring in tons of pollution from mills processing sugar cane and alcohol.

Resources

Most of Latin America's lush tropical forests are being converted into pastureland, subsistence farms, and cash crops. In the seventeen-year period from 1970 to 1987, 50 million hectares of tropical forest in Latin America were destroyed (Hinrichsen 1990).

Panama's Pacific watershed has been logged and the forests replaced with cattle ranches. Between 1980 and 1990, the country lost, on average, around 64,000 hectares of forests a year, a rate that converts to almost 2 percent a year. The only region with intact tropical forests is the San Blas, on Panama's Caribbean coast, controlled and defended by the Kuna Indians.

Ecuador's lush interior tropical hardwood forests are being cleared and the uplands given over to farms and cattle ranches. Ecuador's pastureland has more than doubled in area since the 1960s, at the expense of its forests. There are now only a few remnants of true wilderness left in the country. Many of Ecuador's remaining forests consist of biologically impoverished secondary growth.

Chile and Peru still have extensive tracts of unlogged woodland in the Andes watershed. Each country has around 30 percent of its total land covered by wilderness. But both countries are losing timber faster than they are replacing it. Over the decade of the 1980s, Peru lost, on average, around 270,000 hectares a year, and Chile lost around 50,000 hectares.

As upland forests give way to mining, agriculture, and ranching, soil erosion is one of the inevitable consequences. In Colombia's central highlands around Bogota, deforested hills are being washed into the valleys. Hillsides are slit with huge gulleys carved out by torrential rains. Rivers and reservoirs are silting up.

The Tomine Hydroelectric Dam, 60 kilometers from Bogota, near the village of Sesquile, was faced with closure because of eroded sediment flowing into its 690-million-cubic-meter storage reservoir. The Tomine Electric Authority fought back with a major reforestation campaign. During the late 1980s, some 1.2 million trees—80 percent of them native acacia—were planted on denuded slopes. By 1992, the project had stabilized 1,200 hectares of badly eroded hillsides. "We managed to greatly reduce water erosion rates, and the trees also reduced wind erosion," states a spokesman for the electric authority.

In the uplands of Ecuador, Peru, and Chile—particularly in mining areas—deforested hills bear huge erosion scars, as if some giant had slashed the slopes with a knife, slitting them vertically from top to bottom. Like open wounds, they bleed soils into rivers and streams. In certain parts of Peru and Chile, entire mountains have broken apart during heavy rains, sending avalanches of mud and rock into valleys, burying entire villages. Soil erosion in Peru averages 15 metric tons per hectare per year across the country.

Mangrove forests, the regulators of coastal erosion, are being lost on both coasts. Over the past forty years, Central and South America have probably lost at least half of their original mangrove area (see table 9.2). All but 10 percent of Panama's mangroves are concentrated along the Pacific coast, where they now total around 17,000 hectares. Conversion to brackish water shrimp ponds is continuing despite economic difficulties. Luis D'Croz calculates that the country is losing 1 percent of its mangrove resources every year (D'Croz 1988). "There is some hope that we will still be able to preserve a sizable portion of our mangroves," says D'Croz. "With the costs of opening up new areas getting more expensive, shrimp farming is now less attractive than it was a few years ago."

Colombia's mangroves are largely intact, occupying 280,000 hectares along the Pacific coast—the largest contiguous stand in Latin America. Still, there are fears that as Colombia's Pacific coast becomes more developed, mariculture will accelerate as it has in Panama and Ecuador. Already some 5,000 hectares of mangroves around Buenaventura have been converted into shrimp ponds. Mexico's mangroves have been severely affected by mariculture, agricultural expansion, and

*Table 9.2. Comparative Mangrove Area on the Pacific Coast in
Selected Latin American Countries*

COUNTRY	PACIFIC COAST— MANGROVE AREA 1960S (HA)	PACIFIC COAST— MANGROVE AREA 1990 (HA)
Guatemala	23,000	16,000
El Salvador	120,000	30–45,000
Honduras	50,000	25,000
Nicaragua	80,000	60,000
Panama	50,000	17,000
Ecuador	203,000	161,000

Sources: D. Hinrichsen, *Our Common Seas: Coasts in Crisis,* Earthscan, London,
1990, pp. 90–91; Don Rabadue, "Ecuador's Coastal Resources Management Program," *Intercoast Network,* spring 1993, p. 2; Gordon Foer and Stephen Olsen, eds.,
Central America's Coasts: Profiles and an Agenda for Action, USAID, Coastal Resources Center, Washington, D.C., 1992, pp. 104, 136, 174, and 205; Stephen Olsen,
Donald Robadue, and Luis Arriaga, *A Participatory and Adaptive Approach to Integrated Coastal Management: Ecuador's Eight Years of Experience,* Coastal Resources
Center, USAID, University of Rhode Island, Narragansett, 1994, p. 5.

Note: Most of these figures are based on rough estimates, and some have been
rounded off to the most approximate area, combining several different studies.

urban development. Estimated to cover 1.5 million hectares in 1970, they had been
reduced to no more than 50,000 hectares by 1996 (Spalding, Blasco, and Field
1996). A mere 3 percent of Brazil's Atlantic forests are still standing. The rest—97
percent—has been lost to timber merchants, urban and industrial expansion, and
agriculture (Brooke 1992).

Coral reef resources in the region are not extensive. Most Central American and
Colombian reefs are in the Caribbean not the Pacific. Because of waters rich in nutrients, fisheries in the Pacific region are not reef dependent. The only reefs of any
size are in Panama Bay, around the Galapagos Islands, and in the near-shore waters
of Esmeraldas, Ecuador, and Gorgona Island, Colombia. The black coral resource
near Esmeraldas is being heavily exploited for the tourist trade. Brazil has a few
reefs in the south but little in the way of management.

Fisheries

The cold, nutrient-rich waters swept up the Pacific coast of South America by the
Humboldt Current and its offshore counterpart ensure that the waters off the coasts
of Chile and Peru are teeming with marine life. Local upwellings in the coastal waters of Ecuador, Colombia, and Panama are met by wedges of freshwater brought in
by tropical rivers.

The annual take of fish and shellfish from the west coast increased from 10 million metric tons in 1985 to over 14 million metric tons in 1993, making this region
the second largest fish-producing area in the world after the northwest Pacific. The
deep-water fleets of Peru and Chile landed 8.4 million metric tons and 6 million
metric tons, respectively, in 1993, mostly anchovies, sardines, and Chilean horse
mackerel. As of 1992, 70 percent of the region's commercial pelagic species were
fully exploited, overexploited, depleted, or slowly recovering, along with 60 percent
of the region's demersal species (FAO 1995). As a result, catches are expected to
level off, perhaps drop.

Panama Bay is rich in marine life because of an upwelling of deeper, nutrient-laden water. Nearly 90 percent of the Panamanian commercial fishery is based in Panama Bay and the Gulf of Panama. Three species of white shrimp are caught, amounting to around 7,000 metric tons a year. Many anchovies are also taken, and scallops are harvested by artisanal fishermen working in the less polluted southern part of the bay. But mounting pollution is exacting a toll on the bay's fisheries. "At the moment, we are not getting the maximum sustainable yield from our fisheries because of overfishing and pollution of near-shore waters," states D'Croz.

In Ecuador the fishing industry ranks second, after oil production, as a source of foreign revenue. The total amount of fish and shellfish exported amounts to roughly 300,000 metric tons a year. In 1992, over 145,000 hectares of shrimp farms exported 116,000 metric tons of shrimp worth $525 million. Most of the country's farmed shrimp go to markets in North America and Japan (Robadue 1995).

On the Atlantic coast, total catches amounted to a little over 2 million metric tons in 1993. Brazil accounted for most of that, hauling in close to 800,000 metric tons (FAO 1995). But that fishery, too, is reaching the limits of exploitation. Future increases in the southwest Atlantic fishery are unlikely.

In the meantime, coastal pollution is ruining some traditional fishing grounds. Nearly 500 kilometers north of Rio de Janeiro, in Brazil's state of Espírito Santo, pollution caused by a wood pulp operation has resulted in massive pollution of shallow coastal waters. Chemical discharges by Aracruz Celulose, the world's largest exporter of wood pulp, have turned coastal waters a dingy brown and killed off many species of commercially valuable fish.

"This used to be one of the best fishing areas in the country, but local fisheries have been devastated," says Jaoa Pedro Stedile of Brazil's Landless Workers Movement. "Over 50,000 people in the area used to eat fish every day. Now they eat fish no more; some fishermen have stopped fishing because there are so few fish to catch" (Panos Institute 1995).

A Success Story in Tumaco

The Pacific coast of Colombia is unique: it is virtually undeveloped. From the metropolis of Buenaventura in the north to the small fishing community of Tumaco near the Ecuadoran border, the mangrove coast stretches for over 800 kilometers. It is one of the most unspoiled mangrove wildernesses left on earth, broken only by meandering brown-water rivers that flow lazily into the Pacific (see Hinrichsen 1989). Along the entire mangrove coast only one town of any size appears on a map—Guapi, situated at the mouth of the Guapi River.

Unable to coax much of a living from the land, the people of the mangrove coast turned to the sea. The dark, nutrient-rich waters harbor a wealth of marine life: everything from giant prawns to gray whales. But it is shrimp (*Penaeus occidentalis*) that sustains most of the 5,000 small-scale fishermen who ply the coast in simple dugout canoes or small motorboats. Since there are no large trawlers operating in those waters, the region's fisheries are underdeveloped. Only around 17,000 metric tons of fish and shellfish are taken every year, a mere fraction of the annual potential yield, which has been estimated at between 130,000 and 150,000 metric tons (Hinrichsen 1989).

Colombia's southernmost coastal town, Tumaco, is a study in sharp contrasts. Rickety wooden houses line unpaved streets, but nearly every house sprouts a TV antenna. The town lacks an electricity grid, and running water is piped in from a nearby river untreated when the pump is working. Yet you can make a telephone call to Bogota, New York, or London with ease.

What electricity there is in Tumaco tends to be a family affair. Everyone who can afford a generator has one, but they are fired up only at night and are used mostly to run the family television set. The proliferation of television in rural areas is jokingly referred to as Colombia's "family planning program," but in Tumaco, employment for women is a much better form of contraception. And this rare commodity is being supplied by the town's most famous entrepreneur, Luis Buitrago. At forty-seven, he is Tumaco's biggest commercial success.

Starting with one shrimp boat and six employees, within eight years Buitrago managed to build up a shrimp export business worth more than $1 million a year. Today, his shrimp-and-fish-processing factory employs 180 people, mostly young women, and 12 men work on two shrimp boats. He runs his expanding business empire out of Tumaco with his younger brother, Alvaro, forty-four. Most of the large shrimp are exported to the lucrative United States market under the Pacific Pearls brand name.

"It was really difficult to get started," explains Buitrago. "Twenty years ago my brother and I were just simple fishermen struggling to make ends meet. Our family was poor by any standards. But we saved our money and borrowed from the bank to buy our first shrimp boat."

In those days, the Buitrago brothers sold their shrimp catches to a processing factory in Buenaventura. But transport distances were too great and the profits too thin. Luis decided to set up his own processing factory in Tumaco and export directly to the United States. The factory also cleans, packs, and freezes a variety of other fish, which are then flown to Cali in central Colombia and sold to fish vendors.

The economy of the entire region has improved. "This coast used to be underdeveloped in everything except mosquitoes," jokes Luis, "but now we have established a thriving fishing industry." Buitrago buys most of his shrimp directly from small-scale artisanal fishermen, who ply the coastal waters in dugouts. He has even supplied most of them with nets. This system not only spreads the income around but also ensures that fishing pressures do not build up to such an extent that the shrimp resource is overexploited.

"Our future plans include buying a big tuna boat and competing with the American and Japanese deep-water fleets that presently take all the tuna off Colombia's Pacific coast," states Luis. "The government hasn't bothered to develop the fishing potential of this coast at all," he observes, "but we have big plans."

El Niño

The disruptive, and destructive, weather patterns that periodically sweep through the Pacific have come to be known collectively as El Niño, which translates as "the child." When this weather pattern struck the west coast of South America in 1972, the anchovy fishery, worth millions of dollars, collapsed. Years of overharvesting had contributed to its demise.

El Niño can perhaps best be understood as a sort of ocean drought. Climate-controlling ocean-atmospheric interactions are disrupted on a large scale. Trade winds and ocean currents that normally flow across the tropical Pacific from the

Americas to Asia abruptly reverse course. This switch causes drought in normally wet areas of Indonesia and Australia, interrupts the Indian and East African monsoons, and brings storms to normally calm Pacific islands and the west coast of the Americas (Pearce 1993).

Temperatures climb and surface waters become much warmer than normal. Warmer surface temperatures cause more nutrients to be sedimented out of the water column, depleting coastal waters of the food needed to sustain marine life. With food sources scarce and water temperatures becoming unbearably warm, cold-water fish go elsewhere.

Those creatures who cannot leave learn to take advantage of the warmer waters or perish. During the El Niño of 1982–83, coral reef communities in the Gulf of Panama simply died off as water temperatures soared to 31 degrees Centigrade. Tropical crabs, on the other hand, were able to expand their breeding ranges farther south, thanks to the warmer waters.

That El Niño was one of the worst on record. Heavy rains drenched Colombia, Ecuador, and Peru. In northern Peru, rainfall was 340 times greater than normal. Some rivers carried over 1,000 times their normal flow. The widespread flooding that ensued took a terrible toll on crops, livestock, roads, bridges, schools, and homes. In Ecuador, 40,000 families were made homeless by flood waters. Unable to catch anything in the warmer surface waters, tens of thousands of fishermen across the eastern and southern Pacific were left idle. Shrimp, however, benefited from the abundant rainfall and warmer waters, reproducing in great numbers. The El Niño, in part, prompted the boom in shrimp farming in Ecuador because of the abundance of larvae.

Another El Niño struck in 1990 and lasted through 1995, the longest on record. Some researchers warn that El Niño events could become more severe as the world's climate warms.

Management

In November 1981, the five countries of the southeast Pacific region—Panama, Colombia, Ecuador, Peru, and Chile—met in Lima and collectively approved the Convention for the Protection of the Marine Environment and Coastal Areas of the South-East Pacific. At the same meeting, they approved an Agreement for Regional Cooperation to Combat Pollution of the South-East Pacific due to Oil Hydrocarbons and other Noxious Substances in Cases of Emergency. The Lima meeting was the product of four tough years of preliminary negotiation carried out under the auspices of the Permanent Commission for the Southeast Pacific (CPPS) and the region's five governments in cooperation with the United Nations Environment Programme's Oceans and Coastal Areas Programme Activity Centre (OCA/PAC), headquartered in Nairobi, Kenya. Subsequently, in 1983, another important protocol was added to the convention: the Protocol for the Protection of the South East Pacific Against Pollution from Land-Based Sources. They were all duly ratified and came into force in 1987.

One of the first things the CPPS did was to set up a network of collaborating agencies in each of the five member states. Next, a research program was worked out and approved in an effort to assess the state of the marine environment and to pinpoint pollution hot spots in need of immediate attention. "When we started this program, we had no information at all on the state of our coastal waters and marine resources," says Jairo Escobar Ramirez. "We are still upgrading our database, but a solid beginning has been made." The pollution monitoring program now has its own network and its own name: CONPACSE. Baseline pollution studies and assessments have generated a wealth of data and scientific expertise: Over sixty technical reports have been completed by some forty-two regional institutions.

As of January 1993, CPPS had supervised the training of more than 1,100 experts in marine pollution evaluation, environmental impact assessment, and analytical techniques. "We have two phases to the program," states Ramirez, "to provide basic information on the state of the region's marine resources, and to promote sustainable land-use planning."

In the late 1980s, a priority action plan was worked out for the region. It includes three areas: the control and treatment of municipal and industrial wastes; the development of artisanal fisheries; and the promotion of sustainable mariculture and aquaculture industries.

In 1992–93, the CPPS adopted two new protocols: on environmental impact assessment in marine and coastal areas, and on the prohibition of the transfrontier movement of wastes and harmful substances in the southeast Pacific. It is also the first of UNEP's regional seas programs to have its own Plan of Action for Marine Mammals. However, the implementation of those protocols has been slow, and funding has not been commensurate with commitments.

"We have managed to educate the experts," notes Ramirez. "Next we have to educate the public about the need to conserve coastal resources. We have already launched a program in Buenaventura to teach high school students about the marine environment and its importance to them."

Ecuador's Experience

One country in South America has managed to forge ahead: Ecuador. With technical help from the Coastal Resources Center of the University of Rhode Island and aid money from the U.S. Agency for International Development, Ecuador created five special-area management plans for its four mainland coastal provinces.

One of the most important and immediate challenges facing the country's coastal areas, as discussed in chapter 3 (p. 33) was the unregulated conversion of its mangrove wetlands and salt flats into shrimp ponds. By 1991, the country had lost 42,000 hectares of mangrove forest and over 45,000 hectares of salt flats.

In some estuaries, nearly all mangroves were wiped out, precipitating the collapse of capture-shrimp fisheries, the bulk of which are harvested by small-scale, artisanal fishermen. Poor housekeeping practices by the shrimp farms polluted near-shore waters with excessive nutrients. In addition to the overuse and contamination of coastal resources, conflicts among activities such as fish processing and tourism plagued many areas. With the populations of coastal towns and cities exploding as well, triggering a general crisis in municipal services, Ecuador's coastal areas were turning into zones of contention.

It was apparent that some form of coastal governance was needed to bring order out of the economic anarchy that dominated the country's coastal provinces. Ecuador elected to introduce well-defined management plans, addressing specific issues, for limited sections of coastline. Although these five special-area management programs cover a mere 8 percent of the country's coastline, each was carefully designed with extensive input from local communities and both regional and national authorities.

"We took the two-track approach to Ecuador's coastal problems, working from the top down and the bottom up to create broad-based support for the management plans," points out Don Robadue, a marine management specialist working with the CRC. According to Robadue, the management strategy in each area had five primary objectives: (1) to evolve a workable program that would not exhaust the limited financial resources of the country; (2) to devise a program that would create income for local people and "chains of enterprise"; (3) to build up the capacity of indigenous institutions and local citizen groups to implement and maintain management plans; (4) to create a "climate of transparency" within the governing institutions involved in the program; and (5) to promote the equitable distribution of resources.

After nearly a decade of hard work, the five management plans have been implemented by government decree. New shrimp ponds must now be built farther inland, behind mangrove swamps, on old agricultural land. A more effective system for regulating the housekeeping practices of shrimp farms, so that they pollute less and produce healthier shrimp, has also been developed. A "ranger corps" was created under the auspices of the Ecuadoran navy to patrol coastal waters in an effort to prevent the decimation of more mangrove forests, especially those near the border with Colombia, and to work closely with local management committees on common problems.

Captain Walter Neito Bueno, in charge of the Esmeraldas Ranger Corps, thinks the new coastal management system established in Ecuador is working well:

> The management plan has a strong sense of civic responsibility, and it is important that the Navy be involved in it. Conserving natural resources is not just a task to perform, it is essential for the country, just as it is vital that agencies and user groups work together to enforce environmental laws. The ranger corps is one of the best projects in Ecuador, in part because it is a patriotic activity and is building a sense of citizenship. Nothing could be more important.

Robadue believes that the five management plans in operation can easily be replicated in other coastal areas, since they cover the entire range of coastal zone issues affecting the country. "Most important," notes Robadue, "is that we created a process of coastal governance, one based on community involvement on one hand and national political commitment on the other."

Brazil's Experience

On the east coast, only Brazil has launched a national coastal zone management program. In May 1988, a National Coastal Zone Management Plan was voted into law. It has three main components (Herz 1990; Ministry of the Environment 1996):

- creation of a National System for Information on Coastal Management (SIGERCO), encompassing a geo-information, satellite-linked database and an online computer network connecting all seventeen coastal states;

- implementation of a zoning system for the Brazilian coast under the direct supervision of the states but working together with the federal government;

- introduction of decentralized management plans and monitoring systems tailored to fit the primary needs of each coastal state.

Soon after the legislation was passed, ten of the country's seventeen coastal states and three coastal territories began to develop prototype coastal management strategies.

The national law set the boundaries of the coastal zone at 2 kilometers inland from the mean high tide level and 12 nautical miles seaward. Those boundaries created a coastal zone governance area of approximately 200,000 square kilometers (2 million hectares). However, that effort proved too ambitious and complicated to implement. Moreover, the 1996 budget of the entire program amounted to no more than $2 million, far too little to do the job as set out in the 1988 law.

Brazil is now trying to beef up its planning and enforcement capabilities in coastal areas. In 1996, the federal government was busy restructuring the entire fisheries sector in an effort to manage dwindling commercial stocks.

"One of the key elements in successful coastal management programs is to start small and work up slowly," says Stephen Olsen, director of the Coastal Resources Center at the University of Rhode Island. "This way you can create a system built on successes rather than failures."

Conclusions

- All of Latin America needs to introduce sound urban planning systems, if for no other reason than to take account of growing coastal urban populations. Urban planning could take pressure off degraded areas and help to manage remaining resources. But above all, it could help make the region's sprawling cities liveable, not just for the rich, but also for the majority of urban dwellers.

- Resource management plans that take account of the needs of coastal communities and commercial interests need to be developed and implemented. A *modus vivendi* needs to be found that will permit growth but curtail its excesses. Since it is unlikely that regional plans will prove viable, each country with a coastline must forge its own management strategy and find ways to make it work.

- Regional solutions may be necessary in order to evolve management plans for fisheries that work over the long run. The Permanent Commission for the Southeast Pacific (CPPS), a region-wide organization now based in Santiago, Chile, should be pressed into service to assist in regulating fisheries, particularly near-shore, artisanal fisheries.

- If coastal management issues are to become permanent fixtures in the political arena, then broad public constituencies must be created to demand that coastal areas be managed sustainably.

CHAPTER 10

The South Pacific

Water World

The South Pacific's twenty-four island states and territories, including Australia and New Zealand, consist of over 10,000 islands spread out over 41 million square kilometers of ocean. Disregarding Australia, New Zealand, and Papua New Guinea, only 2 percent of the area is land. Many of the islands are so small they have to be exaggerated to be seen on a map.

The Pacific's archipelagos are isolated from each other by hundreds of kilometers of sea. These vast distances are made wider by ethnic and cultural divisions that split the Pacific into three distinct parts: Micronesia, Melanesia, and Polynesia. Even within these major groups, island cultures are often at odds with each other over political, economic, and social issues. Over a third of the world's languages are spoken in just four Melanesian states—Papua New Guinea, the Solomon Islands, New Caledonia, and Vanuatu. In Papua New Guinea alone, some seven hundred languages are spoken.

"The picture postcard image of the South Pacific is, in many ways, false," says Dr. John Pernetta, a consultant for UNEP's regional seas program based in England. "Behind the facade of palm trees waving gently in the breeze is the stark reality of grinding poverty set against a backdrop of some really basic environmental problems."

The region is beset with problems: widespread destruction of mangrove forests and coral reefs around population centers; runaway urbanization; overfishing; deforestation of upland and coastal forests; soil erosion; misuse of pesticides; pollution of rivers and drinking water supplies; improper disposal of industrial and municipal wastes; lack of sewage treatment facilities; and biological impoverishment.

As Western-style economic development intruded into the region, many island communities found themselves undermined by the very development that promised a better life. Limited island resources are capable of supporting limited populations. Too many people pushing upon coastlines ruin coastal resources and economically important fisheries. Too many farmers in the hills bring about deforestation and erosion of fragile tropical soils, compounding the problems of the coasts. Too much big development, without regard for local conditions or needs, often creates more problems that it solves.

Geography

The Pacific is the largest expanse of ocean in the world, an area nearly twice the size of the former Soviet Union. The distances are daunting. The traveler who wants to go from Guam to Tahiti has to cover 8,000 kilometers. It is 11,000 kilometers from Pitcairn to Belau (Palau), the equivalent of flying from Tromso, Norway, to Cape Town, South Africa. From Papeete, the capital of Tahiti, the nearest continental landfalls are Sydney, 6,000 kilometers west; San Francisco, 6,600 kilometers to the northeast; Panama, 8,200 kilometers east; and Vladivostok, 7,500 kilometers to the northwest.

As in the Caribbean, much of the waters in the Pacific are nutrient poor. In the surface area of the tropical Pacific, the thin, warm, upper layer of water becomes quickly deprived of nutrients because of photosynthesis and the removal of organic matter as a result of sedimentation. In this subsidence zone, no upwelling of the richer, deeper water is possible. Hence, the most extreme conditions of oligotrophy can be found (oligotrophy is the environment in which nutrient concentrations are low and organic production small). These nutrient-poor areas are often called ocean deserts. The very clear water, colored a deep blue, has virtually no suspended particles. And there are few living organisms. Predictably, fisheries are concentrated around island shelves, coral reefs, and current convergence zones.

The Pacific is not only wide, it is also very deep, averaging around 4,000 meters, twice as deep as the Caribbean. The various island chains and archipelagos are sometimes bounded by very deep trenches. The most impressive ocean trench in the world is the Marianas Trench in Micronesia. Its deepest part, the Challenger Deep, plunges over 11,000 meters below the surface.

Population

The total population of Oceania is 29 million people, 18 million of them in Australia. With such a small population over a vast expanse of ocean, human impacts on most island states might be expected to be low. The opposite is true. Perhaps only 500 or so of the South Pacific's 10,000 islands are actually inhabitable, the rest being too remote and inhospitable or lacking in freshwater and cultivable soils. Consequently, population densities tend to be high, up to 410 people per square kilometer. And they are getting worse, as island birth rates continue to average nearly 20 for every 1,000 people, double the rates of western Europe. If Australia and New Zealand are excluded, population growth averages well over 2 percent a year across the region. The population of the Marshall Islands is growing by 4 percent a year, enough to double the current population of 100,000 in just seventeen years (see table 10.1).

As in the rest of the world, the overwhelming majority of the people live along or close to the seacoast, even on the continent of Australia and the larger continental islands of New Guinea and New Zealand.

The interior of Australia is mostly barren, inhospitable wasteland. The majority of Australians live on just a few percent of the continent's vast land area, predominately along the eastern coast from Townsville in Queensland, south to Melbourne in Victoria. Over 80 percent of Australia's population is coastal, with 10 million people living in five big coastal cities and their suburbs: Sydney, Melbourne, Adelaide, Brisbane, and Perth.

Table 10.1. Length of Coastline and Population Data
    for Seventeen States and Territories in Oceania

COUNTRY	COASTLINE (KM)	POPULATION 1997 (MILLIONS)	POPULATION GROWTH (%/YR)	% URBAN
Australia	25,760	18.5	0.7	85
Micronesia	6,112	0.1	3.0	26
Fiji	1,129	0.8	1.8	39
French Polynesia	1,147	0.2	2.0	57
Guam	153	0.2	2.4	38
Kiribati	1,143	0.07	1.9	—
Marshall Islands	370	0.1	4.0	65
New Caledonia	1,249	0.2	1.7	70
New Zealand	15,134	3.6	0.8	85
Palau	1,519	0.02	1.4	69
Papua New Guinea	5,152	4.4	2.3	15
Solomon Islands	5,313	0.4	3.4	13
Tonga	419	0.1	0.7	—
Vanuatu	2,528	0.2	2.9	18
Samoa	403	0.2	2.3	21
American Samoa	116	0.057	3.8	—
Cook Islands	120	0.01	1.1	—

Sources: World Population Data Sheet, 1997, Population Reference Bureau, Washington D.C., 1997; *World Resources 1994–95,* Oxford University Press, New York, 1994, pp. 354, 357; Graham Bateman and Victoria Egan, eds., *The Encyclopedia of World Geography,* Barnes and Noble Books, New York, 1995, pp. 489, 491; Alliance of Small Island States, *Small Islands, Big Issues,* (AOSIS) Washington, D.C., 1997, pp. 51–133.

During the last ten to fifteen years, the growth of secondary coastal towns has been dramatic. Many of them more than doubled their populations in a decade, and new coastal towns have been created where previously there were only a few holiday cottages. This growth has been fueled by two major migration flows: one from the interior of the country to coastal cities and towns, the other from the colder southern coast to the warmer climes of northern New South Wales and southern Queensland.

Nearly the entire population of the smaller island states and territories of the South Pacific—close to 6 million—is also made up of coastal dwellers (Hinrichsen 1990).

With their integration into the world economy, the character of many of the smaller island states and territories has changed dramatically. As rural economies have collapsed, migrants from the interior have swollen city and town populations on the main islands. Three-quarters of all remaining Tahitians, for instance, live in and around the capital city of Papeete. Similarly, 95 percent of the population of American Samoa—about 47,000—live and work on the main island of Tutuila, most of them in the capital of Pago Pago. Increasing population density results in the spread of shantytowns and slums, social dislocation, and environmental degradation (Peau 1991).

Urbanization is a trend that the South Pacific shares with virtually every other region of the world. But small islands, with limited space and resources, can least afford the damaging effects of crowded coasts. On many of the main islands, land prices have soared along with population growth. On Tutuila, a one-quarter acre of land now sells for over $20,000, spawning legal battles over land ownership and tenure (Peau 1991).

Not all migrants have moved to urban centers on larger islands. Many have emigrated to the developed rim countries: the United States, Canada, New Zealand, and Australia. Emigration explains the negative growth rates registered on Pitcairn and Tonga. Auckland, New Zealand, is now the largest Polynesian city in the world; in the early 1980s it already had over 60,000 Polynesians of Pacific origin. Similarly, there are twice as many Niueans in Auckland as there are on Niue and more Cook Islanders than on any of the Cook Islands. Honolulu has more than 20,000 Pacific islanders, including 14,000 Samoans, and Vancouver, Canada, is home to more than 12,000 Indo-Fijians.

Fiji, one of the few developing countries in the region with a comprehensive family-planning program, has seen its growth rates drop from 3 percent in the 1970s to 1.8 percent in 1997. In general, however, the region is characterized by high birth rates and low death rates. Pacific populations are growing rapidly, especially in Melanesia and parts of Micronesia. Much of the growth in Polynesian populations is siphoned off through emigration to the rim countries.

The Marshall Islands serve as a reminder of what can happen if rampant population growth goes unchecked. Because those islands lack resources and employment opportunities, most of the archipelago's population is concentrated in two urban centers: the main city of Majuro, and on the island of Ebeye, residence of the Kwajalein workforce, where U.S. nuclear missile systems and radar are tested. The population density of Majuro is now over 2,188 per square kilometer; that of Ebeye is 25,000 per square kilometer. Visitors have described both places as cramped and horribly polluted. Water rationing has been imposed in an effort to conserve what little freshwater the islands have. In dry months, water is routinely supplied for only one hour every second day, making flush toilets useless. Without landfills, most household garbage and human wastes end up in shallow coastal lagoons.

The problems confronting the Marshall Islands, though extreme, are not unique. Chronic water shortages currently plague over 60 percent of the region's islands, forcing many of them to impose "water hours," certain times of the day or night when water is available for household use. On American Samoa the government–supplied drinking water system is operating at close to 85 percent of its sustainable capacity.

Pollution

The worst problem facing 90 percent of the South Pacific's island states is the disposal of raw sewage and domestic wastes. Except in Australia and New Zealand, there are few sewage treatment plants in working order in the entire region. Nearly all sewage and assorted municipal wastes end up in lagoons or shallow coastal waters (Dahl and Baumgart 1983).

In the interior of the bigger islands, sanitation facilities are virtually nonexistent. Freshwater supplies are often contaminated with household slops and human feces. Intestinal diseases are becoming epidemic. Solid wastes are also dumped in coastal mangroves or lagoons, where they create breeding grounds for disease organisms. Since few islands have developed solid waste management plans, supervised landfills are rare.

The coastlines around nearly every urban center in the Pacific are clogged with untreated sewage, municipal wastes, and household garbage. "One of the main problems is that too many people in the Pacific have acquired a throw-away mentality," states Professor David Mowbray from the University of Papua New Guinea. "In the old days, the people only discarded biodegradables; now they are throwing away everything from plastics and tin cans to car shells."

Industrial pollution is a limited problem in the coastal waters of Australia and New Zealand, but five island states are badly affected: Fiji, New Caledonia, Guam, Papua New Guinea, and American Samoa. The water in Suva Harbor, Fiji, for example, is toxic from untreated domestic and industrial effluents. Pollution from food-processing plants, oil storage depots, a cement factory, chemical plants, and households concentrates in the harbor because it is surrounded by a barrier reef, which restricts the mixing of harbor water with the open ocean.

Even in Sydney, partially treated sewage clogs coastal waters, creating health hazards. Until 1992, when the city began pumping its municipal wastes 1 mile offshore through a new system of pipelines, some 977 million liters of minimally treated sewage and wastewater were dumped directly into near-shore waters every day. Despite the longer pipelines, wastes come back on the tides, polluting beaches at Bondi, Malabar, and North Head (Beyer 1989).

The unsafe disposal of pesticide residues and other hazardous chemicals is another problem confronting many South Pacific islands. The trouble is that many pesticides, such as DDT, that are banned or severely restricted in developed countries are still widely used in the region.

There have also been disasters involving toxic chemicals. In August 1983 a barge carrying 2,700 drums of cyanide destined for the gold and copper mine at OK Tedi capsized near the mouth of the Fly River on the south coast of Papua New Guinea. The drums, each containing 102 liters of the poison, were never recovered, and the effects of the spill on estuarine ecology are still not known. In another incident at OK Tedi, the effects were all too evident. When a worker at the ore treatment plant left a valve open by mistake, 270 metric tons of cyanide were released into the Fly River. In the aftermath, thousands of fish and crustaceans and scores of saltwater crocodiles perished.

Two of Papua New Guinea's major rivers, the Fly and the Jaba, are routinely polluted with sediments and heavy metals from mining operations. Before the Bougainville Copper Mine was closed in the early 1990s, the Jaba River was so full of sediments and heavy metals from the mine—about 135,000 metric tons of mine tailings were dumped into it every day—that its slate gray waters were biologically dead. Researchers claimed that wading into the river to take samples was like inching through moving mud. Once the toxic contents of these rivers flow into coastal waters, little is known of their fate or their effects on marine organisms.

Close to 60 million metric tons of polluted slurry has poured into the Fly River from the OK Tedi mine every year since 1984. Local villages complain that the toxic sediment released by the mine has killed wildlife, clogged navigation channels, and ruined their riparian way of life. In a precedent-setting case, local communities filed suit against the holding company, BHP of Australia, and in June 1996, the court ruled in favor of the communities. Under the terms of the settlement, BHP will pay around £300 million (about $500 million) in compensation, including $32 million to relocate ten villages and over $320 million for pollution control. For the first time in decades, there is real hope that those two major river systems will eventually recover (Anderson 1996a; Osborne 1995).

As more people migrate from island interiors and distant atolls to population centers, coastal pollution worsens. Many of the new arrivals find themselves confined to overcrowded shantytowns and squatter settlements, usually built on the most degraded land. The Koki Settlement in Port Moresby, capital of Papua New Guinea, built on wobbly stilts over the remains of a mangrove forest, is one of many examples. Now that there is no forest to blunt storm damage, the poor people of Koki take the brunt of the storms themselves.

A Nuclear-Free South Pacific

There are fears that radioactive waste from nearly four decades of French nuclear testing—carried out on Mururoa and Fangataufa, two remote islands in the Tuamotu group in French Polynesia—poses a threat to human health and the environment. Since the French have consistently refused to disclose any data on the amount of fallout their bombs have caused, or on the amount of radiation leaking from their nuclear dump sites, health and environmental risks can only be guessed.

"Nuclear fallout engendered by the 41 atmospheric tests made at Mururoa and Fangataufa between 1966 and 1974 is still with us (mostly absorbed in our bodies), and the 124 underground tests made since 1975, instead of diminishing the health threats, have added several new sources of radioactive pollution," states Bengt Danielsson, explorer and historian who lives and works in Papeete, Tahiti.

Despite condemnation by virtually every country in the region of the continued testing of nuclear weapons in French Polynesia, France ignored the call for a "nuclear-free zone" in the South Pacific until January 1996 (Anderson 1995). On January 29, 1996, French President Jacques Chirac went on national television to annouce the "definitive end of French nuclear testing" in the Pacific (Whitney 1996). France's about-face on nuclear testing has opened the door for more positive relations with a number of South Pacific states, including New Zealand and Australia.

Resources

Largely because of growing populations, widespread poverty, and limited natural resources, other than the sea, most of the Pacific's populated islands are facing increasingly severe resource constraints. Upland forests have been cleared for subsistence agriculture or to make way for mining operations. Some coastal mangrove forests have been whittled away by urban, port, and industrial expansion, while others have been converted into tourist resorts. Freshwater reserves, often limited to a thin lens sitting atop brackish, or salt, water, are polluted by untreated municipal wastes and agricultural runoff. And coral reefs around population centers suffer from sedimentation and pollution.

In general, land degradation is serious and getting worse. Heavy logging in the Gogol Valley of Papua New Guinea has resulted in increased erosion and landslides. Across the entire country, somewhere between 22,000 and 290,000 hectares of tropical forest are lost every year. On Fiji, upland forests have been cleared for mono-cropping and subsistence agriculture. Small-scale farmers who strip away hill vegetation are frequently rewarded with the erosion of their garden plots during tropical storms. More often than not, their topsoil is washed down onto their neighbor's fields.

Overall, some 60 percent of the South Pacific's islands suffer from soil degradation. "The problem is, no one is taking land degradation as a serious environmental problem," says John Morrison, a lecturer at the University of Woollongong, near Sydney, Australia.

Mining for nickel, copper, cobalt, lead, zinc, chrome, and gold has left huge scars across New Caledonia's landscape. Most of the island's tree cover has been stripped away. Rivers in the southern part of the island have turned orange-red, swollen with soil scoured from open-pit mines. Tons of fine-grained laterite clay carried into coastal waters from mining and logging in the hills killed mangroves and filled in estuaries. Today, practically no undisturbed mangrove stands are left on the main island.

Fiji has destroyed over 4,000 hectares of mangrove forests for the expansion of agricultural land, mostly to increase sugar cane production on the larger islands. "The irony of the situation," explains Padma Narsey Lal, a Fijian researcher at the East-West Center in Honolulu, "is that, for the most part, mangrove land destroyed in the name of agricultural development has still not been put to use; and where it has been, yields have been very low. This is due in part to acid sulfate conditions in the soils and the absence of any one authority responsible for the administration of coastal lands." With over 840 islands in the Fijian archipelago, most of them uninhabited or only sparsely settled, the nation still has some 50,000 hectares of mangroves in fairly good condition. But it is sheer isolation that has saved them.

Although Papua New Guinea and the Solomon Islands have extensive areas of mangrove swamps, they are under threat from upland logging, mining, and coastal development. Much of the Solomon Island's 64,000 hectares of mangrove forests face increasing pressures from commercial logging operations and coastal land clearance for agricultural and urban expansion.

Coral Reefs

Some 70 percent of the South Pacific's coral reef resources—which total around 77,000 square kilometers (15 percent of the global total)—are thought to be in good to excellent condition. However, this leaves 30 percent, roughly 25,000 square kilometers, under immediate threat, mostly from sedimentation, pollution, overfishing, and coral mining (Jameson, McManus, and Spalding 1995). The most pristine reefs are those far removed from population centers and agricultural and industrial activities: the Great Barrier Reef of Australia and inaccessible coral atolls around Papua New Guinea and the Caroline and Cook islands.

Coral communities are under increasing stress around developed islands (see table 10.2). Some of Fiji's coral reefs on the two main islands, Vitu Levu and Vanau Levu, have silted over, victims of upland deforestation, the destruction

Table 10.2. Main Threats to Coral Reefs in Thirteen South Pacific Island States and Territories

COUNTRY/TERRITORY	SPECIFIC AREA	HUMAN-INDUCED IMPACTS ON CORAL REEFS
American Samoa	Pago Pago Harbor	Overfishing, industrial pollution
Belau (Palau)	Main islands	Erosion and sedimentation from coastal development
Cook Islands	Urban center, Rorotonga	Construction, sand mining, land reclamation
Fiji	Suva urban area	Soil erosion from logging and upland farming; disposal of mine tailings, sand dredging; municipal and industrial pollution
French Polynesia	Society Islands	Overfishing, dredge and fill, sand mining, raw sewage, industrial pollution
Guam	Urban center, Apra Harbor	Raw sewage, thermal discharges, port construction, soil erosion and sedimentation
Kiribati	Urban center, Tarawa	Urban pollution, coastal construction, raw sewage, overfishing
Marshall Islands	Majuro, Ebeye	Mining, port and harbor construction, raw sewage
Solomon Islands	Main islands	Upland mining and logging, coastal development
Tonga	Main islands	Municipal pollution, causeway construction, use of dynamite and poisons, overharvesting of shellfish
Tuvalu	Funafuti Atoll	Municipal pollution, overfishing
Vanuatu	Port Vila on Efate	Coastal construction, raw sewage, coral collection
Western Samoa	Main islands	Agricultural pollution, mining, raw sewage, overfishing

Source: Stephen Jameson, John McManus, and Mark Spalding, *State of the Reefs: Regional and Global Perspectives*, NOAA, Washington, D.C., May 1995, pp. 20–22.

of mangrove forests for agricultural development, and the expansion of ports and coastal towns. In the water around Suva, Fiji's capital, coral reefs have been snuffed out by eroded sediment from logging and upland farming, municipal and industrial pollution, and dredging for sand.

On the Solomon Islands, large-scale deforestation and mining activities have ruined near-shore shellfish beds and coral reefs. The pace of destruction threatens to get even worse: There are now over fifty mining companies prospecting for minerals. The situation is even worse on the Mariana Islands. A coral reef survey conducted in 1996 revealed that over 90 percent of the archipelago's coral reefs were in poor to fair condition, containing less than 50 percent live coral cover (Grigg and Birkeland 1996). In French Polynesia, three-quarters of the population of 200,000 live and work on Tahiti, the main island. The island's coral reefs are under intense pressure from overfishing, dredging, coastal construction, sand mining, and chronic pollution from raw sewage and industrial wastes.

Although the South Pacific's states are beginning to recognize the urgent need to conserve and manage coral reefs, progress is slow. As of the end of 1995, only ten countries and two territories (Guam and American Samoa) had set up marine and coastal protected areas. Australia heads the list with 184 sites protecting over 13 million hectares of coastal land and near-shore waters (including the Great Barrier Reef). Other countries with marine protected areas include New Zealand with 32 sites; the United States (Hawaii, Guam, and American Samoa) with 31; Papua New Guinea with 11; the Kingdom of Tonga with 6; Vanuatu and New Caledonia with 5 each; Palau with 4; Western Samoa with 3; and the Cook Islands with 2.

There is some hope that more countries in the region will set aside marine protected areas, areas that protect not only coral reefs but seagrass beds and mangrove swamps as well. The South Pacific Regional Environment Program (SPREP) is now administering a $10 million grant for biodiversity conservation from the Global Environment Facility. SPREP is devoting a large portion of that money for conservation efforts that include coastal and marine areas.

Fisheries

Traditionally, South Pacific islanders have derived about 90 percent of their animal protein from the sea, and many continue to do so. Most fishing is artisanal and small scale, confined largely to reefs and protected lagoons. Around 100,000 metric tons are taken in coastal waters every year, on average. Over 80 percent of the catch is for subsistence. A wide variety of techniques are used, depending on the region and the type of fish to be caught. Fishing craft range from the simple rafts and dugout canoes with single outriggers used by Polynesians and Melanesians, to the larger, double-outrigger canoes favored by Micronesians and Papuans. Fish traps are widely used in Melanesia.

Artisanal fishing is highly evolved and generally based on principles of sustainable yield. By obtaining seafood from different habitats—fish from lagoons; shellfish and seaweed from exposed reefs; crabs, lobsters, and mollusks from mangrove swamps; land crabs from the coast; and freshwater prawns from rivers—Pacific islanders are able to diversify their food sources, thereby reducing the risk of overexploiting a particular species or ecosystem.

Throughout some areas of the Pacific, particularly Melanesia, fishing villages specialize in different species of fish and shellfish. Since people from one village who want to fish in the waters of a neighbor must first get permission, overharvesting is usually controlled. When stocks are low, limits are placed on the amount of fish outsiders can take.

In New Caledonia, Kanak fishing villages—established on tiny coral islands between the extreme northern end of the main island and the Belep Islands—have evolved some unusual ways to ensure the sustainable use of their coastal resources. Clans are organized into independent management units called *Kavebu*. Besides being a social organization, each Kavebu has a well-defined land and marine territory, within which members can harvest freely to satisfy their own needs. Kavebu are barred from using each other's territory unless permission is granted.

"Generally, their system works very well," says social anthropologist Marie Preston, who has studied these remote villages for more than four years. "The islands are dry, so they don't have much agriculture. And their fishing is almost exclusively subsistence. Problems arise when outsiders try to fish in their Kavebu." A Tahitian who wanted to net fish within a coral lagoon controlled by a Kavebu was sent packing. In another case, weekend anglers on a professional fishing boat from Noumea had all their gear confiscated by angry Kanaks after they tried to fish in a Kavebu without obtaining permission.

Around some islands, subsistence fisheries are in trouble. The Cook Islands and Palmerston Island face the unexplained decline of parrot fish on their coral reefs. On Tarawa in Kiribati, fish stocks are declining because of overfishing. Reef fish around Belau, on the other hand, have a potential sustainable harvest of between 2,000 and 11,000 metric tons a year, comparable to the offshore tuna catch.

There is a major impediment to proper fisheries management in the South Pacific: As the region's small island states and territories evolve into modern economies, their traditional fisheries management systems are crumbling. Worse, as those traditional methods of coastal governance—based on clan and community enforcement—collapse, they leave no viable enforcement mechanisms in place.

American marine biologist Dr. Robert Johannes, who has devoted two decades to the study of traditional Pacific fishing societies, explains the dilemma:

Throughout Oceania the right to fish in a particular area was controlled by a clan, chief, or family. Generally this control extended from mangrove swamps and shorelines across reef flats and lagoons to the outer reef slope. It would be difficult to over-emphasize the importance of some form of limited entry such as this to sound fisheries management. Without some control on fishing rights, fishermen have little incentive not to overfish since they cannot prevent others catching what they leave behind. This is a central tenet of fisheries management science.

However, under modern conditions the government must assume the sole responsibility for placing and enforcing fisheries conservation measures. This is a difficult and expensive responsibility under the best of circumstances; it is close to impossible in most tropical artisanal fisheries. Typically, there are far more boats, more distribution channels (both subsistence and market) to monitor and regulate than in high latitude fisheries of similar sizes. And there is usually much less money and expertise available with which to do it (Johannes 1982).

In some areas of the Pacific, fishing is being transformed from a subsistence activity into big business. As the island states and territories of the South Pacific struggle to modernize their feeble economies, a number of indigenous commercial tuna-fishing operations have developed. The Solomon Islands, American Samoa, and Fiji all have their own tuna fleets, which compete with the deep-water fleets from the United States, Japan, Korea, and Taiwan. In 1990, deep-water fleets landed just over 1 million metric tons of tuna from the region, worth around $1 billion.

Since the advent of 200-mile EEZs, South Pacific states charge access fees for deep-water foreign fishing fleets to harvest within their territorial waters. As a result, Kiribati, the Solomon Islands, and Tuvalu now earn more from fisheries than from any other source of income.

Biodiversity on Land

Pacific islands are wonders of biological diversity. Geographical isolation has made them that way. Islands that were once connected to continental land masses, such as New Caledonia and New Guinea, are now a refuge for species that were driven to extinction long ago by evolutionary processes on the continents. Hence, most Pacific islands, including coral atolls, have a large number of plants and animals found nowhere else on earth. For instance, 2,474 of New Caledonia's 3,250 plant species are endemic. Papua New Guinea has some 20,000 species of vascular plants and 2,000 species of ferns—most of which are endangered by habitat loss and competition from introduced species (Dahl 1984).

Isolation alters evolutionary processes in plants and animals. It produces a great cornucopia of species, but they are often unable to cope with invasions of non-native species. So long as islands remain isolated, that is no problem. However, when Europeans began colonizing the Pacific in the eighteenth century, they introduced many new, highly competitive species like cats, rats, dogs, and pigs. The new arrivals annihilated indigenous populations of fauna and flora, which had not developed ways of dealing with predators and their diseases or with habitat competition. Destruction of forests and coastal wetlands killed off others.

"Today, there are probably more endangered species per person in the South Pacific than anywhere else in the world," says Arthur Dahl, former deputy director of UNEP's Oceans and Coastal Areas Programme Activity Centre in Nairobi. For instance, there are fifty-four endangered bird species, one for every 92,000 inhabitants of Oceania. Equivalent figures show one endangered bird for every 400,000 people in Australia and New Zealand, and one for every 670,000 people in the Caribbean, the other region noted for its extinctions. So far only around fifteen countries in the region have set aside parks and protected areas to safeguard their natural heritage.

Management: The South Pacific Regional Environment Program

Environmental issues have been on the South Pacific political agenda since 1969, when the South Pacific Conference first voiced official concern over the state of the region's resources. But it was thirteen years before action was taken. After a long and complicated process, twenty-two island states and territories (minus Australia and New Zealand) met in 1982 at Rarotonga, capital of the Cook Islands, and endorsed the South Pacific Regional Environment Program Action Plan. Because funding mechanisms were in place to finance the initial part of the action plan, no trust fund was formed. Funds to cover program activities were contributed on a case-by-case basis, with most of the initial money coming from UNEP's overburdened Environment Fund.

In the beginning the program was hampered by a general sense of inertia in the face of overwhelming problems. SPREP's priority programs suffered from a lack of political commitment, as well; the region's governments failed to give it the political muscle it needed to get off the ground. In November 1986, however, the program received a needed boost: twelve independent Pacific states, along with four other nations with territories in the Pacific, adopted the Convention for the Protection of the Natural Resources and the Environment of the South Pacific Region. The convention commits signatories to "prevent, reduce and control pollution in the region from ships, land-based sources, any exploitation or exploration of the seabed, atmospheric discharges, all forms of dumping and the storage of toxic and hazardous wastes." Following the spirit of the Rarotonga declaration in 1982, the states also agreed to prohibit the dumping and storage of radioactive wastes in the convention area and to "prevent, reduce and control pollution that might result from nuclear tests in the region." Known as the SPREP Convention, it came into force in 1990.

Another notable advance came in June 1993, when the region's governments moved to turn SPREP into an official intergovernmental, regional organization. By the end of 1996, ten member countries had ratified the Agreement Establishing SPREP, just enough for it to enter into force.

The South Pacific Regional Environment Program has proved to be an effective forum for tackling regional environmental concerns. Also, by forming a cohesive block on key issues, such as nuclear weapons testing, those countries now have a much larger impact in international forums like the United Nations.

Conclusions

- Most resource management problems can be traced to the breakdown of traditional artisanal management systems that evolved over centuries. Most traditional systems had built-in safeguards to ensure limited access to resources while underscoring the importance of community management. It would be useful to review those traditional resource-based management systems to see if some pieces could not be adapted to modern management methods.

- Fisheries urgently need action. Again, traditional management systems, which worked well for centuries, could be examined for their potential revival and application in more comprehensive coastal area management plans. Traditional management was based on community involvement and involved closed seasons, especially during breeding and egg laying, and limited access. The challenge is to adapt an essentially local process of governance to larger-scale problems, affecting much wider areas.

- Planning for future needs will be easier if governments take account of population growth and distribution while forging resource management plans. The Pacific would make an excellent proving ground for pilot projects that attempt to mesh modern demographic techniques and GIS technologies to draw up sustainable development strategies, on both a country and a regional basis.

- Finally, regional cooperation through SPREP, the Alliance of Small Island States (AOSIS), and other forums should be given proper support by the region's governments. AOSIS, which had 42 members as of June 1997, is now lobbying hard for tougher deadlines on reducing CO_2 emissions into the atmosphere, along with a number of other critical resource issues. Only by joining together can these islands have any impact on the global stage. Considering the complexity of issues confronting them, a solid, unified position would have benefits in many areas, particularly resource management.

The Northwest Pacific

Sea Dragons Ascending

Among the countries of the northwest Pacific, the economies of China, Hong Kong, Taiwan, South Korea, and Japan, known as the five dragons—have been in the fast lane for decades. Between 1988 and 1994, China's industrial output, for example, grew by an average of 18 percent a year in real terms. Shenzhen, the special economic zone next to Hong Kong, grew by 30 percent in 1993, one of the highest rates in the world.

Explosive economic growth has not resulted in better resource management, however. With the sole exception of Siberia, coastal areas in the northwest Pacific are under tremendous development pressures as they absorb the brunt of new economic development in China, South Korea, Taiwan, and Japan.

Population

It is not hard to see how the region's resources have been ravaged. The collective population of the northwest Pacific—of China, Hong Kong, Japan, North and South Korea, Taiwan, and Pacific Russia—amounts to over 1.4 billion, making it the most populous coastal region on earth. China's 1.2 billion inhabitants constitute 22 percent of the global population (see table 11.1).

The majority of the region's population is concentrated in a wide swath along coastlines and river valleys (see figure 11.1). Close to 60 percent of China's population lives in twelve coastal provinces, along the Yangtze River Valley, and in two coastal municipalities, Shanghai and Tianjin. Along China's 18,000 kilometers of continental seacoast, population densities average between 110 and 834 per square kilometer. In some coastal cities, such as Shanghai, China's largest, with 17 million inhabitants, population densities average just over 2,000 people per square kilometer (Yeung and Hu 1992; China's Population and Development 1992).

Virtually all of Japan's 126 million inhabitants are coastal. No one in the country lives more than 120 kilometers from the sea. Along the country's Pacific coastline, population densities average more than 500 people per square kilometer. But in the heavily populated urban districts of Tokyo and Yokohama, over 11,500 people per square kilometer is common.

The Korean Peninsula's population of 70.2 million (both North and South Korea) is virtually all coastal. Both countries' populations are also overwhelmingly

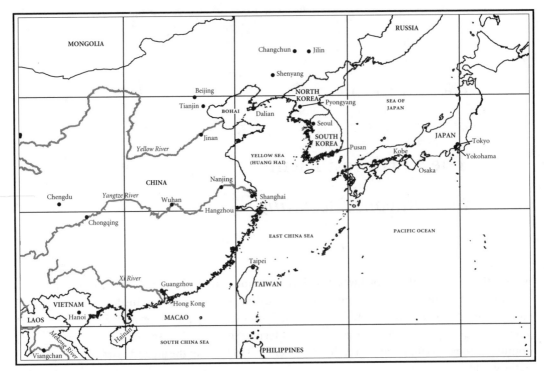

Figure 11.1 *Above:* Geopolitical Map of the Northwest Pacific Region. *Opposite:* Estimated Population Densities for the Northwest Pacific Region, 1995.

Source: Waldo Tobler, Uwe Deichmann, Jon Gottsegen, and Kelly Maloy, *The Global Demography Project,* National Center for Geographic Information and Analysis, Department of Geography, University of California, Santa Barbara, 1995.

Table 11.1. *Length of Coastline and Population Data*
for States in the Northwest Pacific Region

COUNTRY	COASTLINE (KM)	POPULATION 1997 (MILLIONS)	POPULATION GROWTH (%/YR)	URBAN %
China	18,000	1,236.7	1.0	29
Hong Kong*	733	6.4	0.7	100
Japan	13,685	126.1	0.2	78
North Korea	2,495	24.3	1.8	61
South Korea	14,800**	45.9	0.9	74
Taiwan	—	21.5	1.0	75

Sources: World Population Data Sheet, 1997, Population Reference Bureau, Washington D.C., 1997; *World Resources 1996–97,* Oxford University Press, New York, 1996, p. 354; Seoung-Yong Hong, "Assessment of Coastal Zone Issues in the Republic of Korea," *Coastal Management,* vol. 19, no. 4, October–December 1991, p. 391.

Notes: The Pacific coast of Russia is not included, as disaggregated data are not available.

*Hong Kong reverted to China on July 1, 1997.

**South Korea's coastline is often grossly underestimated at only 1,318 kilometers. However, if one includes about 3,200 islands off the west and south coast, the total comes to over 14,000 kilometers.

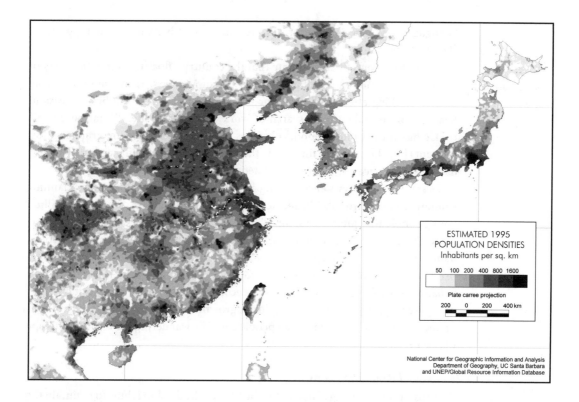

ESTIMATED 1995
POPULATION DENSITIES
Inhabitants per sq. km

50 100 200 400 800 1600

Plate carree projection

200   0   200   400 km

National Center for Geographic Information and Analysis
Department of Geography, UC Santa Barbara
and UNEP/Global Resource Information Database

urban, with many people living in the region's larger cities and outlying suburbs: Seoul, Inchon, Pusan, Taegu, Pohang, Masan, Ulsan, Chonju, and Kwangju in South Korea, and Pyongyang, Sinuiju, Wonsan, Hamhung/Hungnam, and Chongjin in North Korea. Coastal population density in South Korea averages over 600 people per square kilometer, nearly one and a half times larger than the population density for the country as a whole. And the picture is similar across the demilitarized zone in North Korea.

Hong Kong's 6 million residents, Macao's 500,000, and Taiwan's 21 million are mainly coastal zone dwellers. In Hong Kong's more crowded quarters, population densities reach 20,000 per square kilometer.

Only Russia's Pacific coast is sparsely populated. The entire Russian far east contains no more than 8 million people spread out across 6.3 million square kilometers of Siberian wilderness, less than the population of Moscow.

China's Floating Population

Since 1990, China has been beset by the movement of millions of peasants from poor, rural areas into cities and towns. The overwhelming majority of the migrants end up in coastal cities and special economic zones, those areas with the fastest economic growth rates.

Most large coastal cities now have 1–2 million migrants living in shantytowns, overcrowded makeshift dormitories, and public places. Every night Beijing's main railroad station is filled with poor migrants seeking work; with no housing, they sleep on the floor. Shanghai's official population of 17 million is now combined with 3 million rural migrants attracted from all over the Yangtze River Valley. Most have come looking for work on the mammoth development projects in east

Shanghai's new "special enterprise zone" on the east bank of the Huangpu River (Faison 1995).

Chinese demographers estimate that the country's floating population may be as high as 100 million, a population the size of Mexico's. Heavy migration continues, mocking attempts by interior provinces to limit the number of peasants heading for coastal cities in search of work and a better life. "What began as a trickle has turned into a flood," observes one Beijing-based demographer. Most migrants have been able to find work, though some end up toiling eighteen hours a day for subsistence wages on unregulated construction sites.

Despite the country's encouraging economic prospects, China's State Planning Commission sees trouble ahead. According to its projections, around 44 million young people will enter the job market over the next five years. At the same time, some 20 million workers in unprofitable state-run enterprises will "become redundant" and 120 million surplus rural workers will head for coastal cities and enterprise zones to seek work. In all, according to the commission, 180 million new urban jobs will have to be created. If the economy keeps on growing as it has, that may pose no problem. If China's explosive growth begins to slow, however, millions of unemployed or underemployed people could be wandering the urban landscape (Tyler 1994b).

China and Population Growth

Perhaps the most vexing problem for the Chinese leadership is how to maintain the momentum of low fertility rates and slower population growth. China's family-planning program is widely considered a resounding success: Over the past twenty years the country's total fertility rates—the average number of children a woman is likely to have over the course of her reproductive life—decreased from over six in 1970 to around two in 1995. Despite that accomplishment, the country's huge population base produces an extra 13 million people a year. Chinese family-planning officials want to reach zero population growth by early next century. But with rural fertility rates still averaging three or four children per woman in some areas, that may prove difficult.

Professor Tian Xueyuan, director of the Institute of Population Studies of the Chinese Academy of Social Sciences in Beijing thinks the country's rapid urbanization could be a blessing in disguise. "Data show that once rural migrants settle in big cities, their fertility rates drop rapidly," he says. "The current problem is that our huge floating population is not being serviced properly because many of them migrate from city to city in search of work." Services simply cannot keep up with them.

From a resource perspective, continued population growth poses serious risks for China's future. "We already face serious resource limitations in all areas," states Qu Geping, the outspoken administrator of China's National Environmental Protection Agency (NEPA). "We should have 0.6 hectare per person, but in fact we have only 0.1 hectare per person. Our water resources are capable of sustaining only 650 million people, half of our current population. And we are losing prime agricultural land to degradation and rapid urbanization."

Qu is afraid that if trends continue, the country will have trouble feeding itself. About one-third of the country's arable land is suffering from excessive water erosion. Every year China loses some 5 billion metric tons of valuable topsoil. "What we desperately need are integrated policies to manage both land and water resources more sustainably," maintains Qu.

Pollution

China's Coasts in Crisis

Scientists have estimated that along China's 18,000-kilometer coastline (32,000 kilometers if all offshore islands are included), municipal wastes flow into coastal waters at the rate of 50–60 million metric tons *per day*, in other words, 17–21 billion metric tons a year. As of 1993, only 16 percent of the total amount discharged received any form of treatment. That same year, the country's industries reportedly discharged 36 billion metric tons of industrial effluents into rivers, streams, and coastal waters. How much of that waste actually enters the sea from the interior of the country is not known (UNEP 1992).

Shanghai pumps 4.5 million metric tons of untreated and partially treated wastewater into its rivers and coastal areas every day. Concerned about the state of its coastal waters and the deteriorating quality of its drinking water (nearly all of which comes from the polluted Huangpu River), Shanghai's municipal authorities launched an impressive cleanup program in the early 1990s. By 1995, the city had invested in five sewage treatment plants, which now provide secondary treatment for 15–20 percent of the city's municipal wastes. Most still goes raw into the city's rivers.

A $200 million World Bank loan allowed Shanghai to invest a modest $6 million in monitoring and pollution control in the upper watershed of the Huangpu River. Another $10 million was spent on retrofitting heavily polluting industries with wastewater treatment plants. But most of the money is being used for infrastructure improvements and the construction of the city's special tax-free enterprise zone.

Until 1993, Hong Kong emptied 2 million metric tons of mostly untreated industrial and municipal wastes into its coastal waters every day. Much of the industrial waste consisted of highly toxic acids, alkalis, and solvents. On average, scientists have measured 8,800 milligrams per kilogram of copper and 9.6 milligrams per kilogram of cadmium in Kowloon Bay. The government's recommended levels are 25 milligrams per kilogram for copper and 0.4 milligrams per kilogram for cadmium.

In 1993, the Hong Kong government began collecting industrial wastes free of charge and disposing of them in a $167 million waste treatment facility built by Waste Management International's Hong Kong subsidiary, Eviropace Ltd. The plant, designed to treat organic and inorganic chemical wastes, is expected to cut considerably the amount of industrial effluents, including heavy metals, dumped into the colony's coastal waters.

Encouraging as that development is, sewage treatment remains rudimentary. In 1989, the government announced a "strategic sewage-disposal scheme," which, when completed in 1997, will have cost around $3 billion. Even so, the city's sewage will receive only partial treatment with chemicals instead of complete secondary (biological) treatment. The resulting sewage sludge will then be pumped through a 30-kilometer pipeline into the sea off China's coast.

151

Guo Jinghui and He Qiang, two engineers from the Environmental Engineering Department of Tsinghua University in Beijing, have calculated that it would cost China 34 to 44 billion yuan ($4.2 to 5.5 billion) just to build and maintain minimal wastewater treatment facilities for the country's coastal cities and industrial centers. So far, only forty of the country's five hundred major cities have two-stage sewage treatment plants. Guo and He claim such expenditures "are beyond the capacity of the Chinese government at present" (Guo and He 1993).

The Bo Hai and Huang Hai Seas

Wedged between the Chinese mainland, the Korean Peninsula, and southern Japan, China's two semi-enclosed coastal seas, the Bo Hai (Gulf of Chihli) and the Huang Hai (Yellow Sea), cover about 460,000 square kilometers and contain 18,000 cubic kilometers of water. Despite these figures, both bodies of water are relatively shallow: Their average depth varies from 38 meters in the north to 46 meters in the south. But into these two seas pours a huge amount of municipal, industrial, and agricultural waste, along with massive sediment loads and pollutants brought in by sixty-two river systems in China.

On the Chinese side of these two seas, over 100 million people live in coastal population densities of up to 834 people per square kilometer. The region contains 142 industrial cities, including 40 ports. These cities boast 4,500 major industrial complexes, including chemical and petrochemical plants, pulp and paper mills, food-processing facilities, cement and metallurgical plants, and numerous manufacturing concerns (UNEP 1992).

China's provinces along the coast of the Bo Hai and Huang Hai collectively dump around 744 million metric tons of mostly untreated sewage water into coastal waters every year. When all sources of municipal pollution are considered, including the amount of sewage and municipal wastes brought in by river systems, the total figure amounts to around 7.6 billion metric tons per annum (UNEP 1992).

Thirteen large port cities alone dump 200 million metric tons of industrial waste into coastal waters every year. The drainage area (27,000 square kilometers) of just one river system, the Liao, carries 1.3 billion metric tons of untreated industrial effluents annually into the coastal waters of the Bo Hai. In the summer of 1986, parts of that river died from massive amounts of industrial and municipal pollution. Nearly every aquatic organism was killed for tens of kilometers, precipitating the first instance of China's notorious "stink rivers." That same year, the Xiao Qin River was overwhelmed with untreated sewage water, killing 100,000 metric tons of shellfish in the lower reaches of the river. The economic loss was assessed at roughly 20 million yuan ($2.5 million) (UNEP 1992).

Four major rivers in South Korea—the Han, Kum, Yongsan, and Naktong—transport 1.6 million metric tons of municipal and industrial effluents into coastal waters every day. All but one of those rivers, the Naktong, flow into the Yellow Sea. Only a quarter of South Korea's municipal wastes and a third of its industrial wastes receive any form of treatment before being pumped into rivers and coastal waters (Hong 1991). The amount of wastes discharged by North Korea into its coastal waters is not known, but little is thought to be treated before being disposed of.

The near-shore waters around South Korea's industrial cities of Inchon, Masan, and Ulsan are heavily polluted with untreated industrial and municipal wastes.

And chronic red tides in Chinhae and Masan bays have decimated mariculture operations, causing millions of dollars' worth of damage annually.

Japan's Coastal Waters: Some Improvement

As of 1997, only about 50 percent of the Japanese population was connected to a sewage treatment plant. The wastes of 62 million people go into rivers and coastal waters untreated. Japan has introduced industrial wastewater treatment plants in many urban areas, but roughly 15 billion metric tons of partially treated toxic chemicals and other industrial wastes are pumped into surface and coastal waters every year.

Some progress has been made in cleaning up the Seto Inland Sea, once the most polluted body of coastal water in the country. In the 1970s, the government passed several pieces of legislation governing the Inland Sea, including setting up protected areas and greenbelts and promulgating standards to limit the amount of pollutants that can be discharged by industries into the sea. In 1993, Japan passed similar legislation designed to limit the amount of nitrogen and phosphorus entering some ninety semi-enclosed bays and inlets around the country's 13,000 kilometers of coastline. The major flaw in the legislation is that it covers only discharges by industry, not by municipalities and agriculture. Japanese environmentalists argue that nutrient levels will not drop by much until the government builds a lot more sewage treatment plants with at least two stages of treatment.

Resources

The region's coastal and lowland forests have been decimated by urban expansion and the growth of agricultural land and aquaculture and mariculture activities. Along the Pacific coast of Russia, evergreen forests are being devastated by largely unregulated logging companies from Russia, Japan, and Korea. Timber across some 12,000 square miles (30,000 square kilometers) is being felled as fast as it can be cut by ninety forest enterprises (Linden 1995).

The population of Siberian tigers, the largest of all the cats, has been annihilated by poachers. In 1990, wildlife experts estimated there were five hundred of those magnificent animals in the region's coastal forests. Today, no more than two hundred are thought to survive in isolated pockets. With a complete tiger carcass bringing in $25,000, there is little hope for the animal unless existing laws can be enforced and poaching halted (Hinrichsen 1993).

China has also overexploited its woodlands and wildlife. Only 13 percent of the country is covered by forest, most of it in the far western provinces. China has stripped away some 40 percent of Tibet's forests, worth an estimated $54 billion. In the 1950s, Hainan, China's island province, had 20,000 square kilometers of primary tropical forest; by 1987, only 3,800 square kilometers remained (National Environmental Protection Agency 1992).

The country is fighting back, however. In an effort to halt the desertification of highly eroded and degraded land in the Loess Plateau and other dry northern areas, authorities launched the Shelter-Belt (San Bei) Afforestation Project, which has replanted some 9 million hectares of denuded drylands with fast-growing tree species. By 1992, the country had also replanted 5.5 million hectares of coastal lands with trees in an effort to stabilize shorelines.

Progress in one area is often offset by problems in another. The country's coastal wetlands have been seriously damaged by development. Only around 3 million hectares of coastal wetlands remain, and most of those are under immediate threat from expanding agricultural and industrial activities. Of the country's 708 protected areas, only 16 include wetlands, 2.2 percent of the total protected area (Xias et al. 1996). Since 1949, around 400,000 hectares of mudflats in Jiangsu, Zhejiang, Fujian, Guangdong, and Liaoning provinces have been converted into agricultural fields.

Fisheries

In the last five years, aquaculture and mariculture enterprises have exploded along China's coast. Shandong Province has even launched a "Blue Programme" to further exploit its marine resources. By 1993, giant fish farms in offshore waters covered 170,000 square kilometers, an area larger than the province itself. As of 1993, Shandong had three hundred sea farms, eight hundred marine food-processing plants, and ten enterprises growing marine algae. Collectively, those facilities produce and process more than 2.2 million metric tons of seafood a year, worth roughly $2 billion. Exports of farmed fish and shellfish amounted to 90,000 metric tons in 1993, bringing in $300 million (*Window* 1993).

China continues to cash in on the rising demand for fish and shellfish in Asia. Over the past six years, with government help, around 470,000 hectares of land, much of it coastal, has been converted into fish farms. In 1993, the country's aquaculture and mariculture industries produced 17 million metric tons of seafood (FAO 1995).

At the same time that fish farming is expanding rapidly, China's coastal and deep-water fleets harvest more fish and shellfish than those of any other country. In 1993, China's fishing fleet brought in 10 million metric tons of fish and shellfish, compared to Japan's 8 million metric tons, South Korea's 3 million metric tons, North Korea's 2.5 million metric tons, and Taiwan's 2 million metric tons. Overall, the northwest Pacific region is heavily overfished. The UN Food and Agriculture Organization reported in 1992 that 100 percent of the region's commercial fish and shellfish stocks were either "fully fished, overfished, depleted or recovering" (FAO 1995).

The northwest Pacific's catch has been declining steadily since 1990, when over 25 million metric tons were taken by the world's deep-water fleets. FAO blames the declines mostly on wild fluctuations in the abundance of Japanese pilchard. But, in fact, nearly all commercial stocks in the northwest Pacific have reached or are rapidly approaching their upper limits of exploitation.

By 1993, some commercial varieties of fish, such as pollock, were so scarce that Russia demanded a three-year moratorium on fishing off its Pacific coast. When its neighbors refused to support a ban in order to allow stocks to recover, Russian coast guard vessels stepped up patrols in the Sea of Okhotsk and in the southern Kuril Islands, occupied by Russia at the end of the Second World War.

In September 1995 Russian coast guard vessels opened fire on several Japanese fishing trawlers fishing illegally inside Russia's territorial waters in the Soya Strait, off the northern tip of Hokkaido. One of the Japanese captains was wounded, and the boats were confiscated by the Russians. Rear Admiral Sergei Skalinov, chief of staff of the border guards, which oversees the coast guard, told the *International Herald Tribune* that he would use force again if necessary to repel "those who seek others' wealth and ignore the rights and responsibilities of border guards."

Management: Toward Regional Governance

Under the auspices of UNEP's Oceans and Coastal Areas Programme Activity Center in Nairobi, representatives from China, North and South Korea, Japan, and Russia met in the Siberian port city of Vladivostok in 1991 in an effort to agree on a regional action plan. After three years of tough negotiations, an action plan for the northwest Pacific was launched at a meeting in Seoul, South Korea, in September 1994. UNEP is providing interim services as the secretariat for the action plan until a coordinating office can be established in the region.

The action plan's initial goal is to bring coastal pollution under control. But little progress has been made, as the five member states disagree over the area of sea that should be under joint management. So far, "the East China and South China seas lie outside the parameters of the discussion," says Yu Yong-quan, director of NEPA's Sea Division in Beijing, "and this is a serious limitation." Even if the agreement covers only the Sea of Japan and the Bo Hai and Huang Hai, coastal managers point out that it is a beginning, a process that can be built upon and expanded later.

Conclusions

- The entire region needs rational and enforceable fisheries management programs. Perhaps the best hope for this is through a regional initiative.

- More use should be made of the Asian Urban Information Centre formed by Kobe, Japan, with funding from the United Nations Population Fund in an effort to provide other coastal cities with modern planning tools and expertise. Coastal urban planning is particularly crucial, since these areas contain the greatest concentration of people and industrial infrastructure.

- Concerted efforts are needed to manage remaining coastal wetlands and estuaries. The best way forward may be through UNEP's regional seas initiative.

- Finally, all of these countries need to invest in sewage and industrial wastewater treatment plants.

Southeast Asia

A Sea of Islands

Southeast Asia's seas are littered with islands of all shapes and sizes. From the air, the region looks like a garden of skerries. Indonesia's 5,000-kilometer-long archipelago contains over 14,000 islands; the Philippines has 7,100. And there are hundreds in the Andaman and South China seas. Disregarding Canada and Greenland, Indonesia and the Philippines boast the longest coastlines in the world. Indonesia's exceeds 54,000 kilometers. If stretched out in a straight line, it would encircle the earth twice.

The interplay of land and water through the ages has shaped the lives of Southeast Asians as surely as it has shaped the humid, watery lands they inhabit. Indonesia and the Philippines—the two biggest island archipelagos in the world—are defined by water. Malaysia is a thin peninsula caressed by two bodies of water. Vietnam and Cambodia curve around the bottom of Southeast Asia like a fetus, nurtured by the sea. Singapore is an island city-state, while Brunei is a tiny kingdom on a larger island, and Myanmar (formerly Burma) stretches out along Southeast Asia's western flank. Only Thailand has some land mass to spare in its northern interior, hard against the golden triangle.

Southeast Asia's coastal states form one biogeographical unit, separating Asia from Australia and the Pacific from the Indian Ocean. This vast region covers nearly 9 million square kilometers, or 2.5 percent of the world's ocean surface.

The prime location of these island and peninsular states has forged the region into one of the world's major trading crossroads. Seasonal trade winds allowed the "spice islands" to be visited by Arabs from the west, Chinese traders from the east, and Europeans from both directions. While trying to find a shortcut to the East Indies—meaning the Indonesian and Philippine archipelagos—Christopher Columbus accidentally stumbled upon North America. Not knowing where he was, Columbus called the islands of the Caribbean the "Indies" and their inhabitants "Indians."

In large measure because of its location, Southeast Asia has been frequently colonized. The Philippines were Spanish, then American, while Indonesia remained Dutch until 1949, when it gained independence after a short war. Malaysia, Singapore, and Burma were all British. Vietnam and Cambodia belonged to French Indochina. Only Thailand was never under foreign domination.

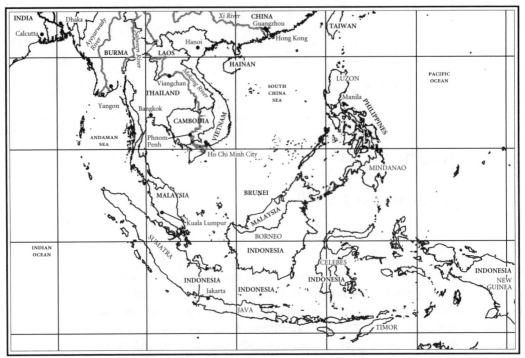

Figure 12.1. *Above:* Geopolitical Map of Southeast Asia. *Opposite:* Estimated Population Densities for Southeast Asia, 1995.

Source: Waldo Tobler, Uwe Deichmann, Jon Gottsegen, and Kelly Maloy, *The Global Demography Project*, National Center for Geographic Information and Analysis, Department of Geography, University of California, Santa Barbara, 1995.

Today, the region finds itself on a major oil route. All Middle East crude destined for Japan, China, Hong Kong, and South Korea must pass through either the narrow Strait of Malacca or the Lombok Channel. One of the reasons the United States maintains a strong military presence in the region is to protect those vital shipping lanes. Not surprisingly, the economies of those states are in high gear. Thailand, Malaysia, Singapore, Indonesia, Vietnam, Brunei, and recently the Philippines are experiencing rapid economic growth. Only Myanmar and Cambodia are lagging behind.

Malaysia has been galloping along for eight years with annual economic growth rates of 8 percent or more. Singapore and Thailand chalked up similarly impressive records: In 1995 both economies grew by close to 9 percent. Indonesia, once extremely poor, is now well on the road to economic recovery, with a growth rate in 1995 of 7.5 percent. In 1970, close to 60 percent of all Indonesians lived in poverty; by 1990, that figure had been whittled to just 15 percent. Even the Philippines's sluggish economy grew by 5.3 percent in 1995 and over 6 percent in 1996.

However, without coastal management plans in place and working, the brunt of unplanned development and economic expansion is being borne by the coasts. Without sustainable management of resources, coastal economies are unlikely to be sustainable, either. Coral reefs are being pillaged for profit, reducing tourism dollars. Mangroves are being destroyed to make room for fish ponds and agriculture, reducing offshore shrimp and fish catches. Upland forests have been decimated in the name of development, sending tons of eroded sediment into rivers

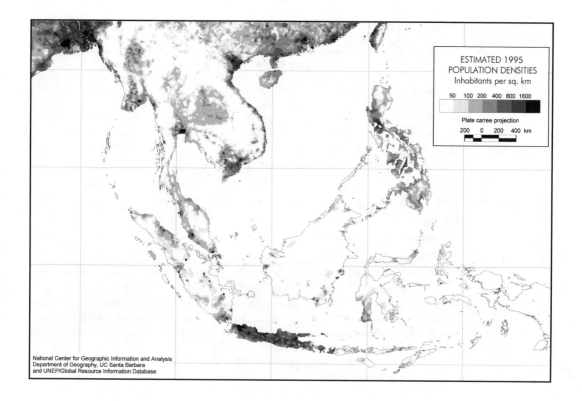

National Center for Geographic Information and Analysis
Department of Geography, UC Santa Barbara
and UNEP/Global Resource Information Database

and coastal waters, ruining habitats and killing off marine life. In many places the link has not been made between what happens on land and what happens in the sea. Roughly half of Southeast Asia's coasts are at severe risk from development pressures (see figure 1.3 on pate 10).

Population

In 1997, Southeast Asia's collective population—that of Brunei, Cambodia, Indonesia, Malaysia, Myanmar, the Philippines, Singapore, Thailand, and Vietnam—amounted to 496 million people. Of that 85 percent, or 400 million, live along a coast or within 200 kilometers of one. Southeast Asia has the highest percentage of coastal dwellers in the world (see figure 12.1 and table 12.1).

There are a number of reasons for the region's spectacular economic advances, but two have to do with demographics: first, the creation of highly successful population and reproductive health programs in all countries except Cambodia; second, increased emphasis on educating girls and women.

Population growth and total fertility rates (TFRs) have been falling steadily for over two decades. By the end of 1970, Singapore, Thailand, Indonesia, Malaysia, the Philippines, and Myanmar had all launched population and family-planning programs (Malaysia's program dates from the 1950s). As of the beginning of 1997, annual population growth rates for Thailand, Indonesia, Myanmar, Vietnam, and Singapore were all below 2 percent a year. The Philippines's rate was 2.3 percent in 1997 and Malaysia's 2.2.

In both Thailand and Indonesia the average number of children women have over the course of their reproductive lives has fallen dramatically: from six children

*Table 12.1. Length of Coastline and Population Data
for Nine Southeast Asian Countries*

COUNTRY	COASTLINE (KM)	POPULATION 1997 (millions)	POPULATION GROWTH (%/YR)	TFR 1997	CPR 1997 (%)
Brunei	161	0.3	2.3	3.4	—
Cambodia	443	11.2	2.9	5.8	—
Indonesia	54,716	204.3	1.7	2.9	55
Malaysia	4,675	21.0	2.2	3.3	48
Myanmar	3,060	46.8	1.9	4.0	17
Philippines	22,540	73.4	2.3	4.1	40
Singapore	193	3.5	1.1	1.7	65
Thailand	3,219	60.1	1.1	1.9	66
Vietnam	3,444	75.1	1.6	3.1	65

Sources: World Population Data Sheet, 1997, Population Reference Bureau, Washington, D.C., 1997; *World Resources 1994–95,* Oxford University Press, New York, 1994, p. 354.

Notes: Length of coastline is often a crude estimation and frequently fails to include offshore islands. For example, Malaysia's total coast line, including all islands, has been calculated at 48,000 kilometers. Indonesia's has been estimated at over 80,000 kilometers. Figures here are from the World Resources Institute for the purpose of consistency.

TFR = total fertility rate (average number of children a woman is likely to have over the course of her reproductive life).

CPR = contraceptive prevalence rate (percentage of women using some form of contraception).

per woman in 1970 to fewer than three by 1995. And those rates continue to drop as more women stay in school longer, many of them entering the job market. Contraceptive prevalence rates remain high: 66 percent of Thai and 50 percent of Indonesian women use some form of contraception.

Indonesia's population growth rate would be manageable if the population was distributed more evenly over the country. However, nearly two-thirds of Indonesia's population—about 130 million people—live on the main island of Java, occupying just 7 percent of the country's land area. Like the rest of Southeast Asia, Indonesia is going through an urban revolution. The coastal cities of Jakarta and Surabaya (on Java) grew by 3–4 percent a year during the decade of the 1980s, enough to double their populations every seventeen years.

Even more impressive growth rates were recorded for some of Indonesia's secondary cities. Pekalongan, Tegal, and Probolinggo, all located on Java's overcrowded north coast, grew by over 5 percent a year during the 1980s, a population doubling time of less than fourteen years. Tanjung Karang, on Sumatra's south coast grew by over 8 percent a year during the 1980s. As a result of urbanization pressures, economic growth, and Indonesia's policy of siphoning off people from the overcrowded islands of Java, Bali, and Lombok and sending them to underpopulated Kalimantan, the east Kalimantan city of Samarinda holds the country's all-time growth record: over 4 percent a year, on average, since the 1930s.

"Our population is increasing in coastal zones," says Dr. Emil Salim, former minister of state for population and the environment in Indonesia. "By 2010 our population is projected to be over 240 million, and most of these people will be living in coastal areas," continues Salim, currently an advisor to the government on environment and population issues. "These trends pose immense challenges for Indonesia's future, challenges we will have to confront. By the turn of the century, we should have evolved policies to better manage coastal zones."

Meanwhile, Indonesia's coastal cities are being overrun with migrants from the countryside. Many of the recent migrants are small-scale, subsistence farmers, forced off their land by larger landowners or lack of access to capital. Others are environmental refugees. Hill farmers in central Java, encouraged to cultivate every hectare of land that could be put to the plow, started farming steep slopes. Without proper terracing and soil conservation measures, many saw their topsoil and their livelihoods carried away by the monsoon rains. An increasing number of these "economic and environmental refugees" are ending up in coastal cities, living in squatter settlements and slums. If they have jobs at all, most likely they are under-employed and underpaid. Population densities for Java, Bali, and Madura average close to eight hundred people per square kilometer.

In war-ravaged Vietnam, coastal populations are growing by two-tenths of a percentage point faster than the country as a whole: 2.5 percent a year, compared to 2.3 percent for the rest of the country. Along the Gulf of Tonkin in the north, population densities average between 500 and 1,100 people per square kilometer. Comparable figures for the south coast range between 100 and 2,000 people per square kilometer. The country's two main cities, Hanoi and Ho Chi Minh City (formerly Saigon), are growing by 3–4 percent a year. In parts of Hanoi, population densities reach 35,000 per square kilometer (Warfvinge and Pham Dinh Huan 1991).

There are three main reasons for the rapid increase in Vietnam's coastal populations: a booming tourism trade, said to be growing by 10 percent a year; increased fisheries efforts, which now provide jobs for 13 percent of the coastal population; and industrial development, particularly oil exploitation.

Thailand's capital city, Bangkok, contains over 10 percent of the country's total population. The city generates 45 percent of the country's wealth, handles 95 percent of all imports and exports, and boasts an average per capita income over twice that of the rest of the country.

Nearly 60 percent of the Philippines's population lives in some 10,000 coastal barangays (counties) and large urban areas. Close to twenty of the country's twenty-five cities with more than 100,000 inhabitants are located on the coast, and no one in the country lives more than 150 kilometers from the sea (Country Report, Philippines 1993).

Pollution

Nearly all of Southeast Asia's river systems are polluted with organic wastes, industrial poisons, raw sewage, plastics, and junk. All of Metro Manila's rivers are essentially biologically dead. The bays of Jakarta and Manila are so polluted that low oxygen levels have asphyxiated most marine life. The murky waters of the Chao Phraya estuary in the Gulf of Thailand can no longer sustain shellfish like clams and oysters. Artisanal fishing for shrimp has ceased. Those lacking motorboats have had to abandon fishing altogether; they are unable to catch anything worth eating in the gulf's near-shore waters.

The wastes of close to 10 million people and 10,000 industries end up in Manila Bay every day; just over 10 percent of the population is served by sewers. "The bacteria count in Manila Bay is over one hundred times above safe standards," says Celso Roque, undersecretary of the Department of Environment and Natural Resources in Manila. Efforts to get industries to voluntarily reduce toxic effluents have floundered; the city has no money to entice them and no way to enforce water quality standards.

Manila's 2 million poor, many living in makeshift hovels and squatter settlements along the bay, pay a heavy price for this pollution. Their children swim in the bay, they draw drinking water from its canals, and they must eat what little they can catch in it. Disease rates have soared, and infant mortality in some poor waterfront communities approaches 100 per 1,000 live births.

The fishing communities on Jakarta Bay, just north of Indonesia's capital, pay a similar price for their poverty. Over the past thirty years, millions of tons of industrial and municipal wastes have been discharged into the bay every year without treatment of any kind. Scientists have found PCB concentrations in green mussels as high as 1.32 ppm. At that level, a 5-gram mussel would contain 6.6 micrograms of PCBs, a dangerously high amount (Harger 1993).

One village struggling against the effects of pollution and unplanned urban growth is the small fishing village of Muara Angke. "I've lived in this village for more than thirty years," says eighty-five-year-old Sho Boen Seng, the town's unofficial head. "In the early 1960s you could catch as much fish in two hours as it now takes one or two days to bring in. Today, we have to travel out of the bay, three to six hours each way in order to fish. You can still catch some fish in the bay itself, but no one wants to eat them."

Pollution and sedimentation are the main culprits. Parts of the bay are becoming eutrophic because of the heavy input of nutrients from untreated sewage and industrial wastes. Sedimentation from dredging and landfilling operations have also helped to obliterate vital near-shore spawning areas for shellfish like shrimp and clams. "All corals have been stripped out of the bay," reports Seng. "And our mangroves are all but gone, thanks to landfilling and urban expansion."

Like all of Muara Angke's 3,000 residents, Seng and his wife and daughter live in a simple house made from breeze (or cinder) blocks and concrete with a tin roof. Seng's main work these days consists of trying to get the city government to pay more attention to the needs and concerns of the village and its people. In 1980 the government banned trawlers from Jakarta Bay in an effort to give small-scale fishermen a better chance to make a living. It was a good gesture but came too late to have much impact on fish catches. The bay's shallow waters were already so degraded and overfished that trawlers would have left anyway.

"Still, we do get some illegal trawlers coming into the bay to fish, mostly from Taiwan and South Korea," points out Seng. "But the worst practices are those of the dynamite fishermen. In the thousand islands region of the bay, they have exterminated most of our reefs, even those eight hours away by boat. And with the reefs have gone our fish."

Seng was instrumental in helping to set up a local environment committee in an effort to improve living conditions and also get fishermen to adopt more sound fishing practices. "I did get many of them to start using larger mesh nets, in order to reduce the by-catch of juveniles and smaller species," says Seng. "But we have had no luck with improving our access to clean drinking water and sanitation."

The village still lacks potable water. Its wells are polluted. Everyone must buy water from vendors who sell it door to door in bottles and buckets. And there is no sewage system. All wastes go into unlined canals that channel them directly into coastal waters.

After a decade of lobbying them, municipal authorities still have not built a cold storage facility for the village. "There is one in nearby Muara Daru," says Seng, "but not here." The village's women continue to toil long hours in the sun preparing the fish for drying on long pallets. What isn't dried must be sold the same day in the market or it will quickly rot.

Seng is now trying to introduce an environmental education course into the local school system so that the younger generation will act to protect their environment. "Who knows, perhaps our children will be able to reverse the mistakes of the past," he says. "Thirty years ago we had tree-lined canals and rivers you could swim in. Now look at it," he says, pointing to a black, garbage-filled canal. "It is all ruined."

Since the 1970s, the industrial city of Samarinda, on Kalimantan's east coast, has grown explosively as a result of oil and gas extraction. Fifteen-percent of Indonesia's oil and gas reserves are concentrated in the area, with another 30 percent located a short distance inland. Every day, approximately half a million barrels of oily water, with an average oil concentration of 25 milligrams per liter of water, are discharged untreated into the waters of the Makasar Strait, the body of water separating east Kalimantan and Sulawesi, by three oil companies (Dahuri 1994). A huge fertilizer plant near Samarinda also dumps untreated ammonia wastes into the sea. Massive fish kills in coastal waters have been reported periodically since the late 1980s.

If pollution were not bad enough, the Mahakam River Delta is silting up due to massive logging operations in the interior of Kalimantan. In 1989, authorities had to dredge 2 million cubic meters of sediment out of the delta to keep shipping channels open.

Thailand fares little better. Roughly 3 million poor people crowd along the four major rivers that run through Bangkok: the Chao Phraya, the Thachin, the Maeklong, and the Bangpakong. These four rivers deposit roughly 10,000 metric tons of raw sewage and municipal wastes into the Gulf of Thailand every day.

Although Thailand's capital has drawn up water-quality guidelines and has initiated a Master Sewerage Plan for the city, half of its 9 million citizens would still not be covered by a sewage treatment network. The World Bank has calculated that the cost of connecting all of Bangkok's residents to sewage lines would be $1.4 billion. An alternative system using the city's many refuse-clogged canals is now underway.

Vietnam's coastal waters are also clogged with untreated municipal and industrial wastes. Most of the country's urban areas lack adequate waste treatment facilities. In Hanoi, for instance, a third of the residents still depend on double-vault latrines, each one shared by thirty people on average. The situation is even worse in Ho Chi Minh City. There 16,000 dwellings along the city's canals and waterways,

163

containing over 100,000 people, have no sanitation facilities at all (Warfvinge and Pham Dinh Huan 1991).

Resources

The loss of mangrove forests in Southeast Asia may be the most extensive anywhere. Although the region still has roughly 5 million hectares of mangroves, that figure is thought to constitute less than half of their original area. In the 1930s, mangrove forests covered between 500,000 and 1,000,000 hectares of coastal land in the Philippines. Today, only 100,000 hectares remain, and most of those stands consist of secondary growth (Spalding, Blasco, and Field 1996; Country Report, Philippines 1993).

Indonesia contains the world's largest area of mangroves: some 4.25 million hectares, close to 30 percent of the world's total. However, 20 percent of the country's remaining mangroves—roughly 1 million hectares—have been designated for use as "production forests," a euphemism for logging.

Even mangroves set aside as conservation areas have been decimated by the desire for more agricultural land and the rapid expansion of mariculture operations. Java once had about 50,000 hectares of mangrove forests protected as reserves; little of that area remains. On the eastern half of Java, mangroves used to cover 1,600 kilometers of coastal land; today, no more than 800 kilometers remain, much of it degraded.

Mangroves have fared little better in the rest of the region (Jameson, McManus, and Spalding 1995):

- Since 1960, Thailand has lost half its mangrove resources; only around 196,000 hectares remain. About 64 percent of the country's mangrove area has been exploited for mariculture. Today, the undeveloped Andaman coast is the only place in the country with healthy, relatively undisturbed mangrove forests.

- Malaysia's mangrove forests have also been whittled away over the years. They now occupy no more than 650,000 hectares; at least one-third has been lost to development, especially logging, the construction of fish and shrimp ponds, and the development of rubber and palm oil plantations. Most of the country's remaining stands are located on the Malaysian half of the large island of Borneo.

- At the time of its founding in 1819, 13 percent of Singapore's land area consisted of dense mangrove swamps. Now there are virtually no mangrove stands except a few remnants on the north coast of the main island and a few isolated patches on its offshore islands.

- By the end of the Vietnam War in 1975, defoliation, napalming, and bombing had completely destroyed around 124,000 hectares of the mangrove forests in southern and central Vietnam. More than 50 percent of the mangrove swamps in the Mekong Delta and in the Cape of Camau (at the very southern tip of the country) were lost. Currently, the country has around 250,000 hectares of mangroves left, a 50 percent loss since 1945. Between 1981 and 1994, about 500,000 hectares of mangrove swamps were converted into shrimp farms (Kemf 1990; Lindén 1995).

- Cambodia has no more than 6,000–10,000 hectares left, much of it degraded from years of civil war.

- Myanmar has just over 344,000 hectares, mainly in the Irrawaddy River Delta. Much of the country's mangrove resources have been degraded by fuelwood collectors, charcoal makers, and timber merchants.

The loss of so much of Southeast Asia's mangrove forests has badly affected the livelihoods of millions of small-scale commercial and artisanal fishermen. Offshore shrimp catches have fallen dramatically, as have landings of groupers, jack, mullet, snappers, and other valuable species.

Southeast Asia's coral reefs are thought to be in even worse shape. Thirty percent of the world's coral reefs, representing all known genera of corals and all morphological types of reefs, are found in a huge lopsided triangle of tropical ocean stretching from the Philippines southeast through Irian Jaya to northern Australia, then westward through the Indonesian archipelago to Borneo. Throughout most of that area, coral colonies face numerous deadly threats. They are dynamited and poisoned by fishermen, mined for lime (for use in cement production), smothered by eroded sediment, damaged by boat anchors and coral collectors, and snuffed out by pollution from industries, agricultural activities, and municipalities.

At least 70 percent of the region's 200,000 square kilometers of coral reefs—amounting to about 140,000 square kilometers—are under threat. Most will be lost within ten to twenty years unless workable management plans can be put in place (Wilkinson et al. 1992).

When the Philippines carried out a survey of 632 reefs in 1981–82, researchers discovered that two-thirds of them were in poor to fair condition, with less than 50 percent live coral cover. A reevaluation of the country's coral resources in 1991, coordinated by the Marine Science Institute of the University of the Philippines in Manila, found that only 5 percent of the country's reefs were in excellent condition (with more than 75 percent live coral cover); 70 percent of the 735 sites visited were degraded, many beyond recognition (table 12.2).

The Philippines isn't the only country whose coral resources are under threat from fishermen using dynamite and poisons. In Indonesia, Malaysia, and Thailand, subsistence fishermen routinely resort to dynamite in a desperate effort to put food on the table. Others use sodium cyanide in small squirt bottles to stun fish, capturing them alive for the lucrative aquarium trade or for restaurants, mostly in China, Hong Kong, Taiwan, and Japan, specializing in live fish. The poisons often cripple or kill many other kinds of sea life and impoverish the reef communities by killing off coral polyps, the tiny animals that build reefs.

A recent report authored by American marine biologist Dr. Robert Johannes and Michael Reipen, a fisheries economist in Wellington, New Zealand, claims that the use of cyanide is now so widespread in Southeast Asia, that the global epicenter for coral diversity is threatened with destruction (Dayton 1995).

Tremendous loads of industrial and municipal pollution from Bangkok and Pattaya have destroyed the coral reefs in the northern part of the Gulf of Thailand. Reefs have also been damaged off the west coast island of Phuket, victims of overfishing, coastal development—particularly the construction of numerous tourist resorts—and the release of untreated sewage into Phangnga Bay. In all, over 60 percent of Thailand's coral resources are damaged (see table 12.2).

*Table 12.2. The Status of Southeast Asian Coral Reefs
Based on National Surveys in the Early 1990s*

COUNTRY	NO. REEFS SURVEYED	EXCELLENT 75–100%	GOOD 50–75%	FAIR 25–50%	POOR 0–25%
Indonesia	124	5.6	30.6	33.9	29.8
Philippines	735	5.3	25.2	39.2	30.3
Singapore	65	1.5	36.9	47.6	16.9
Thailand	168	—	36.0	33.0	30.0

Sources: C.R. Wilkinson, "A Regional Approach to Monitoring Coral Reefs: Studies in Southeast Asia by the ASEAN-Australia Living Coastal Resources Project," Proceedings of the 7th International Coral Reef Symposium, June 1992, Guam, p. 3; Michele Lemay and Lynne Hale, *A National Coral Reef Strategy for Thailand, Vol. 1: Statement of Need,* Thailand Coastal Resources Management Project, University of Rhode Island, USAID, 1991, pp. 16–17.

Notes: Percentages refer to amount of live coral cover. Coral reefs are degraded if they have less than 50 percent live coral cover. There is no percentage available for the excellent category for Thailand, since good and very good were combined into one estimate.

In Jakarta Bay, Indonesia, researchers had to travel 25 kilometers offshore to find viable coral reef communities. The rest had been done in by heavy pollution, dynamite fishing, and sedimentation.

Fisheries

Southeast Asia's fisheries are in deep crisis. "Nearly all Asian waters within 15 kilometers of land are considered overfished," says Dr. Ed Gomez, director of the Marine Science Institute at the University of the Philippines in Metro Manila.

Since Southeast Asians derive most of their protein from the sea, fishing and various mariculture operations have an added importance. Filipinos and Vietnamese get at least 60 percent of their animal protein from fish; Indonesians get 63 percent. "You have to keep in mind that the average Asian eats rice and fish with nearly every meal," points out Dr. John McManus, reefbase project leader working with the International Center for Living Aquatic Resources Management in Manila.

Furthermore, millions of artisanal and commercial fishermen make their living from the sea or through fish farming and pond cultures. Java has at least 7.5 million artisanal fishermen operating in shallow coastal waters or raising fish and shrimp in small ponds. Vietnam has close to 1 million small-scale fishermen, many of them turning to mariculture and aquaculture. Some are also resorting to dynamite: In 1992, fishermen operating near Haiphong caught 2,000–3,000 metric tons of fish using explosives (Lindén 1995).

But as more people fish for a living, resource conflicts escalate and sometimes turn ugly. Gun battles occur in the coastal waters of Indonesia and the Philippines between trawlermen and artisanal fishermen. Malaysia's gun battles over fish ceased in the early 1970s, with no clear victors.

Although most trawlers are now officially banned from the near-shore waters of the Philippines and Indonesia, many violate the law, slipping in under cover of darkness to exploit coralline fisheries. Still others bribe the coast guard and fish with impunity in broad daylight. Because of their destructive fishing gear, trawler fleets have been called the vacuum cleaners of the sea in Southeast Asia. Decades of bottom trawling in the Arafura Sea, in eastern Indonesia between the Tanimbar Islands and Irian Jaya, has plowed up the seabed, turning it into a desert.

Fishermen in Malaysia's states are allowed to fish only in their own waters, but there are many violators. Fish stocks are assessed regularly by the Fisheries Research Institute in Penang. Based on such assessments, licenses to fish are allocated to each state. "The problem is, too many poor fishermen are operating without licenses, and exploitation is getting out of control in some areas," explains a fisheries expert.

In an effort to come to grips with depleted fisheries and ruined coastal resources, Malaysian authorities are trying to reduce the number of coastal fishermen. Some 50,000 are expected to be turned into fish farmers or retrained for some other occupation. Expanding the area of fish farms in Malaysia is likely to backfire, however, since most of the new sites will be carved out of mangrove swamps, hurting offshore fisheries.

As stocks continue to collapse, taking more fishermen off the sea may be the only course of action for a number of Southeast Asian countries. Dr. Daniel Pauly, a fisheries biologist and director of the Coastal Resource Systems Program with the International Center for Living Aquatic Resources Management (ICLARM) in Manila explains the dilemma:

> The problem is that absolute fish densities decrease offshore, while the shrimp:fish ratio decreases even faster. This implies that in Southeast Asia, at least, attempts to disentangle the artisanal inshore fisheries from the trawl fisheries using traditional legislative or administrative means (exlusive zones, bans on inshore fishing for trawlers, and others) usually fail to work because the offshore stocks simply cannot support lucrative commercial trawl fisheries (Pauly and Thia-Eng 1988; Pauly, Silvestre, and Smith 1989).

The Lingayen Gulf on the northwest coast of Luzon, the Philippines, is a classic example of overexploitation, aggravated by habitat destruction from blast fishing and sedimentation from mining operations. Near Bolinao, a fishing town with 50,000 inhabitants, fish stocks were so depleted in the late 1980s, complains Dr. Liana McManus, a marine biologist working with the Marine Science Institute in Metro Manila, "that we couldn't even assess how to optimize fisheries in this part of the Gulf, because we couldn't find enough fish to carry out a proper study."

Many of the gulf's coral reefs had already been reduced to rubble by 1988. "In 1983 I didn't hear any underwater explosions [from dynamite fishing] while diving near Bolinao," recalls Dr. Helen Yap of the Marine Sciences Institute. "But in 1988 when I returned to do research, I couldn't count the number of explosions I heard during one dive. They were happening all the time."

Conflicts occur regularly in the Lingayen Gulf between trawlers and the 20,000 small-scale fishermen who work the gulf. Although trawling is not permitted within 7 kilometers of the coast, many trawlers lay out their nets 2–3 kilometers offshore. Whereas twenty years ago, fishermen could fish twenty-five days a month during the peak season, they now go out one week a month and are lucky to catch anything worth eating. As resources ran out, more of the gulf's poor fishing villages turned to illegal and destructive fishing methods.

By the early 1990s the situation was desperate. During that time, the Marine Sciences Institute in cooperation with the U.S. Agency for International Development and ICLARM worked out a rescue plan for the gulf. However, USAID, having funded an initial five-year program designed to identify major coastal zone problems and develop appropriate management plans, was unable to provide follow-up funding to actually implement the recommendations. The Marine Sciences Institute and ICLARM, working in close collaboration with local community groups in the gulf, managed to forge ahead anyway, though at a much slower pace (McManus et al. 1992).

The goal of the Lingayen Gulf Coastal Area Management Plan, according to Ed Gomez, is "social and intergenerational equity in the use of coastal resources, poverty alleviation, enabling legal arrangements, and collective advocacy." Specific projects focus on the rational use of land and water resources through zoning regulations, the rehabilitation of degraded coastal ecosystems, livelihood development, law enforcement, and the provision of "infrastructure and support services." The plan covers 2,100 square kilometers of the gulf and includes eighteen coastal municipalities, one city, and five noncoastal towns with estuarine fish ponds.

As a subproject of the gulf plan, the Marine Science Institute is overseeing the development and implementation of the Bolinao Community-Based Coastal Resources Management Project. The 50,000 people who live in that municipality are heavily dependent on marine resources for their livelihoods.

According to Liana McManus, among the major components of the management program for Bolinao are: community organization, environmental education, resource management, livelihood development, and networking and advocacy.

Recently, one more threat to the Bolinao reef system surfaced in the shape of a new open-pit mine. "The current plans for the mine, which could cover at least 2 square kilometers, site it up-current from the reef," says John McManus. "Scientists believe that during typhoons, the sediments deposited during the operation will be washed onto the reef, killing the remaining corals." To make matters worse, the mine would generate only two hundred jobs, nearly one hundred of them requiring outside experts. "Thus, it appears that the livelihoods of 20,000 people could be jeopardized for the sake of one hundred jobs, which will last for less than a decade of mine operation," concludes McManus.

Ultimately, salvaging coastal resources will depend largely on whether local stakeholders can work out and implement management strategies themselves with help from local or national authorities. Once communities feel that they can manage their resources in a more sustainable fashion, resource recovery can be very rapid.

Capiz Province on Panay, one of the poorest in a country filled with poverty, provides an example. Of its 680,000 residents, over 60 percent make their living from the sea, directly or indirectly. Half live below the poverty line, set at 2,500 pesos a month (about $100) (Hinrichsen 1994b).

As in much of the rest of the Philippines, the island's fisheries have been devastated over the past thirty years from chronic overfishing and the use of dynamite, poisons, and fine-mesh nets. Much of Capiz's coastal zone has been converted into fish and shellfish ponds, 20,000 hectares in all. The province's natural fish farms—

mangrove swamps—have been converted into fish ponds and rice paddies, harvested for timber, or filled in for urban expansion. Only around 1,000 hectares remain.

In order to tackle the pervasive poverty that grips these coastal fishing villages and reduce family size, the governor of Capiz launched a multifaceted program to raise income levels, improve the status of disadvantaged women, and promote better resource management. Technical assistance was provided by the Philippine Department of Agriculture in cooperation with the Manila office of the UN Food and Agriculture Organization. Funding came from the United Nations Population Fund, as part of its country assistance program for the Philippines.

After three years, there are thirty women's groups in Capiz, involving close to 1,000 women. Most of those groups are now at the center of community development initiatives, including improved health care, family planning services, and better resource management. Collectively, their enterprises have generated savings of over 200,000 pesos.

There is one widespread social factor that may mar attempts at the rational management of coastal resources: landlessness. Most rural households in Southeast Asia do not own the land they work. Around 78 percent of Filipino and 85 percent of Indonesian peasants own no land. When the soil wears out, or they are unable to pay rent, these poor families are often forced off the land. In many cases, their only recourse is to head for the nearest coast.

Once these migrants arrive at the coast, they usually add to the problems of the fishing communities already there. "Migrants are not restrained by webs of family and informal ties, nor do they have, as old fishermen sometimes do, a small plot of land to resort to when catches are poor," explains Dr. Daniel Pauly. "Hence, they are usually among the first to use fishing techniques such as excessively fine-mesh nets, dynamite, cyanide, and bleaches. In the Philippines and Indonesia these techniques now constitute the major fishing gear in coralline areas."

Poverty is at the root of many of the problems afflicting Southeast Asia's fisheries. "Solutions to fishing problems will be forthcoming only when the central issue, poverty itself, has been resolved," observes Dr. Pauly.

Conservation

Tiny Apo Island, off the southern coast of Negros, may be only a pinprick on a map, but it is big news for fishing communities in the Philippines and elsewhere as they struggle against great odds to manage and conserve marine resources. By preserving as a sanctuary only 8 percent of the colorful 106-hectare coral reef that fringes the island, this poor fishing community has managed to rebuild dwindling stocks of edible fish and shellfish (Hinrichsen 1997).

"Before the sanctuary was set up, it was difficult to catch fish, but now it is much easier," explains Jesus Delmo, president of the Apo Island Marine Management Committee. "In the mid-1980s, we had to travel as far as Mindanao to fish, but not now. We can catch as much fish as we need around our own island."

The only activities permitted in Apo's coral reserve are scuba diving and snorkeling. No spearfishing is allowed. "Once when some Japanese scuba divers tried to spearfish in our reserve," says Delmo, "we confiscated their gear and sent them packing as a lesson."

Before the reserve was established in 1986, the island's one hundred fishing families were using dynamite and poison, as well as hooks and lines, to put food on the table. They even employed the old *muro-ami* technique, which involves scores of children free-diving down to the reef face, pulling a large fixed net behind them. In order to frighten the reef fish out of their coral refuges and into the net, the coral was bashed with big wooden mallets. The fish usually obliged by fleeing into the net, but the coral was pulverized in the process. The waters around Apo were beginning to resemble a desert, devoid of life.

Once the sanctuary was established, with help from the marine laboratory at nearby Silliman University in Dumaguete City, it took only two years for fish stocks to recover from decades of overharvesting. Today, the sanctuary not only acts as a feeding station and nursery for fish, it also generates hard cash for Apo Islanders. The Marine Management Committee charges each scuba diver and snorkeler who visits the reef a modest user's fee, the equivalent of around $2.00. The money is used to police the reef and to educate islanders about the economic importance of healthy reefs. "Eventually, we hope to be able to build a coral reef information center here on the island," says Delmo.

Apo Island's reserve is now visited nearly every day by scores of divers from around the world. Here they can still see schools of jack, dozens of species of parrot fish, coral trout, and giant groupers. The marine laboratory at Silliman University is culturing giant clams in tanks and transplanting them to the reserve after they are deemed viable enough to survive. After the clams reach maturity, Apo Islanders will be allowed to harvest a certain number of them.

"We are now thinking of extending the sanctuary to cover more of our reef," says Conversion Regalado, one of the women members of the Marine Management Committee. "We may adopt a system similar to the one operating on the Great Barrier Reef of Australia, which divides the reef into different user zones. This way we can better control human activities and also limit diving to certain areas."

Not far from Apo Island, on the east coast of Negros, another rescue operation is underway: Knee-deep in mangrove mud, Wilson Vailoces is all smiles. At sixty-three he is as energetic as his grown-up sons. Wilson is examining the growth of a luxurious stand of red mangroves, trees he planted himself more than a decade earlier along the tidal flats in front of his house. His trees are over 10 meters high and still growing, stabilizing the shoreline, anchoring sediment, and providing valuable habitat for hundreds of species of fish and shellfish (Hinrichsen 1994b).

Wilson's mangroves do not stop at his property line. Thanks to his pioneering efforts, they now extend over 4 kilometers up the coast, covering more than 100 hectares. As of 1995, there were over 100,000 mangrove trees in Bindoy Municipality, with more being planted every month. Wilson has planted around 10,000 trees himself, helping to reforest a coast that was virtually devoid of mangroves a decade ago.

"We had cut down all the mangroves and reduced most of our reefs to rubble by dynamite fishing," explains Wilson. "We destroyed in one generation what it took nature a hundred years to build."

By the mid-1980s, with the entire regional economy collapsing from overfishing and the destruction of critical habitats such as mangroves and coral reefs, Wilson decided he had to rebuild the region's resources if there was to be any future for his children and grandchildren. He began his mangrove rehabilitation project with help from the marine laboratory at nearby Silliman University. After his neighbors saw how important mangroves are as natural fish farms, they followed his lead.

But Wilson didn't stop with mangroves. He also built and sank over 1,000 artificial reefs made from bamboo, rubber tires, or concrete. Designed as large pyramids, they provide excellent cover and homes for a host of marine life. At the same time, he began working with local farmers in an effort to convince them to use less harmful pesticides on their fruit crops (particularly mangos). And he persuaded the Department of Environment and Natural Resources to encourage subsistence hill farmers working steep slopes to plant trees around their fields in order to reduce soil erosion into coastal waters.

"Great damage has been done to coastal fisheries from landslides and siltation due to deforestation in the hills," explains Wilson. "Coral reefs have been buried and mangroves inundated. But we have now turned this situation around."

Renewing resources is only part of the battle. Mangroves, coral reefs, and associated fisheries must be managed properly and protected from poaching. Today, Wilson is vice president of the local fishermen's association, as well as president of the Bindoy Sea Watchers Organization. In order to ensure that fishermen do not employ illegal and destructive techniques, such as the use of dynamite and poisons, to catch fish, he has recruited a fifty-member force to patrol the waters north of Bais Bay. Offenders are apprehended and turned over to the proper police authorities for fines or prosecution.

Like his mangrove project, which spread up the coast of Bindoy, his sea watchers organization is cloning itself. More and more villages on the east side of Negros are forming seagoing patrols in an effort to safeguard remaining resources.

Wilson's work is not done, but the region is well on its way to recovery. Fisheries and the fecund coastal ecosystems they depend on are now being managed not simply exploited. "Before, we were slowly killing ourselves," says Wilson. "No longer. We are rebuilding our resources for the future. It's all we have."

Singapore's Singular Success

In 1995, Singapore officially entered the ranks of the world's developed economies. With an average annual per capita income of over $20,000, Singapore has actually been part of the "developed" world since the early 1980s. The only thing lacking was the UN label.

Singapore is now widely regarded as having one of the best urban environments in the world. It is cleaner than Tokyo, safer than Stockholm, and greener than Paris. Its 3 million residents enjoy the highest standard of living in Asia after Japan. The city's economy is robust and has been growing by over 8 percent a year for a decade.

Singapore's waterfront used to be an unsightly scene. Floating scum and garbage covered the harbor and collected around boats. Pig and chicken farms discharged their wastes directly into the Singapore River. Boats dumped their untreated wastes into the harbor. Food vendors washed out their stalls on the streets; the slops washed into the Kallang Basin. All of this pollution eventually collected in the city's near-shore waters.

In 1977 the government decided to rid the city of this eyesore and public health menace. The catchments of the Singapore River and the Kallang Basin were cleaned up. Ten years and $100 million later, the entire waterfront has been restored. It is now one of the city's premier tourist attractions.

The job was difficult and time consuming. Sewage lines had to be built to accommodate some 12,000 unserviced homes and industries. A multimillion dollar two-stage sewage treatment plant was built. Over 1,000 pig and chicken farms were uprooted and moved. Around 5,000 street vendors were relocated and given new stalls with proper drains connected to the sewer system. The harbor was dredged (Sien 1992; Sien and Ming 1989). "Once we removed the main sources of pollution, the rivers cleaned themselves," points out Kuan Kwee Jee, former deputy director of the Ministry of the Environment.

Management: The East Asian Seas Action Plan

Asian governments have traditionally viewed environmental protection as a luxury they cannot afford. Athough most governments in the region have established ministries or departments of environment, they are usually relegated to the sidelines, forced to operate on shoestring budgets with too few qualified personnel. Programs are routinely swallowed by red tape, and coordination with other ministries is wishful thinking.

But attitudes toward environmental and resource management are changing, however slowly. When five of the region's governments asked UNEP to help them set up a regional seas action plan in the early 1980s, it readily agreed. Confronting regional problems as a region seemed to make good sense.

After a series of preliminary meetings, the East Asian Seas Action Plan was adopted in April 1981 at an intergovernmental meeting in Manila. The action plan aims to provide a framework for a comprehensive and environmentally sound approach to coastal area development. In order to oversee its implementation the Coordinating Body on the Seas of East Asia (COBSEA) was established. This group, with representatives from each participating government—originally the Philippines, Indonesia, Malaysia, Singapore, and Thailand—has overall authority to identify priorities and determine the budget. In May 1993, at the request of the region's governments, UNEP established the Regional Coordinating Unit for the East Asian Seas Action Plan (EAS/RCU), based in Bangkok.

The Coordinating Unit also manages the region's trust fund, which was set up to cover the costs of the program. Initial contributions totaled around $100,000 a year. In recent years, however, more money has been allocated for regional activities under the action plan. UNEP continues to augment the trust fund for COBSEA-approved activities.

At the sixth meeting of COBSEA, in 1987, government representatives proposed a long-term strategy for managing the region's marine and coastal resources. At the same time, UNEP helped launch the Association of South-east Asian Marine Scientists (ASEAMS), in order to involve the marine science community and provide a mechanism for interregional cooperation. ASEAMS now functions as an independent scientific advisory body to both COBSEA and the executive director of UNEP. Since its inception in 1987, ASEAMS's membership has grown to 150. The association held its first symposium in 1989, followed by two more in 1991 and 1995.

"Before the initiation of the program, the countries of the region had no integrated activities to address the problems of the marine environment," points out

Ed Gomez. "Through this action plan, scientists and environmentalists concerned with the protection of the marine environment have been able to broaden their focus from a national to a regional perspective."

Although COBSEA has moved slowly from the science phase to the policy phase, 1994 marked the beginning of a new era for regional cooperation. "In October 1994 a high-level meeting of plenipotentiaries on the East Asian Seas Action Plan was held," explains Reza Amini, coordinator of the EAS Action Plan in Bangkok. "This meeting was very significant in that it approved a revised action plan for the region and broadened considerably the number of participating states, adding Australia, Cambodia, China, South Korea, and Vietnam."

National Coastal Management Efforts

While regional strategies are forging ahead with renewed vigor, a few national efforts have continued. Until recently, Malaysia's efforts in coastal zone management were confined to one special-area management program in Johore, the state near Singapore. However, an environmental unit has now been set up in the Economic Planning Unit of the prime minister's office. Based on the experience gained from the Johore project, which was sponsored by the U.S. Agency for International Development, the Economic Planning Unit has drafted a National Coastal Resources Policy.

Meanwhile, under the Danish Fund for Development and the Environment (DANCED), Malaysia is undertaking three pilot projects that focus on integrated coastal resources management; one in the peninsular state of Penang and one each in the east Malaysian states of Sabah and Sarawak (on the island of Borneo, shared with Indonesia). "The experience gained from these pilot projects will be taken into account in fine tuning the National Coastal Resources Policy," explains Kim Looi Ch'ng, who works for UNEP's Regional Coordinating Unit for the East Asian Seas Action Plan in Bangkok. "In addition, the Malaysian government, under a loan from the Asian Development Bank, is working to strengthen its capacity to manage shorelines by concentrating on collecting and analyzing GIS data, monitoring coastal erosion, and modeling physical processes in near-shore areas."

The Fisheries Department has even established and implemented a National Policy on Marine Parks and Marine Protected Areas. So far, twenty-one offshore islands have been placed under protection as marine sanctuaries.

Thailand can also point to an important achievement: the development and implementation of a national coral reef conservation strategy, with assistance from the Coastal Resources Center of the University of Rhode Island and the U.S. Agency for International Development (Lemay and Hale 1991, 1993).

The CRC began working with local authorities on Phuket Island in 1986 in an effort to help them develop a realistic management plan to safeguard the island's coral resources, one of the area's main tourist attractions. The rapid development of resorts with little regard for environmental impact was undermining the tourist-based economy.

One of the first things the CRC team did was to hold workshops and town meetings in an effort to get all stakeholders together. Next, they built approximately twenty permanent boat moorings to reduce anchor damage to the coral and launched information campaigns aimed at local tour operators and businessmen.

An agreement worked out between the Patong Sanitary District (the local governing body), the private sector, and the national government provides for patrol and maintenance of the mooring buoys. Local businessmen contributed close to $25,000 over two years for education and information campaigns. Once local communities could see the value in preserving coral reefs, they got behind a management plan. CRC was also careful to work with national authorities in Bangkok in order to create a positive atmosphere for implementing the proposed management plan. Resorts on Phuket now have to build on-site sewage treatment plants to reduce nutrient pollution and take steps to lower the amount of sediment flowing into coastal waters from their activities.

In March 1992, a National Coral Reef Strategy for the entire country was adopted by the Thai cabinet, and $2 million was appropriated for its implementation, based on the Phuket model. The Coral Reef Strategy has six main policy objectives: (1) to manage coral reefs according to their different ecological and economic values; (2) to reduce degradation of coral reefs by increasing the effectiveness of existing laws and regulations; (3) to build and maintain public support for the management strategy; (4) to make revisions to existing laws and administrative directives in order to make the strategy practical; (5) to monitor and evaluate progress in accomplishing the objectives; and (6) to support management efforts through ongoing scientific research.

Conclusions

- There is a clear need to develop ways to distill the experience of community-based resource management programs and apply it to provincial and country-wide initiatives.

- The international donor community, including the World Bank, should require that all new development initiatives have appropriate plans for treating wastewater. And major cities in the region should include wastewater management components in all new infrastructure projects.

- Enforcing regulations already enacted to keep trawlers from fishing too close to the coast would bring immediate relief to coastal communities as they struggle to manage dwindling stocks of fish and shellfish.

- The region's governments need to actively foster the management of resources by local communities and action groups. The Philippines is in the process of granting coastal communities a "contract of stewardship" over coastal resources. These contracts typically last for twenty-five years or more and are responsible for some remarkable success stories. Uneven as the process is, it remains a promising initiative and one the rest of Southeast Asia could emulate.

South Asia

A Sea of People

South Asia, which includes Bangladesh, Sri Lanka, the Maldives, India, and Pakistan, contains more poor people than any other region on earth, with the exception of parts of sub-Saharan Africa. The average per capita income for India, Pakistan, and Bangladesh is less than $400 a year. The Maldives and Sri Lanka average a little over $500 a year per person. Deepening poverty continues to undermine development efforts.

All along South Asia's coasts, human numbers overwhelm resources. Mangroves are being cut down for fuelwood, tannin, and building materials. Coral reefs face assaults from coral mining and from sedimentation due to coastal construction, soil erosion, and dredging. Coastal forests, like most of the region's forests, have been replaced by agricultural land and towns. Overfishing has reached crisis proportions. Practical management strategies, designed to safeguard coastal resources and livelihoods, have yet to emerge anywhere except in Sri Lanka.

Geography

South Asia's geography is dominated by the Indian subcontinent, which slices into the Indian Ocean like an inverted shark's fin. Consequently, South Asia's seas are divided into four distinct parts: the Arabian Sea, the Bay of Bengal, and the Andaman and Laccadive seas. The continental shelves vary in width from a few hundred meters to several hundred kilometers. Upwellings of nutrient-rich water in the Arabian Sea make it a valuable fishing ground.

Throughout South Asia, land and sea are locked in a perpetual tug-of-war, much like the region's inhabitants. Seasonal monsoons and cyclones routinely swamp the low-lying coasts, inundating thousands of square kilometers of land. At the same time, millions of tons of soil are being stripped off agricultural land. So much is being lost every year—about 25 billion metric tons in India alone—that new islands are regularly formed in the huge deltas of the Ganges and Brahmaputra rivers in the Bay of Bengal. Indian scientists have calculated that the 25 billion tons of lost soil contain nutrients equivalent to those in all of the chemical fertilizers applied yearly throughout the entire country. The islands formed by this soil are often washed away during tropical storms but are soon replaced by others. This give and take of land and sea makes coastal areas difficult and dangerous places to live, but millions of the region's poor people have no choice but to occupy every little sand island (or char) that rises above the waves (Harrison 1991).

As a result of widespread deforestation and soaring erosion rates, most of the countries in South Asia experience devastating floods during the monsoon seasons. In India, annual flood losses are more than fourteen times what they were in the 1950s when the country had most of its upland forests intact. In a typical year, 46,000 Indian villages may be flooded during the monsoons, affecting close to 10 million hectares of land and inundating crops worth roughly $60 million. Between 1973 and 1987, annual floods in Bangladesh destroyed 1.7 million metric tons of food crops.

Population

The collective population of Bangladesh, India, Sri Lanka, the Maldives, and Pakistan amounts to just over 1.2 billion people, making this region the second most populous on earth. India, the world's second largest country in terms of population, has close to 1 billion people (table 13.1). If its growth rate continues at around 2 percent a year, it will surpass China as the world's most populous country sometime during the first half of the twenty-first century; by then it could have a population of 1.5 to 2 billion. As figure 13.1 illustrates, the bulk of the country's population is concentrated along coastal areas and in the wide Gangetic Plain. Demographically speaking, one of India's main population challenges will be to foster planned growth along its overcrowded coastline.

All coastal areas in South Asia are heavily populated. India's nine coastal states contain nearly 420 million inhabitants (Population Reference Bureau 1994). Demographers calculate that of this number roughly 330 million live on a coast or within 150 kilometers of one. Most of Sri Lanka's 18 million people are coastal. So too is the majority of Bangladesh's population and all of the Maldives's 300,000 inhabitants. Pakistan's largest city, Karachi, sits near the broad delta of the Indus River, on the Arabian Sea. South Pakistan contains close to 30 percent of the country's population, about 40 million people; more than one-quarter of them live in the greater Karachi area. In terms of population, Pakistan is the world's seventh largest country and Bangladesh the ninth largest.

Karachi now has 10 million inhabitants in a city designed for no more than 3 million. The municipal government has completely broken down, unable to provide adequate services and infrastructure for a population that is growing by half a million a year. Most newcomers from the impoverished countryside join the 4 million people who live in slums and squatter settlements. In some areas, people are crammed five to a room, and 20 percent of infants do not live to see their first birthday (Pearce 1996).

With 1.2 billion people living on a little more than 5 million square kilometers of land, South Asia's average population density is high, more than 200 per square kilometer. India's population density now averages close to 300 people per square kilometer. Along the region's coastlines, densities swell to between 400 and 600 people per square kilometer.

Kerala, a sliver of a state with 30 million residents, wedged into the southwest tip of India, has one of the highest population densities in the country. So many people live in the state that it has been described as one continuous village. Even so, the

Table 13.1. Length of Coastline and Population Data
for Five South Asian Countries

COUNTRY	COASTLINE (KM)	POPULATION 1997 (MILLIONS)	POPULATION GROWTH (%/YR)	POPULATION 2025 (MILLIONS)	TFR 1997
Bangladesh	580	122.2	2.0	180.3	3.6
India	12,700	969.7	1.9	1,384.6	3.5
Maldives	644	0.3	3.6	0.6	6.2
Pakistan	1,046	137.8	2.8	232.9	5.6
Sri Lanka	1,340	18.7	1.5	23.4	2.3

Sources: *World Population Data Sheet, 1997,* Population Reference Bureau, Washington, D.C., 1997; *World Resources 1996–97*, Oxford University Press, New York, 1996, p. 269.

Notes: TFR = total fertility rate. Population for 2025 is based on the UN medium-term population projection.

density of the fishing population along the coast is twice that of the rest of Kerala, about 1,000 people per square kilometer.

Like urban areas in the rest of the developed world, cities and towns in South Asia are growing rapidly. Many of the region's largest cities are coastal: Bombay, Calcutta, Ahmadabad, Karachi, Madras, Dacca, and Colombo. At current growth rates, cities like these will double their population in seventeen years or less. In 1990, nearly 5 million of Bombay's 13 million residents were living on the streets or in overcrowded slums and squatter settlements. By the turn of the century, up to 75 percent of Bombay's population could be living in makeshift, inadequate housing.

In Calcutta, close to 4 million people crowd into some 3,000 slums and squatter settlements. Household slops and wastes are dumped into open latrines and drainage ditches. Respiratory ailments and rampant intestinal diseases afflict about three-quarters of slum dwellers. Most sickness is blamed on the acute lack of clean drinking water and access to sanitary facilities. Every morning, one of the most common sights across South Asia is that of scores of people defecating into drainage canals or by stagnant streams, while a few meters away others bathe, brush their teeth, and wash their clothes.

With urban populations continuing to expand, municipal authorities often find they are unable to provide adequate public services. "It seems as though we take two steps backwards for every one we take forwards," laments one Indian civil servant in Bombay.

Even good intentions often backfire. The Calcutta Metropolitan Development Authority has been trying since the early 1970s to provide a sanitary latrine for every twenty-five slum dwellers, a potable water tap for every one hundred people, and proper drains. Even where this modest goal has been achieved, the facilities are often poorly maintained. Drains are clogged, latrines overflow, and taps run dry. Funds appropriated to maintain these services often disappear into the pockets of corrupt local officials or unscrupulous contractors.

In Male, the capital of the Maldives, some 50,000 people live cheek-by-jowl on 6 square kilometers of land. The city has no public sewer system. Instead, people use the Gifili system, digging a small hole for every use. The island's freshwater lens, only a few meters deep in loose, sandy soil, sits precariously atop saltwater.

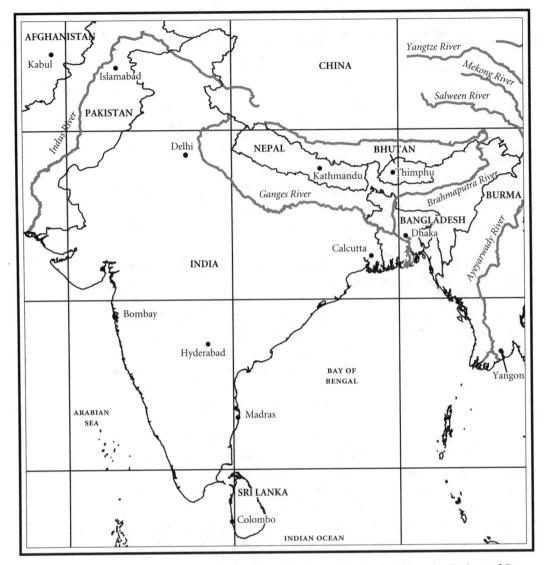

Figure 13.1 *Above:* Geopolitical Map of India, Bangladesh, and Sri Lanka. *Opposite:* Estimated Population Densities for India, Bangladesh, and Sri Lanka, 1995.

Source: Waldo Tobler, Uwe Deichmann, Jon Gottsegen, and Kelly Maloy, *The Global Demography Project,* National Center for Geographic Information and Analysis, Department of Geography, University of California, Santa Barbara, 1995.

Not only is the island's limited supply of freshwater contaminated with fecal matter, which seeps into it from all the latrines and shallow holes, but overuse is causing saltwater to intrude into the freshwater lens. Consequently, diarrhea and intestinal disorders are epidemic, and freshwater availability is becoming a serious problem. The freshwater lens, 20 meters thick in 1973, was no more than 6–8 meters thick one decade later. Presumably, it is even thinner now.

The World Health Organization estimates that no more than half the region's population—about 600 million people—have access to clean water and adequate sanitation. But that figure is thought to be misleading. Even where sewage lines exist, their contents are often pumped untreated into coastal waters or into rivers and streams. Eventually, much of the wastes of 1.2 billion people find their way into the region's near-shore waters.

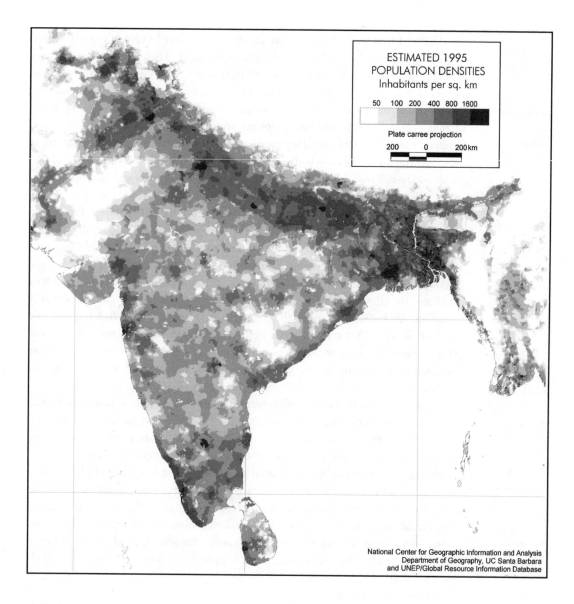

ESTIMATED 1995
POPULATION DENSITIES
Inhabitants per sq. km

50 100 200 400 800 1600

Plate carree projection
200   0   200km

National Center for Geographic Information and Analysis
Department of Geography, UC Santa Barbara
and UNEP/Global Resource Information Database

Pollution

Proper sewage treatment in South Asia is practically nonexistent. At best, cities
have one-stage treatment plants. All of Bangladesh's sewage, for instance, is flushed
directly into the sea or into rivers and streams. Only a tiny fraction of municipal
and industrial wastes from India and Pakistan receives any treatment before being
dumped into coastal waters or pumped into rivers and canals. India's coastal population produces over 4 billion cubic meters of sewage per year; much of it ends up in
coastal waters with little or no treatment. Bombay, for example, discharges roughly
365 million metric tons of untreated or partially treated sewage and municipal
wastes every year into the Arabian Sea. Calcutta dumps close to 400 million metric
tons of raw sewage and other wastes into the Hooghly Estuary every year through
361 outfalls. The estuary, one of the many mouths through which the Ganges
reaches the sea, is said to be biologically dead. Karachi adds another 175 million
metric tons of sewage and industrial wastes to the Arabian Sea annually (Harrison
1992; Hinrichsen 1990).

Much of that waste is not dispersed by currents but lingers in coastal waters, poisoning marine life and endangering human health. High bacteria counts from untreated wastes are found in nearly all near-shore waters along South Asia's urbanized coastline. Shellfish beds near Madras are contaminated with municipal wastes, making them unfit for human consumption. Industries in Colombo, Sri Lanka's capital, discharge heavy amounts of untreated wastes into the city's canal system, equivalent to 4,000 kilograms BOD of every day. The Manora Channel in Karachi Harbor reeks from hydrogen sulfide gases, its black, bubbling waters filled with industrial poisons and untreated sewage. Although Karachi has two single-stage sewage treatment plants, their capacity is constantly overwhelmed.

Pollution of near-shore waters from chemical pesticides and fertilizers is a serious problem and getting worse. Agricultural communities along India's 13,000 kilometer-long coastline spread more than 5 million metric tons of fertilizers on their fields every year and spray crops with over 55,000 metric tons of pesticides. Residues from those chemicals leach into canals and rivers, which carry them into shallow coastal waters. Pakistan's Indus River, for example, is full of pesticide residues. Sri Lanka uses so much fertilizer on its fields, up to 120 kilograms per hectare of land, that ground water reserves throughout the island are polluted with nitrates. In some areas, water wells are so saturated with nitrates and other agricultural chemicals that runoff from them kills nearby vegetation (Hinrichsen 1990).

Industrial pollution is limited to major urban centers, but untreated industrial wastes pose an increasing threat to the health of coastal areas. India's coastal industries dump some 390 million cubic meters of effluents into coastal waters every year. In the Bangladeshi city of Cittagong, a petroleum refinery's wastes contaminated the city's water supply, making it unusable. Water now has to be brought in by tanker and pumped from neighboring areas. In Sri Lanka, industrial effluents from food-processing plants and slaughter houses have killed fish in the Kelani River and turned the Valachchenai Lagoon dark brown. The waters off Bombay, besides being clogged with municipal effluents, contain elevated levels of heavy metals. There are widespread fears that as the region becomes more industrial, pollution will accelerate, compounding the problems posed by untreated or partially treated municipal wastes.

India's fourteen major rivers are all polluted. In all, they carry around 50 million cubic meters of sewage into coastal waters every year. The Yamuna, a tributary of the Ganges, picks up a daily load of 200 million liters of raw sewage and 20 million liters of assorted industrial wastes as it passes through New Delhi. The Meghna in Bangladesh (a branch of the Ganges) and the Indus in Pakistan also ferry millions of tons of pollutants to the sea (Harrison 1992).

The Ganga Action Plan

By the mid-1980s, the Ganges (or Ganga, as the Indians call it) was full of raw sewage, toxic wastes from industries, and, along most of its 2,525-kilometer length, decomposing bodies. For centuries, the burned remains of both rich and poor were laid to rest in India's most sacred river. In recent years, however, wood to burn bodies has become too expensive and hard to get. The poor cannot afford the costs. At one Indian city, Varanasi, about 10,000 half-burned bodies were put into the river every year in the late 1980s, along with 60,000 carcasses of cows, dogs, and buffaloes. (See Sampat 1996; Shankar 1992; D'Monte 1996.)

The Ganga Action Plan, launched in 1986, was the government's response to the condition of the river. The Ganga is the longest and most important of India's rivers. Twenty-five of the country's largest cities, containing in excess of 100,000 inhabitants each, are found along its banks. These towns and cities generate around 1.4 billion liters of sewage per day. The Ganga Action Plan aims to divert and treat 870 million liters of this sewage each day. After nearly a decade, however, only around half of that amount was being diverted, through settling ponds and new sewage treatment plants.

More success has been achieved in reducing industrial pollution to the river, but industries contribute no more than 20 percent of the river's pollution load. By the end of 1992, out of sixty-eight industries targeted for action, ten had been closed down, eight were being prosecuted, forty-three had installed treatment plants, and seven were in the process of installing treatment plants.

Varanasi is responsible for up to a quarter of the pollutants dumped in the river in the state of Uttar Pradesh. The city's sewage system, dating from 1917, was designed to service no more than a tenth of its current population. Parts of the system have been inoperable since 1920. Coliform counts in the Ganga at Varanasi routinely exceed 100,000 per 100 milliliters of water; the World Health Organization's recommended maximum is 10 per 100.

The number of corpses put in the river has not been appreciably reduced. Although the government built more electric crematoria along the river, most of them remain inoperable. Of the four erected in the state of Uttar Pradesh, for instance, only one was in working order in 1996, and in Bihar state no more than one in nine was operational. Consequently, thousands of half-burned bodies still end up in the river. In response, the Indian government has introduced specially bred snapping turtles to feast on the corpses in the river.

The first phase of the Ganga Action Plan has cost the government over $120 million. The states are expected to bear the recurring costs of operating and maintaining sewage treatment plants and monitoring factory effluents. However, the three main states involved in the cleanup—Uttar Pradesh, Bihar, and West Bengal—have said they cannot afford to pay. At the end of 1995, the action plan seemed to be stalled, with only about half of its overall objectives achieved.

Oil Pollution

South Asia is located on the world's oil superhighway. Every year roughly 250 million metric tons of Middle East crude pass through the Arabian Sea and the Maldives and go on to the Far East. Oil spills occur, but the real worry, as elsewhere, is the routine discharge of dirty ballast waters, bilge slops, and tank washings from oil tankers and other ships plying the shipping lanes in the Indian Ocean. UNEP has estimated, very roughly, that every year some 5 million metric tons of oil is spilled into the Arabian Sea and 400,000 metric tons into the Bay of Bengal, all of it from routine shipping operations (UNEP 1987).

Despite attempts at coming to terms with hydrocarbon pollution, the countries of this region have not been able to agree on regionwide mechanisms to combat the problem. Nor do they have an oil spill contingency plan.

Resources

Healthy mangroves and coral reefs are essential components of coastal ecosystems in the Indian Ocean. If managed properly, they can provide employment, as well as food, fiber, fodder, and building materials. But everywhere these valuable ecosystems are overexploited for short-term profit.

The region's mangroves have been pillaged for decades. India has lost half of its mangrove stands since 1963. No more than 670,000 hectares are thought to remain, much of it degraded. The 20,000 square kilometers of mangrove swamps known as the Sundarbans, an impressive coastal wetland between Calcutta, India, and Bangladesh, thought to be one of the largest single stands of mangroves in the world, have lost at least one-third of their original area. The main reason they have not been further exploited is because the area is inhabited by man-eating tigers. Mangroves in the rest of South Asia are found mostly in small, isolated patches. No more than about 10,000 hectares remain in Sri Lanka. Pakistan has around 340,000 hectares of coastal salt marshes and wetlands, but only about 160,000 hectares consist of true mangroves, most of them located in the delta of the Indus River (Spalding, Blasco, and Field 1996; Jameson, McManus, and Spalding 1995).

In recent years, Bangladesh has replanted around 120,000 hectares of coastal land with mangroves in an effort to provide storm buffers for the land and protect coastal settlements from cyclones, which regularly slam into the Bay of Bengal.

The region's coral resources—concentrated around the republic of the Maldives, Sri Lanka, and the Nicobar, Andaman, and Laccidive islands belonging to India—also suffer from widespread degradation. As a result of decades of exploitation, no more than 50 percent of South Asia's reefs—9,000 square kilometers—are thought to be in good to excellent condition, containing more than 50 percent live coral cover. The only reef colonies on India's continental coast are located in the Gulf of Kutch, on the Arabian Sea, but coral growth is poor and the reefs heavily mined for lime.

Fifty years ago, Sri Lanka had extensive areas of coral reefs. Today, as mentioned in chapter 3, most of them have been mined for cement production or dug up and used as road fill and building materials. In the mid-1980s, the country was mining over 13,000 metric tons of coral a year from living reefs and collecting another 5,000 metric tons of coral debris from beaches (Rajasuriya et al. 1995).

On the main island of Male in the Maldives, coral reefs have been virtually eliminated along the north coast, due to decades of extraction for cement production. Every year some 2,000 cubic meters of living coral is excavated for use as building material. A report written for the United Kingdom's Overseas Development Administration in 1984 concluded that at present rates of extraction, the island's coral resources would be totally exhausted by 2014. Researchers now feel the island's coral reefs may be gone by the turn of the century.

Elsewhere in the Maldives, however, coral reefs are in excellent condition. Of the 2,000 islands in the Maldive archipelago, only about 200 are inhabited, and only 25 have more than 1,000 inhabitants, accounting for 53 percent of the population. Still, marine biologists fear that without proper management plans in place, even remote coral reefs could be overexploited for the tourist trade. So far, 70 islands have been set aside exclusively for tourism (Kelleher, Bleakley, and Wells 1995).

Fisheries

The coastal waters of the Arabian Sea and the Bay of Bengal are rich in fish and shellfish, thanks to upwellings and favorable currents. The five countries of South Asia, and foreign fleets operating in this part of the Indian Ocean, harvest 3–4 million metric tons of fish and shellfish every year. Offshore fleets pull in lucrative catches of skipjack and yellowfin tuna, this being one of the few areas of the world ocean where catches have not fallen off in recent years. There has even been a 50 percent increase in the landings of panaeid shrimp since 1990 (FAO 1995).

India and Pakistan both have deep-water fishing fleets. India has over 20,000 mechanized trawlers, in addition to nearly 100 deep-sea fishing vessels. These bigger boats account for the lion's share of fish landed in the region, but thousands of tons of fish are taken by small-scale commercial and artisanal fishermen as well. In many areas, particularly off India's south coast, conflicts have erupted between trawler men and subsistence fisher folk. In recent years, as near-shore fish stocks plummet, those conflicts have become more desperate. In June 1993, artisanal fishermen from a port near Trivandrum, in the state of Kerala, torched fourteen trawlers, seized four others, and held one trawler captain hostage. In retaliation, trawlermen took four fishermen hostage. The contest ended with no deaths (Panos Institute 1995).

For the 10 million or so subsistence fishermen along India's coastline, making a living from the sea is becoming more difficult and dangerous every year. Trawler fleets often fish within 3 kilometers from shore, robbing artisanal fishermen of badly needed income and protein. "When India opened its offshore waters to foreign fishing fleets, they opened a wound," claims one old-timer from Madras. "Nothing will come of this policy except more confrontations and more deaths."

One positive development has occurred: As a result of the government's neglect, India's fishing communities were forced to become better organized. In November 1994, poor fisherfolk from the coastal states of West Bengal, Orissa, Kerala, Tamil Nadu, and Goa put away their nets and went on a highly publicized three-day strike. In January 1995, after reviewing the fishermen's complaints, the Indian government decided to freeze all foreign licenses. At last report, only thirty-four foreign fishing vessels were operating in Indian waters.

The regional Bay of Bengal Program, funded by multinational agencies under the supervision of the UN Food and Agriculture Organization, has been running for over a decade. One of its principal aims is to improve the management of small-scale coastal fisheries for participating countries, including India, Bangladesh, the Maldives, and Sri Lanka. The program has received mixed reviews; critics claim that more attention needs to be paid to the poorest fishing communities.

Turning fishermen into fish farmers is thought to be a long-term solution to the region's growing fisheries crisis. Yields from cultivated fish and shellfish operations continue to expand year after year. In 1993, India harvested 1.3 million metric tons of pond-raised fish and shrimp; Bangladesh farmed 230,000 metric tons, and Sri Lanka produced close to 100,000 metric tons (FAO 1995). There is clearly room for expansion, especially for the export market. The problem is that expanding aquaculture activities will probably mean cutting down more mangroves or destroying coastal wetlands. That will undermine near-shore capture fisheries, especially shrimp, which depend on mangroves for critical stages in their life cycle.

Conservation

South Asian governments are beginning to recognize that their coastal and marine resources need to be protected, but so far no more than fifty protected areas have been established in the entire region. India has designated sixteen sites on the mainland; another fourteen are found on the Andaman and Nicobar islands. Bangladesh has three coastal sites under some form of protection, but a management system is nonexistent. Sri Lanka has identified twenty potential sites for the creation of marine parks and sanctuaries, but so far only two have been officially declared: the Hiddaduwa Marine Sanctuary, formed in 1979, and the Bar Reef Sanctuary, established in 1992. Both, however, suffer from a lack of management capacity. The Hiddaduwa sanctuary now consists mostly of sandstone and dead coral; less than a quarter of the reef community is still alive. As of 1995, the Maldives, which have perhaps the most pristine coral reefs in the central part of the Indian Ocean, have not declared any marine or coastal protected areas . However, several areas are under consideration (Kelleher, Bleakley, and Wells 1995).

Management: An Action Plan for South Asia

UNEP assisted South Asia with the preparation of a draft action plan in the late 1980s, but it was not until March 1995 that the region's governments formally adopted the Action Plan for the Protection and Management of the Marine and Coastal Environment of the South Asian Seas Region, at a high-level meeting in New Delhi. A secretariat for the South Asia Cooperative Environment Programme (SACEP) has been set up in Colombo, Sri Lanka, to coordinate the action plan. And a network of cooperative centers has been established to oversee the scientific work. Over the years, UNEP has invested half a million dollars in the region, trying to build up the capacity of key national institutions to cooperate on a broadbased plan.

National Efforts at Coastal Management

Only Sri Lanka has implemented a national coastal area management plan. Although limited in scope, it is considered a qualified success (see chapter 3, p. 39). In recent years, the Coast Conservation Department (CCD) has begun a process of delegating more authority for issuing "minor" permits to local authorities. As part of this devolution of political authority from the central government to local agencies, the CCD has organized an extensive training program for local officials. Minor permits will be issued for modest houses, small-quantity sand removal from beaches, and other activities with limited, local impacts. However, one of the country's biggest challenges remains: halting the widespread use of coral reefs for building materials (Lowry and Sadacharan 1993).

In 1971 India created the Central Beach Erosion Board in an effort to reduce coastal erosion. According to Dr. B. U. Nayak of the National Institute of Oceanography in Gao, the board has proposed the establishment of a Coastal Zone Management Authority in each of India's coastal states. The government has yet to act on this request (Nayak, Chandramohan, and Desai 1992).

Conclusions

Increasingly, South Asian countries are beginning to develop the capacity to plan for future population growth, taking into account the resource implications of that growth. Comprehensive, integrated planning is emerging as a real alternative in India, Sri Lanka, and Bangladesh. Ways must be found to encourage economic growth, yet at the same time protect the resource base on which so many of the region's people depend for a living, either directly or indirectly. With India's current drive to decentralize the central government, devolving more authority to its states and districts, there is an opportunity for local authorities to develop community-based coastal management programs.

Of the coastal challenges collectively confronting the region, the following management issues are the most pressing:

- Develop effective national and regional fisheries management plans that can be implemented with the support of both artisanal fishing communities and large-scale, commercial trawler interests.

- Reduce the amount of nontarget species by-catch—averaging between 100,000 and 130,000 metric tons a year—from shrimp trawling operations in the Bay of Bengal, or find ways to harvest and market it.

- Develop effective management plans for coastal ecosystems, particularly mangrove wetlands and coral reefs.

- Reduce the flow of untreated municipal and industrial wastes into near-shores waters. Consideration should be given by the World Bank and other development agencies, both multilateral and bilateral, to funding the construction of two-stage sewage treatment plants in all major coastal cities. There is plenty of scope to capture and recycle wastes from the chemical and petrochemical industries, but the region's governments need help in acquiring new technologies and processes.

CHAPTER 14

The Arabian Gulf

Politics and Oil

Since the 1950s the Arabian Gulf states have been dominated by the international politics of oil. The Gulf War of 1990–91 underscored the region's geopolitical significance. Even by conservative estimates the eight gulf states—Iran, Iraq, Kuwait, Bahrain, Saudi Arabia, Qatar, the United Arab Emirates, and Oman—sit atop the largest hydrocarbon reserves on the planet. Over 76 billion metric tons of recoverable oil lie under and around the gulf, while its natural gas reserves amount to a staggering 32.4 trillion cubic meters. In a region consumed by political, economic, and religious rivalries, large reserves of oil and gas are the only things the countries have in common (WRI 1996).

There is one other common denominator: sand. The entire Arabian Peninsula, which gets less than 10 centimeters of rainfall a year, has been described as a "burning anvil," a scorched, inhospitable land worn by the sun and the wind. Summer temperatures regularly rise above 45 degrees Centigrade, turning cities into ovens. Asphalt roads soften, concrete crumbles, water evaporates in minutes, and buildings are in constant need of repair from heat damage.

The gulf is not very big. It stretches only 1,000 kilometers from the Shatt Al Arab waterway in southern Iraq to the Strait of Hormuz and varies in width from 75 to 350 kilometers. With an average depth of only 35 meters, it is considerably shallower than the Mediterranean or Red sea. But it is blessed by strong tidal currents, which flush out its waters. Its water volume is completely renewed every one to three years. This fortunate hydrobiological feature means that oil and other pollutants from domestic, industrial, and agricultural activities are pushed out into the much larger and deeper Arabian Sea.

Population

In 1997, the gulf states had a collective population of 115 million (including all of Iran's 67 million inhabitants). With the exception of the populations of Iran and Iraq, the overwhelming majority of gulf residents are coastal dwellers. Most of Iraq's population is inland, but the majority of its people are crowded into the Tigris and Euphrates River valleys and the capital city of Baghdad (Population Reference Bureau 1997).

Approximately 15 million people live and work along the gulf's highly developed coastline. Native populations are bolstered by over 5 million foreign "contract workers." The bulk of the region's foreign workers come from other Middle East countries such as Egypt, Jordan, and Yemen. Others journey across the Arabian Sea from India, Bangladesh, Pakistan, and Sri Lanka. In 1990, before the Gulf War, foreign workers made up over two-thirds of the labor force in the Arab gulf states. They made up 86 percent of the labor force in Kuwait, 89 percent in the United Arab Emirates, and 92 percent in Qatar (Omran and Roudi 1993a).

As a result of the influx of foreign workers, all of the gulf states are encouraging high fertility rates among their nationals. The hope is that more natives will mean fewer foreigners in the future. As a consequence of that policy, each woman in the gulf states averages between three and seven children over the course of her reproductive life, among the highest rates in the world (see table 14.1).

"The entire coastline of Kuwait is packed with people," observes Dr. Manaf Behbehani, professor of marine science at Kuwait University. Even before the Gulf War, "the pressures on our coastline were enormous. Now all new construction is along Kuwait Bay, and this urban expansion has ruined tidal flats and wildlife habitat."

Throughout the entire region oil wealth has fueled a building boom—of everything from apartment buildings to industrial complexes. Growth has outpaced government attempts to control it. Although Kuwait City has a master plan for phased development, most cities in the region do not. UNEP reports that some cities have doubled their population in as little as four years due to soaring birth rates and the importation of foreign labor.

According to UN projections, the region's population will more than double within the next thirty years. By 2025, the gulf's Arab states (i.e., all Gulf states but Iran) could be struggling to support over 115 million people, including millions of foreign contract workers.

Pollution

After World War II, with the discovery of vast amounts of oil, the gulf states went from feudal to industrial economies in forty years. One of the by-products of that transformation was severe coastal pollution and loss of wildlife habitat.

All the gulf states have working sewage treatment plants, covering most of their coastal populations. Before the Gulf War, nearly 95 percent of Kuwait's population was served by three-stage sewage treatment plants. Today, over 75 percent of Bahrain's population is covered by sewage treatment networks, as is close to 100 percent of the United Arab Emirates's and 79 percent of Iran's.

In sharp contrast, however, most gulf states have done little to control industrial pollution. Millions of tons of industrial effluents are dumped into the gulf's shallow waters every year with little or no treatment. Moreover, coastal investment in the gulf is estimated to be worth between $20 million and $40 million per kilometer. Over twenty major industrial complexes have been completed or are under construction. So far, few states have well-established, integrated pollution-control programs.

The need for controlling industrial and municipal pollution is receiving increased attention, however. The reason is simple: Most of the region's drinking

Table 14.1. Length of Coastline and Population Data
for the Arabian Gulf Region

COUNTRY	COASTLINE (KM)	POPULATION 1997 (MILLIONS)	POPULATION GROWTH (%/YR)	TFR 1997
Bahrain	161	0.6	2.0	3.0
Iraq	58	21.2	2.8	5.7
Iran	3,180	67.5	2.7	4.7
Kuwait	499	1.8	2.2	3.1
Oman	2,092	2.3	3.4	6.2
Qatar	563	0.6	1.7	4.3
Saudi Arabia	2,510	19.5	3.1	6.4
United Arab Emirates	1,448	2.3	1.8	3.8

Sources: World Population Data Sheet, 1997, Population Reference Bureau, Washington, D.C., 1997; *World Resources 1996–97,* Oxford University Press, New York, 1996, p. 269.

Notes: Coastline for Saudi Arabia includes Gulf and Red Sea coasts. TFR = total fertility rate.

water comes from the sea. The gulf hosts the largest desalination plants in the world. The intake valves in all desalination plants are monitored carefully for dangerous pollutants. A chlorine plant on Kuwait Bay was closed down when scientists discovered it was discharging mercury into coastal waters.

Sometimes masses of jellyfish get sucked into the intake valves of desalination plants and thermal power stations (which use seawater for cooling purposes). Although there is no proof that jellyfish swarms are linked to pollution, circumstantial evidence points in that direction.

Red tides have also increased in recent years, badly affecting fisheries, particularly maricultured species like groupers and sea bream. Although these killer algae have been around for centuries, their outbreaks are aggravated by untreated industrial and municipal wastes, which consume oxygen needed by marine organisms. Throughout the gulf, coastal waters near industries and cities are increasingly oxygen starved.

The Oil Highway

Oil is the real polluter in the gulf. The world's oil highway starts and ends here. Somewhere around 25,000 tankers sail in and out of the Strait of Hormuz every year. They carry 60 percent of all the oil carried by ships throughout the world, a billion metric tons a year. The oil is exported from twenty-five major oil terminals scattered around the region. Most oil is carried by supertankers at least as large as the quarter-million-ton *Amoco Cadiz.*

With all that oil being pumped and transported, the gulf's waters are heavily contaminated with oily residues and tar balls. Roughly 2 million barrels of oil are spilled into the gulf's waters every year from the routine discharge of dirty ballast waters and tanker slops and from the region's eight hundred offshore oil and gas platforms.

"Oil pollution from routine operations is our biggest environmental problem," states Dr. Hosny Khordagui, a research scientist at the Kuwait Institute for Scientific Research (KISR) in Kuwait City. "Oil pollution affects a lot of our marine biota, tainting fish and mussels." Near shipping lanes and offshore oil rigs, petroleum residues 100 times higher than in uncontaminated waters have been found in the tissues of mussels. Fish are also affected but not as severely as filter feeders (clams, oysters, mussels). One KISR scientist notes that "fish so far show no signs of pollution stress from hydrocarbons, but every fish we have ever sampled contains oil in its tissues. Fortunately, the levels are too low to raise health concerns."

With such large amounts of oil being spilled into the gulf every year, visiting research scientists in the late 1980s expected to find vast stretches of sand beaches coated in oil and ecosystems ruined. Such was not the case. "Although some beaches are heavily contaminated with tar, the background levels of petroleum hydrocarbons in sediments and biota are not exceptionally high," states Dr. Olof Lindén, a Swedish marine biologist who has studied the effects of oil on marine ecosystems throughout the world. "This is probably due to the rapid degradation and weathering of oil in the gulf's warm, shallow waters." Lindén made his studies before the Gulf War.

The Gulf War

The Gulf War was a disaster for tiny Kuwait. Kuwait City's infrastructure was left in ruins, its hotels and public buildings destroyed, its power plants and desalination plants sabotaged.

In the final hours of the war, retreating Iraqi troops dynamited over six hundred oil wells and opened the valves at Kuwait's Sea Island storage facility, sending thousands of tons of oil into the gulf. They also are thought to have brought in and sunk seven oil tankers. The results were devastating (see Renner 1991; Canby 1991; UNEP 1992a; WWF 1991):

- The 610 oil fires burned out of control for months, consuming around 5 million barrels of crude oil a day along with 70 million cubic meters of natural gas. The fires were concentrated in seven oil fields, located both north and south of Kuwait City, with the majority centered at the Al Burgan oil field south of the airport.

- Fire storms—described by a *National Geographic* reporter as resembling "flaming tornadoes, twisting and writhing in the wind"—raged from March to November 6, 1991, when the last oil fire was extinguished. The fires were estimated to have released into the atmosphere more than 1 million tons of sulfur dioxide and around 100,000 tons of nitrogen oxides *each month*.

- The sooty pall from the fires blackened the sky over the gulf. Black rain fell in Saudi Arabia and Iran, and 1,500 miles northeast, black snow fell in the mountains of Kashmir.

- United Nations experts sent in to assess the damage estimated that somewhere between 4 million and 8 million barrels of oil had been spilled into the gulf. At one point, an oil slick 100 kilometers long and 30 kilometers wide was reported moving south, but it broke up and dissipated before reaching Bahrain. A major part of the slick was trapped in Musallamiyah Bay near the Abu Ali islands (on the Saudi coast) by booms and hastily built sand banks.

- All told, the oil stained about 560 kilometers of coastline in Kuwait and Saudi Arabia. Along some areas of the Saudi coast, scientists reported that all mangrove stands and 90 percent of salt marshes had been oiled. Up to 30,000 seabirds are thought to have perished, including thousands of cormorants and terns.

The long-term effects of the Gulf War are not known. Immediately after the war, a United Nations interagency task force—fifty experts from twelve agencies and fourteen countries—worked for ninety days to evaluate the effects of the war on gulf ecosystems. Predictably, they found high levels of oil in bivalves and in sediments, but the team was not in the region long enough to evaluate long-term damage (UNEP 1992b). Based on initial damage assessments—damage to the Saudi coast alone was put at $700 million—a $1.2 billion rehabilitation program was proposed by the United Nations, involving nineteen major projects, but the funding never materialized.

John Robinson, head of NOAA's Gulf Program Office, told reporters in 1993, "Two years after the oil spill there are some signs of recovery, but despite all our attempts it is going to be decades before we can fix the situation."

Robinson's team was puzzled by the sudden death of 90 percent of Kuwait's coral reefs. "Reefs in November 1991 appeared in good shape," explained Robinson, "but when our scientists got there in mid-1992 the reefs appeared to be in very bad shape" (World Conservation Union 1992). The scientific community is divided over the causes. The corals may have succumbed to colder water temperatures or reduced salinity due to heavy rainfall. But they may also have died from the sublethal (long-term) effects of oil. No one knows for sure.

At the end of 1992, researchers from the International Atomic Energy Laboratory in Monaco reported that many of the worst hit beaches in Saudi Arabia were clean of oil. Apparently, the gulf's warm waters were able to degrade and weatherize (evaporate) the oil more quickly than scientists thought possible.

Resources

The gulf's mangroves have been particularly hard hit by coastal development. Only around 125 square kilometers of mangroves survive in the entire region: 90 square kilometers off Iran; 10 square kilometers on the coast of Saudi Arabia and Bahrain; and the remainder along the coast of the United Arab Emirates. Nearly 40 percent of Saudi Arabia's coast has been reclaimed for urban, port, and coastal development, destroying 50 percent of the country's gulf coast mangroves (Jameson, McManus, and Spalding 1995).

The coral reefs in the gulf are already pushing their environmental limits. Consequently, they are low in species diversity: fifty-five to sixty species of coral are now found there, compared to over two hundred in the Red Sea. Most communities consist of patch reefs, but a few fringing reefs are found around offshore islands such as Qaru and Kubbar off the coast of Kuwait. Island reefs, in general, are damaged by boat anchors and trawling operations. About 40 percent of Qatar's coral reefs and nearly one-quarter of those around the coast of Bahrain have died, killed by sedimentation and pollution.

The state of the region's seagrasses is not known, but land reclamation and dredging, in particular, churn up bottom sediments, which smother seagrass meadows and choke off coral growth.

Between August and October 1986, thousands of dead and dying dugongs, dolphins, turtles, and fish washed up on beaches along the coasts of Saudi Arabia and Qatar. Scientists don't know what killed them, but industrial pollution and oil are at the top of the suspects list.

There seem to be few genuine conservation efforts in the gulf states. Kuwait was planning to designate several of its offshore islands as nature reserves, to protect marine turtles, but the Gulf War ended those plans. The Gulf of Sulwa region, shared by Bahrain and Qatar, harbors the world's second largest population of dugong, yet nothing has been done to protect their vital near-shore habitats. There are few marine and coastal protected areas in the entire gulf region: Saudi Arabia has two protected coral reefs, Bahrain one, and Iran one. Oman's two protected coral reefs are on the Arabian Sea coast (Kelleher, Bleakley, and Wells 1995).

Fisheries

The region's rich fisheries have not yet recovered to their prewar levels. Gulf fishing fleets used to haul in around 4,000 metric tons of shrimp a year (mostly *Penaeus semisulcatus*). In 1992, the total catch of shrimp was 25 metric tons. By 1993, shrimpers from Bahrain were returning to port with only 1 percent of their pre–Gulf War catch (Pearce 1993).

Two factors are likely to help fish stocks recover: (1) most gulf states have established fishermen's associations that regulate the number of fishermen and boats allowed to exploit fish and shellfish stocks; (2) because many of Kuwait's boats were destroyed during the war, there are fewer fishermen in the gulf's northern waters. Both Kuwait and Saudi Arabia have imposed bans on shrimping from March to September, giving the shrimp time to spawn. No commercial trawling is permitted in near-shore waters, a regulation that helps artisanal fishermen make a better living.

Some gulf states, such as Kuwait, are also developing large-scale fish farming enterprises for the domestic and export markets. At KISR's Mariculture and Fisheries Department, three species of fish—sea bream, grouper, and tilapia—are being cultured in large sea pens. Saudi Arabia is also beginning to develop a mariculture industry in the gulf.

Management

In 1978, the eight governments of the gulf adopted a Convention and Action Plan. Two legal instruments, the Regional Convention (known as the Kuwait Action Plan) and a protocol titled Regional Cooperation in Combating Pollution by Oil and Other Harmful Substances in Cases of Emergency, were ratified and came into force in 1980.

The United Nations Environment Programme acted as interim coordinator for the action plan, until the Regional Organization for the Protection of the Marine Environment began operating on January 1, 1982. The ROPME secretariat was set up in Kuwait, but during the Gulf War it was exiled to Bahrain (it is now back in Kuwait).

In connection with the protocol concerning regional cooperation in combating oil spills, the Marine Emergency Mutual Aid Centre (MEMAC) was established in Bahrain in August 1982. "MEMAC's objective is to help contracting states develop their own national capabilities to combat pollution from oil and other harmful

substances as well as to coordinate information exchange, technological cooperation, and training," says Hamid Shuaib, president of the Kuwait Environmental
Protection Society. "MEMAC took a leading role in combating oil pollution during
the Naw Roz oil field disaster in 1983 and the Assimi oil tanker disaster in 1985." It
also played a key role in coordinating activities in the aftermath of the Gulf War
(Shuaib 1988).

A second protocol on controlling oil pollution—this one designed to combat
pollution from offshore oil and gas platforms—was signed in December 1988. Two
more protocols were ratified in the early 1990s: one dealing with the control of
land-based sources of pollution and one concerning the transport of hazardous
wastes.

During the initial phase of the action plan, the gulf states concentrated on
oceanographic studies and baseline pollution surveys. A major emphasis has been
on making data comparable to information gathered in other regional seas programs. Between 1982 and 1990, more than five hundred technicians received special training in oil and non-oil pollutant sampling, data handling, oceanographic
modeling, marine monitoring and research, and marine pollution prevention. As a
whole, the Kuwait action plan focuses on oil pollution, industrial wastes, sewage
treatment, fisheries resources, and the environmental impacts of coastal engineering and mining.

Because the region has been wracked by wars for a decade, ROPME's programs
got put on hold. In addition, the gulf states have handicapped ROPME with an annual budget of only $1.5 million, enough to pay for basic coordination work but
not much else. Nevertheless, ROPME remains the only truly regionwide forum in
the gulf through which cooperation on marine and environmental issues can be
achieved.

National Efforts at Coastal Management

National efforts at coastal zone management have continued in fits and starts. The
Meteorological and Environmental Protection Administration (MEPA) of Saudi
Arabia, which played a major role in cleanup activities after the Gulf War, has completed detailed coastal surveys in cooperation with the World Conservation Union
for both of its seacoasts. A National Coastal Zone Management Plan was proposed,
but no action has yet been taken to implement it.

Only Oman has been able to plan and implement a comprehensive coastal area
management program. Since Oman, like many developing countries, had no tradition of coastal management, the process took nearly a decade to complete. Rodney
Salm, coordinator of IUCN's Marine and Coastal Conservation Programme in
East Africa, spent eight years in Oman during the 1980s surveying the entire
coastal zone and developing a practical and implementable coastal area management plan (see Salm 1988, 1991, 1992).

"This was an exciting project to work on, since the scientific community knew
virtually nothing about Oman's coastal areas," recalls Salm. "I discovered that the
country's coasts were largely unspoiled and incredibly diverse: enormous sandy
beaches stretching for hundreds of kilometers, sculpted rocky outcrops composed
of limestone and sandstone, and deep fjords slicing inland near the Strait of
Hormuz."

Oman's biological wealth was equally impressive. Along the southern coast,

Salm discovered areas of rich upwellings, supporting great quantities of marine life, including kelp beds and prolific algal growth. He also found more extensive coral reefs than anticipated. "Oman has forty-six genera of coral," he says. "Some of the communities have even adapted to colder water."

One of Oman's real treasures is the largest nesting population of loggerhead turtles in the world, along with hundreds of nesting green and hawksbill turtles. Several huge colonies, each containing 6,000–12,000 nesting turtles, have been discovered at Ra's al-Hadd and Ra's al-Jumayz on Oman's Arabian Sea coast and in the Daymaniyat Islands. Just south of Ra's al-Hadd, tens of thousands of seabirds can be found.

"Despite the pristine nature of many coastal areas, rapid coastal development was beginning to outpace the ability of the government to manage resources sustainably," says Salm. "Coastal construction and urban expansion, the mining of beach sand, oil and tar balls washing up on beaches, abandoned fishing nets on the seafloor, and gill net fisheries were becoming serious problems in some areas. So we proposed a management plan that could be implemented immediately, without need for loads of new legislation or the creation of new institutions."

Salm and his team carefully worked out a coastal area management plan for the entire country based on the idea of multiple-use areas and protected, or conservation, zones. All management activities were tied into the mandates of existing ministries, with the cabinet having overall authority to settle disputes.

This strategy seems to be working. The proposed management plan was adopted by the government of Oman, and the first stage was implemented in 1986. By 1992 Oman had a functioning coastal zone management program. Five areas received immediate attention: (1) beach sand mining has been banned; (2) Environmental Impact Assessments (EIAs) or "no environmental objections" by the Ministry of Regional Municipalities and Environment (formerly the Ministry of Environment and Water Resources) are required for all new coastal development projects; (3) both marine and air pollution are now being monitored and regulated, and a rapid-response oil spill contingency plan has been worked out; (4) the only mangrove forest in the capital area at Qurm has been set aside as a nature reserve; and (5) visits to the Daymaniyat Islands have been prohibited during the nesting season for birds and marine turtles. In addition, three marine sanctuaries were set up (Salm and Dobbin 1987).

Oman's coastal area management plan is considered a model for other countries struggling to conserve resources in the face of mounting development pressures.

Conclusions

- All Coastal management issues need to be given greater priority by the region's governments.
- More research is needed on the sublethal effects of chronic oil contamination, especially the longer-term effects of hydrocarbons on marine plants and animals, particularly coral communities and seagrass meadows.
- It is imperative to plan for future resource and infrastructure needs, taking into account fertility levels and population distribution patterns.
- Since most of the Gulf's population can be classified as "coastal urban," each country needs to develop a workable system of urban planning.

CHAPTER 15

The Red Sea and the Gulf of Aden

Between Two Deserts

The Red Sea is unique. Nowhere else on earth is such a productive sea surrounded by inhospitable wasteland. On a map the Red Sea appears like a giant bayonet wound in the earth's crust—deep, narrow, and long. This gash extends for 2,100 kilometers, from Suez in Egypt to the Strait of Bab el-Mandeb, which connects the sea to the Gulf of Aden. Its average depth is 500 meters, but in many places the bottom plunges to 2,000 meters. The sea is, in fact, an ocean in the making. It was born as a crack in the desert floor some 25 million years ago and has been expanding ever since, by about one inch a year (Doubilet 1993). Since there are no big rivers flowing into the Red Sea, it contains probably the warmest and most saline water of any sea. In the summer, surface water temperatures regularly exceed 30 degrees Centigrade.

Due to the semi-isolation of its bottom waters, the sea is filled with unique and endemic species of fish and plants. Divers who have visited the Red Sea claim that its generous coral reefs are among the most spectacular on the planet. It contains at least 1,000 species of fish and shellfish, many of them reef dependent. Species that are becoming scarce in other seas can still be found in the Red Sea in abundance: the humphead wrasse, coral grouper, Picasso triggerfish, and emperor angelfish, to name a few.

Were it not for the Suez Canal, which opened in 1869, the Red Sea might very well have been bypassed by development entirely. The canal, however, transformed the Red Sea into a major commercial crossroads, providing easy access to Mediterranean and European markets. Some 20,000 merchant vessels and small tankers ply its warm waters every year, sailing to and from the Persian Gulf and Asia. On average, its commercial traffic accounts for 15 percent of all international shipping every year (Hinrichsen 1990).

Since the canal opened, over forty species of Red Sea fish have taken up residence in the eastern Mediterranean. There is even a special name for those fish: Lessepsian migrants, named after Ferdinand de Lesseps, the French engineer who built the canal.

Population

The collective population of the nine countries that border the Red Sea and the Gulf of Aden amounts to slightly more than 150 million people (see table 15.1). However, to calculate human impacts on the Red Sea itself, that figure is grossly misleading. Most of the inhabitants of those countries live and work in other areas: In Egypt, 99 percent of the population lives along the Nile River or the Mediterranean coast; most Israelis live along the Mediterranean coast or the Jordan River; most of Saudi Arabia's population is concentrated along the gulf side. No more than 5 million people are calculated to live along or close to the Red Sea coast and the Gulf of Aden. According to the World Bank, only around 520,000 people live along Sudan's Red Sea coast; in Eritrea, around 1.3 million are coastal out of a total population estimated at 3.6 million in 1997. And most of them live in small fishing villages or the region's few cities. Hundreds of kilometers of the coast are virtually uninhabited (World Bank 1995; Population Reference Bureau 1997).

Despite a thinly settled coastline, the overall population growth rates in the Red Sea region remain high. Leaving out Egypt and Israel, women average between five and eight children each over the course of their reproductive lives (see table 15.1). If those growth rates continue, Djibouti, Eritrea, Jordan, Saudi Arabia, Somalia, Sudan, and Yemen will nearly double their populations in one generation.

With few people in the region, urbanization is not a problem. There are only a few important urban centers on the sea: Suez, Egypt, at its most northern point; Elat, Israel, and Aqaba, Jordan, squeezed together at the end of the deep and very narrow Gulf of Aqaba; Port Sudan, Sudan; Massawa and Assab, Eritrea; and Jiddah and Mecca, Saudi Arabia. Along the Gulf of Aden three more urban areas are found: Djibouti in Djibouti and Aden and Al Mukalla in Yemen.

There are three main reasons the Red Sea coast has never been developed: the heat, the lack of freshwater, and the remoteness from traditional centers of population and commerce.

*Table 15.1. Length of Coastline and Population Data
for Nine Red Sea and Gulf of Aden States*

COUNTRY	COASTLINE (KM)	POPULATION 1997 (MILLIONS)	POPULATION GROWTH (%/YR)	TFR 1997
Djibouti	314	0.6	2.3	5.8
Egypt	1,840	64.8	2.1	3.6
Eritrea	1,094	3.6	2.9	6.1
Israel	14	5.8	1.5	2.9
Jordan	26	4.4	3.3	5.6
Saudi Arabia	2,510	19.5	3.1	6.4
Somalia	3,025	10.2	3.2	7.0
Sudan	853	27.9	2.1	5.0
Yemen	1,906	15.2	3.5	7.2

Sources: World Population Data Sheet, 1997, Population Reference Bureau, Washington, D.C., 1997; *World Resources 1996–97,* Oxford University Press, New York, 1996, pp. 268–69.

Notes: TRF = total fertility rate. Saudi Arabia's coastline includes both Red Sea and Persian Gulf; Somalia's includes Red Sea and Indian Ocean.

Pollution

With only a handful of cities of any size along the sea, municipal and industrial wastes pose few problems to its health. Jiddah has a sewage treatment network, so wastes are not dumped raw into coastal waters. Port Sudan, Massawa, and Assab lack any real treatment facilities, but sewage is swept away by currents and dispersed. Most of the other small towns have no sewage treatment plants, and wastes are dumped into coastal waters, but the effects are thought to be minimal. Except for petrochemicals, there are few industries along the entire coastline.

One blot on the Red Sea is oil pollution. Nearly 100 million metric tons of crude oil are transported through the Gulf of Aden and the Red Sea every year. Most pollution, however, is concentrated around the oil fields in the Gulf of Suez. Tar balls and floating slicks are common in that part of the sea. How much oil is spilled into the Gulf of Suez from offshore drilling and routine shipping is not known, but pollution of beaches is extensive. "Some beaches along the Gulf of Suez and its islands are already oiled beyond recovery," says Dr. Youssef Halim of the Department of Oceanography at the University of Alexandria in Egypt.

Tourism: The Main Threat

Perhaps the most significant threat to coral reefs in the Gulf of Aqaba and elsewhere in the Red Sea is unregulated tourism. Elat and Aqaba are among the most popular destinations in the Red Sea for thousands of recreational divers and snorkelers. Anchor damage to reefs at popular dive sites is becoming a serious problem in the region (*Economist* 1996; Watzman 1995; Welwig 1995).

Localized damage to reefs has also been reported on Egypt's Red Sea coast. Hastily constructed tourist resorts in the towns of El-Ghardaqa, Hurghada, and Quseir have resulted in excessive nutrient pollution and sedimentation of reefs due to coastal construction.

There are fears that once the Red Sea is discovered by adventurers and ecotourists, the region will be overrun with visitors flown in by the jet load. In East Africa, particularly Kenya, some game reserves have been terribly degraded by too many tourists. The same fate could await the Red Sea unless steps are taken to develop its tourist potential in a more rational manner.

Resources

Resource exploitation is not a widespread problem in this area, as it is in virtually every other sea on the planet. Few people are around to exploit anything. Mangroves—pushing their northernmost limits—are troubled more by grazing camels than by people. In a few areas, they are a valuable source of building materials, fodder, and fuelwood. But exploitation is local and limited. The sea's ten species of seagrasses are in relatively good health. Coastal dredging along the Saudi and Egyptian coastlines has ruined seagrass habitat in some areas, but again the effects are localized.

Around 220 species of coral have been identified along the Red Sea's 2,000 kilometers of reefs. The most prolific coral communities are concentrated in the sea's northern and central portions. One of the most impressive reef formations in the sea is the 400-kilometer-long series of reefs lying 10–40 kilometers off the coast of Saudi Arabia. UNEP has estimated the standing crop of fish on Red Sea reefs to be ten to fifteen times as productive as that in the North Atlantic. Overall, the sea's reefs are in good condition, thanks to low population densities, the lack of industrial centers, and the mostly artisanal nature of Red Sea fisheries (Jameson, McManus, and Spalding 1995).

Metalliferous Muds

Metalliferous muds, rich with metals, are found on the deep seabed, midway between Saudi Arabia and the Sudan. When geologists estimated that the muds could contain 2 million metric tons of zinc, 500,000 metric tons of copper, 80 metric tons of gold, and 4,000 metric tons of silver, the two countries formed the Saudi-Sudanese Red Sea Commission in 1974, in a bid to exploit those mineral resources. Despite research investments of over $100 million, mining has not commenced. The technology needed to extract valuable metals from such depths has not developed as fast, nor proved as economic, as previously thought. The commission insists, however, that commercial mining operations will begin sometime before the turn of the century (Hinrichsen 1990).

Fisheries

Fisheries in the Red Sea are mostly exploited by small-scale, artisanal fishermen. Yemen's fishing fleet, for instance, consists of 1,000 small *sambuks* with outboard motors. A beach seine fishery also exists for immature sardines and anchovy. Eritrea's fleet, mostly destroyed during thirty-one years of continuous warfare, is now being rebuilt. Eritrea has about 2,500 fishermen, close to 1,000 of them based in Massawa. Sudanese fishermen operate out of small dugouts or canoes and use hand lines to catch reef fish like rock cod, red snapper, coral trout, and emperors.

Saudi Arabia's Red Sea fishery is not nearly as developed as its gulf-side operations. Artisanal fishermen use small dugouts or planked canoes and fish close to shore. However, the government recently formed a company aimed principally at exploiting the shrimp fishery for export to Japan. New fishing harbors and processing facilities are being built along the Saudi coast.

According to FAO, the total take of fish and shellfish from the Red Sea and the Gulf of Aden is still relatively small—around 150,000 metric tons a year on average. Tuna are not exploited much, but estimates of their potential annual yield have been set at 20,000 metric tons (FAO 1995).

Yemen has developed a deep-sea lobster fishery, which was yielding about 300 metric tons a year by the late 1980s. More recently, catches have fallen off dramatically due to overharvesting. In order to give stocks a chance to recover, Yemen has limited the number of boats and has prohibited the taking of females with eggs. The lobster fishery off the tip of the Sinai Peninsula in Egypt has also collapsed from overexploitation (Hinrichsen 1990).

Farther south, the picture looks more encouraging. Eritrea is on course to develop and implement a rational coastal management plan, including the introduction of a sound fisheries management policy. Independent only since 1991 after a bitter thirty-one-year war of liberation against Ethiopia, Eritrea has revoked the fishing license of an Egyptian commercial operation for overharvesting and taking undersized fish. Currently, the only foreign fleets operating in Eritrean waters consist of a Saudi Fisheries Council bottom-trawling operation and an Israeli trawl fishery for demersal fish. In addition, two companies—one American and one a joint venture between a German firm and Eritrea—are taking live reef fish for the aquarium trade. Early fears that the country's pristine waters would be swiftly overharvested have dissipated, especially since the FAO estimated the annual sustainable yield of fish and shellfish at close to 80,000 metric tons. As of January 1997, Eritrean fishermen were pulling in no more than 3,000–4,000 metric tons a year, consisting mostly of red snappers, coral trout, yellow-tailed jack, and a variety of groupers (Rowntree 1993).

Yossef Kahsay, head of marine development in the Ministry of Marine Resources, based in Massawa, claims that Eritrea has two basic aims in its efforts to manage fisheries and near-shore waters: "First, we want to promote the development of our fisheries resources, especially as a badly needed source of protein in Eritrean diets; and second, we want to generate foreign currency through targeted exports."

Conservation

With limited human impacts in the southern part of the Red Sea, marine turtles are thriving there. Researchers along the Sudanese coast report the highest hawksbill turtle nesting densities ever observed. Thousands are also found on Saudi Arabia's deserted offshore islands. Hawksbill and green turtles nest on deserted beaches along the coast of Yemen. The Ithmun and Sharma beaches are counted among the most important nesting sites in the world for green turtles. Exploitation of turtles is low, and there are plenty of good nesting beaches. Still, no conservation legislation exists to safeguard these important nesting areas.

A few parks and protected areas have been set up in the Red Sea region. Egypt has designated the exquisite coral reefs of Ras Mohammed, off the southern tip of the Sinai Peninsula, as a marine national park. It is managed vigorously. Coastal development is rigidly controlled, and strict guidelines have been enacted requiring on-site sewage disposal plants for all new resorts. On Ras Mohammed's reefs, anchors are not allowed; boats use special anchor buoys. And divers are not allowed to collect shells or fish from any part of the park (*Economist* 1996).

Saudi Arabia has undertaken a survey of important shallow-water marine habitats in cooperation with the World Conservation Union, the Red Sea and Gulf of Aden Environment Programme (PERSGA), and UNEP. The country intends to develop a system of marine protected areas along its relatively unpolluted Red Sea coast.

The Eritrean coast is one of the least populated and least developed in the world. The coral reefs surrounding the Dahlak Archipelago are thought to be some of the most pristine in the southern part of the Red Sea. Dr. Chris Hillman, a marine resources advisor for the British Department for International Development (DFID), has spent three years in Massawa working with the Ministry of Marine Resources in an effort to develop a coral reef management system. "We have surveyed large tracts of Eritrean waters and have identified about 130 species of coral," says Hillman. "But we want to develop a multiple-use management system, patterned somewhat after Australia's Great Barrier Reef Marine Park. We are not interested in locking marine resources away in restrictive reserves."

In recent years, the New York Zoological Society has fielded several scientific expeditions to the Dahlak Archipelago. According to the reports, those islands also contain a wealth of wildlife, including hundreds of species of birds (e.g., pink-backed pelicans, brown boobies, and dozens of species of waders and waterfowl) and relatively rare mammals such as the golden jackal. An Eritrean Coastal Conservation Project has been launched (Hillman 1993).

Management: PERSGA

After several preliminary meetings, the governments of the region met in Jiddah in January 1981 and approved the Action Plan for the Red Sea and the Gulf of Aden. The Regional Convention was adopted a year later by six of the region's states, along with a protocol to combat pollution from oil spills and other harmful substances.

Since many of these countries had no experience in cooperating on marine issues, one of the first priorities was getting member countries to set up a network of scientific institutions capable of coordinating the work. A secretariat for PERSGA was established in Jiddah, originally under the auspices of the Arab League Educational, Cultural and Scientific Organization (ALECSO).

Building institutional support for the action plan was not easy. Yemen, still divided into north and south at that time, did not have a functional marine laboratory. One was built with assistance from UNESCO and the Islamic Development Bank.

Meanwhile, as part of the action plan, Djibouti, Somalia, and Yemen agreed to establish a "subregional response center" for combating oil spill emergencies in the Gulf of Aden. The center was set up with assistance from the International Maritime Organization and other donors.

In 1993, PERSGA expanded its membership by admitting Egypt, bringing the total to seven: Egypt, Jordan, Saudi Arabia, Yemen, Somalia, the Sudan, and Palestine (in lieu of Israel). That same year, Jiddah became the headquarters for a new, reinvigorated action plan, independent of the Arab League.

So far, baseline studies of pollution sources to the marine environment have been carried out and inventories of marine resources completed. There is a wealth of basic oceanographic data on the Red Sea; Egypt has been studying the physical and chemical properties of the sea since the 1930s. As more data accumulate, the countries of the region will be better prepared to implement coastal area management plans.

Conclusions

- More efforts are needed to make PERSGA a functioning organization with an achieveable action plan and sufficient funding to implement it.
- Since ecotourism is likely to be a major source of foreign exchange for Red Sea states, coral reef protection and management plans should be drawn up and implemented as soon as possible, particularly in those states such as Israel, Jordan, and Egypt, that are already beginning to take advantage of this potential bonanza.
- Fisheries management, though not critical now, is likely to be in the future, as more Red Sea states grant fishing licences to foreign fleets. Eritrea's efforts at regulating its fisheries could be a model of rational management for the rest of the region.

CHAPTER 16

East Africa

Exploited Coasts

Not only is virtually every East African country struggling with high population growth rates, most are also deeply in debt. Collectively, the four mainland countries of East Africa—Mozambique, Somalia, Kenya, and Tanzania—plus Madagascar owe $22 billion to banks and lending institutions.

Over the course of the 1980s, Africa's collective GNP fell by 20 percent, or 2.6 percent a year. As of 1993, the average per capita GNP for East Africa, not including the island states, was less than $200. In Mozambique the GNP is barely $80 per person per year, the lowest rate in a continent full of poverty.

Mounting debt, deepening poverty, and loss of resources have crippled the governments of East Africa as they struggle to develop. The region's colonial past has not helped them. During the decades of colonial exploitation, little was done to protect the environment or conserve resources. Somalia was occupied by the Italians, then the British, finally gaining independence in 1960. Kenya was a British colony until 1963. Tanzania was taken over by the Germans under Bismarck in 1898 in order to secure trade concessions, but was ceded to the British after World War I; the country became independent in 1961. Mozambique was Portuguese until 1975 and after fighting a bitter war of liberation.

All of the island states were initially colonized by France. The French annexed Madagascar in 1896, with the connivance of the British, remaining in control until the country became independent in 1960. The Comoros Islands gained independence from France in 1975. The Seychelles were annexed by France in 1756 to support its colony on Mauritius; they passed to the British after the defeat of Napoleon and became an independent state in 1976. Mauritius became an independent republic in 1992. Reunion and one island in the Comoros, Mayotte, remain French dependencies.

Most of the mainland countries have been independent for more than thirty years, yet, like most of the rest of the continent, they have been unable to renovate their antiquated infrastructures and build modern economies. No East African mainland country has implemented comprehensive land-use policies. "Real planning capacity simply doesn't exist," says one Kenyan scientist. "In our rush to develop, we have forgotten resource management completely."

Among some of the most stunning scenery on the planet live some of the world's poorest people. A quarter of East Africa's mainland population—some 20 million

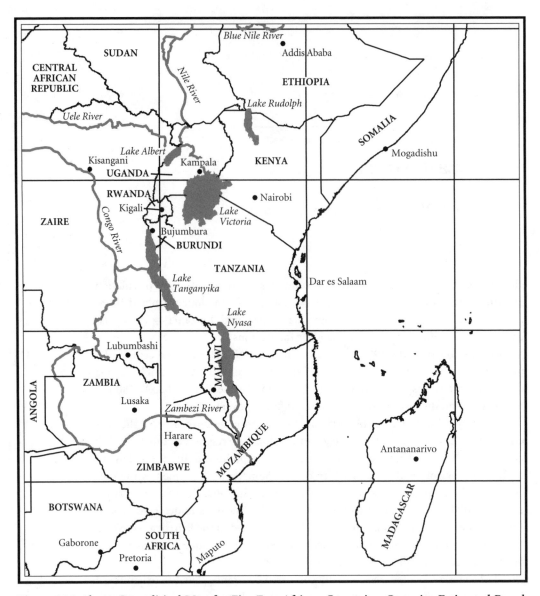

Figure 16.1 *Above:* Geopolitical Map for Five East African Countries. *Opposite:* Estimated Population Densities for Five East African Countries, 1990.

Source: Waldo Tobler, Uwe Deichmann, Jon Gottsegen, and Kelly Maloy, *The Global Demography Project,* National Center for Geographic Information and Analysis: Santa Barbara, 1995.

people—live in severe poverty and suffer from impaired diets and lack of access to basic services such as health care, family planning, clean water, sanitation, education, and adequate housing.

Population

East Africa's population is growing, on average, by around 3 percent a year, enough to double the region's human numbers in twenty-three years. The collective population of East Africa in 1997, including the islands, amounted to just over 100 million people. The World Bank calculates that roughly 16 million East Africans live

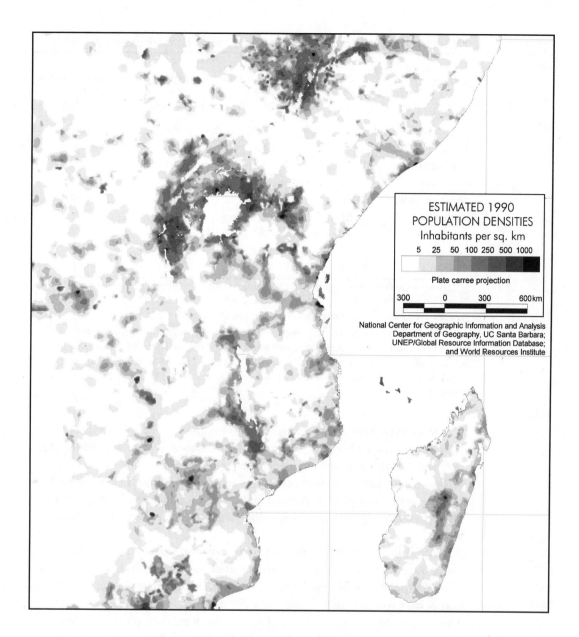

and work in the 60-kilometer-wide coastal zone extending from Somalia to Mozambique (World Bank 1995) (see figure 16.1). The 17 million people living on the islands are all coastal dwellers, since none of them live more than 200 kilometers from a coast. This means that over 33 million East Africans live in coastal areas, roughly one-third of the total population (see table 16.1).

Coastal population densities are not yet as high as those of West Africa, but urban centers are experiencing dramatic growth. In the first half of the 1990s, Dar es Salaam grew by 6.7 percent a year, Maputo by 7.2 percent, Mogadishu by 10 percent, and Mombasa by 5 percent, enough to double their populations in seven to fourteen years. The population of the island states, with the exception of Madagascar and the Comoros, is growing slowly, due in large measure to the introduction of family planning and reproductive health programs.

Table 16.1. Population and Coastal Data for Nine East African Countries

COUNTRY	COASTLINE (KM)	POPULATION 1997 (MILLIONS)	COASTAL POPULATION 1997 (MILLIONS)	POPULATION GROWTH (%/YR)
Comoros	340	0.6	0.6	3.6
Kenya	536	28.8	1.8	2.6
Madagascar	4,828	14.1	14.1	3.3
Mauritius	177	1.1	1.1	1.2
Mozambique	2,470	18.4	6.0	2.7
Reunion	201	0.7	0.7	1.6
Seychelles	491	0.1	0.1	1.4
Somalia	3,025	10.2	4.0	3.2
Tanzania	1,424	29.5	4.8	3.0

Sources: World Population Data Sheet, 1997, Population Reference Bureau, Washington, D.C., 1997; *World Resources 1996–97,* Oxford University Press, New York, 1996, p. 268; *Africa: A Framework for Integrated Coastal Zone Management,* World Bank, Washington, D.C., 1995, p. 17.

Notes: The World Bank's coastal population figures are based on the 1990 censuses, disaggregated for coastal municipalities, counties, and provinces. However, the bank limits its coastal population to within 60 kilometers of the coast. In this book, 200 kilometers is generally used. According to the larger definition, all of Madagascar's population is coastal; according to the bank's definition, only 50 percent is coastal.

The weight of agricultural and grazing pressures on fragile uplands throughout the region has contributed to widespread land degradation and soil erosion. The best land in Kenya has been subdivided again and again. The amount of land available per person has fallen as the country's population continues to grow: from 0.40 hectares per person in 1969 to less than 0.20 hectares by 1990. Poor farmers are forced to use marginal land, which swiftly deteriorates, or else move to cities and towns.

Nairobi, Kenya's capital, grew 600 percent between 1950 and 1980, with most of that growth concentrated in slums and squatter settlements scattered around the fringes of the city. By the year 2000, Nairobi's municipal government is expected to have to cope with 4.5 million people in a city designed for no more than 1 million. According to Nairobi's mayor, John Kingori, the population crush has effectively paralyzed the capital. "Nairobi has been dismantled," he says matter of factly. "Water and sewage service is a great problem. Roads are a huge problem. And we lack the funds [to tackle infrastructure problems]" (Lorch 1995).

Meanwhile, population density along Kenya's coast, which now averages around 200 people per square kilometer, continues to grow. As more people move to the coast in search of better livelihoods, they are often forced to squat on whatever land they can find.

In next-door Tanzania, government resettlement programs have backfired in unexpected ways. The creation of *ujamaa* ("working together") villages created chaos rather than stability. Many of the villages were not sited properly, so their new inhabitants could not make a living from the poor soils. Necessities like potable water and sanitation were neglected. Dissatisfied with the arrangements in these ujamaa, many peasants migrated to coastal cities like the capital, Dar es

Salaam. Huge squatter settlements quickly sprang up around the city, prompting a crisis in municipal services. When government authorities evicted thousands of squatters in the early 1980s, more displaced peasants moved in to take their place.

Another relocation program forced people out of the traditional rice-growing region in the Rufiji flood plain and into the surrounding hills. The peasants, unaccustomed to growing anything in the drier uplands, could barely produce enough food for their own use. Meanwhile, rice production on the flood plain collapsed.

Well-meaning but wrong-headed agricultural policies have contributed to the impoverishment of soils and upland watersheds, aggravating Africa's food crisis. Overgrazing of arid rangelands in the northern areas of Somalia, Kenya, and Tanzania has resulted in the advance of the desert.

Pollution

The overwhelming majority of municipal and industrial wastes that find their way into the western Indian Ocean are untreated. Sewage systems in East Africa serve a minimum of 20 percent of the collective population, no more than 20 million people (Coughanowr, Ngoile, and Lindén 1995).

Even that figure may be an overstatement. Kenya claims to have 150 sewage treatment plants operating throughout the country, but they cover only half the population. Moreover, most of those plants date from the British colonial period and have no more than one stage of treatment. Mombasa's antiquated sewage system is constantly overwhelmed. "The plain fact is, most of Kenya's wastes go untreated into rivers and coastal waters," maintains Dr. Holly Dublin with the World Wide Fund for Nature in Nairobi.

The sewage treatment system for Dar es Salaam has collapsed completely, prompting municipal planners to design a new master sewage plan for the city. To be completed in stages, it will not be operational until the year 2010. Meanwhile, everything is dumped raw into coastal waters. Somalia has no sewage network at all; it relies on pit latrines and shallow septic tanks, which pollute groundwater. By 1990, 60 percent of Mozambique's urban population had access to sanitation facilities, but only 11 percent of rural dwellers did. A cholera epidemic in Maputo in the mid-1980s was linked to the sewage-contaminated waters in the Bay of Maputo.

Madagascar has virtually no sewage treatment networks; wastes are channeled raw into coastal waters. Even though 75 percent of the population of Port Louis, the capital of Mauritius, is connected to a sewage network, the wastes receive only rudimentary treatment before being discharged into coastal waters. Raw sewage damages nearby coral reefs and pollutes beaches.

Although coastal urban areas invariably create hot spots of pollution, another threat to the health of the region's coastal waters is the amount of industrial and agricultural poisons brought into the sea from rivers and streams. Most of Nairobi's municipal and industrial wastes, for instance, are flushed into the Athi River, which exudes a stench that can be detected for kilometers. Some of the wastes eventually reach the seacoast, but their effects on marine ecosystems have not been studied.

Industries in coastal cities dump their wastes directly into the sea. Both the Tudor and Kilindini creeks in Mombasa are badly polluted. Mogadishu Harbor is patrolled by huge sharks, attracted to the raw wastes from a slaughter house that are drained into near-shore waters. Coastal industries in Tanzania dump untreated wastes from soap factories; sisal and sugar mills; and plastics, woodprocessing, and super-phosphate plants into coastal waters. At Dar es Salaam, industrial effluents have turned the waters of the Mzimbazi Creek—which flows through a mangrove swamp—into an anaerobic cesspool bubbling with sulfide gases.

The runoff from pesticides and fertilizers used on agricultural lands is another worry for East African governments. Residues from pesticides, such as DDT, seep into rivers and streams. Madagascar alone used to spray its fields with 20,000 metric tons of pesticides a year, much of it DDT.

Solid wastes are often dumped into coastal lagoons and mangrove swamps. On the main island of Mahe in the Seychelles, solid wastes are dumped into a lagoon known as "stinking corner." Next to it is a marine park. Efforts to clean up the lagoon have failed, as authorities do not know where else they can dump their municipal garbage.

Living on the Oil Highway

East Africa lies on the other half of the world's oil highway, the branch that flows out of the Arabian Gulf west to markets in Europe and North America. In 1990, roughly 135 million metric tons of crude oil were transported into and through the region's waters; though earlier estimations put this figure at around 470 million metric tons a year. About half of the amount is transported by Very Large Crude Carriers (VLCCs), averaging about 200,000 metric tons apiece. On any given day there are about two hundred oil tankers plying the shipping lanes of the western Indian Ocean (Hinrichsen 1983).

With that much oil being transported, oil spills are inevitable. However, most of the oil pollution found in that section of the Indian Ocean is attributed to the routine cleaning of tanks and the discharge of dirty ballast waters and bilge slops. Captain James Ferrari, former principal secretary in the Ministry of Transport for the Seychelles, estimates that roughly 35,000 metric tons of oil are spilled into the western Indian Ocean every year through normal shipping operations. But he admits this may be an underestimation. Its effects, if any, on marine biomass are not known (Hinrichsen 1983).

Resources

One of the main threats to coastal resources remains the widespread degradation of uplands. Deforestation, poor farming practices, and overgrazing of rangelands have resulted in extensive damage to soil cover. Barren soil, exposed to wind and torrential rains, quickly erodes. What doesn't erode is baked hard by the tropical sun. The major rivers of East Africa and Madagascar now run red with sediment pulled out of the watershed. Most of it ends up in shallow coastal waters, where it clogs estuaries and smothers seagrasses and coral reefs.

The Galana-Sabaki River in Kenya brings sediment washed out of the uplands into coastal waters near a marine park at Malindi. During the rainy season the sediment-choked waters around this resort town turn dark brown. The beach, once a major tourist attraction, has expanded seaward by over 500 meters due to sediment deposits.

So much sediment has been dumped at the mouth of the Zambezi River in Mozambique that its delta is sinking under the weight. As a result, the sea has advanced, intruding 80 kilometers up the river, altering the entire ecology of the delta.

By far the worst case is Madagascar, a country that is losing 100,000 hectares of forests a year. Its denuded uplands are the result of excessive fuelwood cutting, slash-and-burn agriculture, and overgrazing by sheep and goats. The most noticeable feature of the island's central highlands is its barren hillsides, stripped of vegetation. Water erosion has gouged out huge craters. Nearly every river on the island is clogged with sediment. So much sediment was dumped into the coastal waters of Mahajanga, that the city's port became completely inoperable. A newer one had to be built at great expense, farther down the coast (Population Reference Bureau 1989; Hinrichsen 1983).

In some upland areas, as much as 300 metric tons of soil is stripped away from each hectare every year. Madagascar's soils tend to be erosion prone, a condition aggravated by tropical cyclones, which can deluge the island with as much as 15 millimeters of rain in fifteen minutes. UNEP has figured that the average loss from erosion in dollar terms amounts to $100–150 per hectare per year, the equivalent of 70–100 percent of the average Malagasy's annual income. The government has counterattacked with soil conservation programs, resettlement schemes, and tough legislation, but the battle is far from over (Hinrichsen 1983).

Coasts are also being decimated directly by nonsustainable harvesting of mangroves and the overexploitation of coral reefs. East Africa's mangroves cover somewhere between 600,000 and 1,000,000 hectares. Accurate surveys have not been carried out. The most extensive growth—320,000 hectares along the northwest coast of Madagascar—supports a thriving shrimp fishery; but Madagascar has lost 40 percent of its mangrove forests. Kenya's 96,000 hectares represent a 70 percent loss since the turn of the century. Remaining stands are concentrated around Lamu, northward through the Kiunga Marine Reserve, along the coast from Diani to the Shimoni area, and around Kilifi, with small pockets on the south coast from Gazi to Vanga. Tanzania's 240,000 hectares are found mostly in the broad Rufiji Delta. And Mozambique's 345,000 hectares cover 40–50 percent of their former area, mostly in the country's central delta region (Jameson, McManus, and Spalding 1995; Semesi and Howell 1992).

The region's coral reefs are suffering from multiple assaults: destructive fishing practices such as the use of fine-mesh nets and dynamite; wholesale mining for use in making cement; and overexploitation by the tourist trade. Up to 20 percent of East Africa's coral reefs have been destroyed, along with 5 percent of its seagrass beds. Remaining reefs are facing intense development pressures. If no conservation measures are taken in time, the region could lose up to 70 percent of its coral resources—10,000 square kilometers—within forty years (Jameson, McManus, and Spalding 1995).

Already, many of Tanzania's reefs have been blasted by dynamite fishermen and excavated by coral miners. On Mafia Island, off the Tanzanian coast near Dar es Salaam, miners excavated over 5,000 cubic meters of coral a year during the 1980s and early 1990s, most of it sold through one commercial dealer. Mozambique's few reefs have been overharvested by fishermen, and many corals have been smothered by silt (Salm 1996).

Some of the worst damage has been done to the coral reefs surrounding East Africa's island states. Coral sands and reefs have been excavated for cement production and building materials on Mauritius and Comoros. During the 1980s, some 500,000 metric tons of coral sand were excavated every year on Mauritius. Similarly, most beaches on Comoros are severely pockmarked by sand mining operations.

Mass tourism has also contributed to the demise of coral communities on Mauritius. Tourist resorts, built in coralline areas, have provided reef access to tens of thousands of tourists from Europe, who unthinkingly tramp across the coral heads at low tide, crushing them, in their search for souvenirs.

There is at least one bright spot in the region: the Seychelles. Apart from those of the four main islands of the Seychelles, where 90 percent of the population resides, the coral resources found around most of this nation's 115 islands, scattered like grains of sand over 1,000 kilometers of ocean, are still intact and in excellent condition.

The magnificent raised coral atoll of Aldabra was made a World Heritage Site in 1983. Not only are its pristine coral reefs protected, but the island is also one of the last refuges of the giant land tortoise (*Geochelone gigantea*), a close relative of the species living on the Galapagos Islands off the coast of Ecuador. The island is also a rookery for millions of nesting seabirds.

Fisheries

Despite the fact that most fishing in East African waters is small scale and artisanal in nature, near-shore fisheries are being overexploited along most of the mainland coast. Between 70,000 and 100,000 fishermen work the reefs, lagoons, estuaries, and near-shore coastal waters of the region. Fishing takes place predominately in shallow lagoons and on reefs; Kenyan and Tanzanian fishermen, for instance, are largely confined to coral lagoons, where they fish on foot during low tide using hand nets. Those with boats—usually outrigger canoes—stay within 5 kilometers of the shore.

On the Comoros, some 4,000 outrigger canoes fish within a couple of kilometers of the coast, using handlines. Madagascar's 4,000 artisanal boats, mostly outriggers, stay on the calmer, western side of the island, where they use hooks and lines and gill nets. Trawling is limited to a few offshore areas, where tuna, sardines, anchovies, Indian mackerel, and billfishes are caught.

According to FAO, in 1993 fish catches in East African waters, including the islands, amounted to around 311,000 metric tons, a doubling of the take since the early 1980s. But FAO admits that that figure is probably a gross underestimation, since most artisanal catches are not recorded (FAO 1995).

Shrimping is big business in Madagascar, where some forty trawlers haul in $50 million worth of the crustaceans a year. On the continent, shrimping is confined to a few important mangrove areas in Kenya, Tanzania, and Mozambique.

Offshore, in deeper waters, distant-water fleets from as far away as China, Korea, and Japan hauled in 3.7 million metric tons of fish and shellfish in 1992, a 25 percent increase over 1990 landings. FAO attributes the rise to bountiful catches of skipjack and yellowfin tuna (FAO 1995).

Fishing technologies are still rather primitive throughout most East African coastal waters. Cold storage facilities are few and far between. Most fish must be sold the same day in local markets or to middlemen who pack them in ice and deliver them to buyers in nearby towns and cities.

But lack of sophisticated gear has not prevented the region from overharvesting and degrading coral reefs. As increasing numbers of people head to the coasts in search of work and new livelihoods, many turn to the sea as a last resort. Unfamiliar with fishing techniques and conservation methods, they often use fine-mesh nets, dynamite, or poison to put food on the table.

A Tale of Three Fishermen

Akida Abushiri and Jackson (not his full name) live in the same fishing village in Kenya, Shariani, yet they are poles apart in their approach to fishing and the use of coastal resources.

Akida's greatest fear is that he will have to give up the sea. He has been fishing in the clear, coralline waters north of Mombasa for over fifty years. But his fishing days may be coming to an end. "More people are fishing now than ever before," says Akida, "so there is less to go around. When I started fishing in this area fifty years ago, there were only five other fishermen. Today, sixty people fish regularly in this part of the reef."

Increased numbers of fishermen are only part of the problem. The poor fishermen of Shariani have no boats, not even simple dugouts. Without boats, they cannot fish outside the reef, where fish are more plentiful. Like most of the older fishermen from the village, Akida uses a net made from nylon or cotton fiber to catch fish, a technique that has been used along this part of the Kenyan coast for centuries. With two of his comrades, Akida wades out to the reef at low tide. When they find a likely spot, usually up to their necks in water, the net is spread out and readied. While one fishermen beats the water with his hands, the other two guide frightened fish toward the net. Any fish taken are quickly placed in fish bags woven from the leaves of the screw pine.

Akida averages about 1,000 Kenyan shillings a month ($18). From this he supports two wives and ten children. Four children have already left home and started families of their own. Since his land is too sandy to grow vegetables, staples like rice and maize must be bought. "Twenty years ago I could save some money in the bank," says Akida, "but not today. One good week is usually followed by several bad ones. So any surplus is quickly used up."

The real threats to coastal fisheries in Kenya, as elsewhere in the region, are the proliferation of illegal fine-mesh nets, which catch everything, even eggs; uncontrolled spearfishing; and the use of seine nets weighted with chains. So far, the use of dynamite and poisons is rare in Kenyan waters, though they are commonly used in Tanzania.

Jackson is a spearfisherman and does not apologize for it. At thirty, he says his employment opportunities are severely limited. He dropped out of school because he wanted to make money now, not wait another four years. He came to the coast from the region around Lake Nakuru and quickly saw the profits to be made from fishing with spearguns. To make the task even easier, he fishes at night with a waterproof flashlight.

Jackson justifies his methods because they work. "We can get 20 kilos of fish a night," he explains. "We only go out for about three hours a night. Day fishing is too much work, and there are too many guys doing it." Usually Jackson fishes with nine other young spearfishermen. They all use crudely fashioned, homemade spearguns. "Night fishing is really easy," he says excitedly. "The fish are resting and don't expect to be taken at night. You can swim right up to them and almost grab them with your hands." Jackson and his friends each make about 4,000 shillings a month from night fishing, four times the amount Akida brings in. And they do not have families to support.

Jackson also collects shells for the tourist trade. That practice has depleted the reef of many trochus and triton mollusks. And night hunting has drastically reduced the number of triggerfish and other predators that control sea urchins, upsetting the natural balance of the reef.

There is a middle way between Jackson's nonsustainable approach and Akida's poverty. South of Mombasa, in the village of Vanga, hard against the border with Tanzania, local fishermen have formed a cooperative society in an effort to manage their resources rationally.

Mohammed Kitawana is one of the officers of the Mwagugu Fisheries Cooperative Society, formed a decade ago. Nearly all fishermen in Vanga, around two hundred, are members of the cooperative. Unlike those in Shariani, Vanga's fishermen all own their own boats and dugouts. They use a variety of techniques including hooks and lines and different kinds of nets, depending on what type of fish they are after. "Some of our fishermen specialize in shrimp and lobsters," explains Kitawana. "Others go after mullet and rabbitfish. But most of the fishermen here harvest whatever edible fish they can catch."

Kitawana is thirty-seven years old and has a high school education. He was born in Vanga and is happy to be able to contribute to the advancement of his village. "It took us almost three years to set up the cooperative," he says. "We had to get government permission and organize all the fishermen. Some were skeptical at first that we could make this work. But today, the cooperative's success is well known. All our members make a comfortable living from the sea." The cooperative sells no less than 500 kilograms of fish and crustaceans a day and often unloads up to 1,000 kilograms.

A few years after the fishing cooperative was established, the government built the Vanga Fish Depot, where the cooperative sells its fish. They also built an office. "Once you have an organization, you can get bank loans and assistance from the government," points out Kitawana.

There is no overharvesting of fish around Vanga, and none of the fishermen resort to illegal techniques like dynamite or poisons. Occasionally, Tanzanian poachers infringe on Kenyan waters, but anyone caught dynamiting is quickly driven off. "If anything, our waters are still underfished because we don't have big trawlers in these waters," notes Kitawana.

Conservation

Over the past decade, the governments of all East African countries except Comoros have taken some solid steps toward the protection and management of marine and coastal sites with high species diversity. As of 1996, there were fifty marine

Table 16.2. Distribution of Marine and Coastal Protected Areas in the East African Region

COUNTRY	MARINE SITES	COASTAL SITES	MANAGEMENT STATUS
Comoros	0	0	none
Kenya	4 marine parks; 6 marine reserves	0	most have enforcement
Tanzania	8 marine reserves	0	no enforcement
Mozambique	1 marine reserve; 1 marine park; 2 game reserves	1 game reserve	no enforcement
Madagascar	1 marine park	1 special reserve	unclear
Reunion	6 fishing reserves; 4 turtle protected areas	0	no enforcement
Mauritius	6 fishing reserves	9 island nature reserves	little enforcement
Mayotte	1 fishing reserve	0	unclear
Seychelles	1 strict nature reserve; 5 marine parks; 5 special reserves; 4 marine mollusk reserves	7 seabird islet protected areas	2 marine parks have a management plan and enforcement capabilities; more are expected

Source: Adapted from table 12.3, "Distribution of Marine and Coastal Sites in Protected Areas in the East African Region," in *A Global Representative System of Marine Protected Areas,* World Bank, Washington, D.C., 1995, p. 83.

protected areas in the region along with eighteen protected coastal sites. Many of them, however, lack proper enforcement capabilities (Kelleher, Bleakley, and Wells 1995).

The Seychelles is one promising exception, with an impressive twenty-one sites listed as protected (see table 16.2). Although only two of the Seychelles's marine national parks are adequately managed, the government is working with the World Conservation Union to improve its management capacity in the remaining three. Despite its limitations, this island nation of 100,000 inhabitants is trying to cash in on the boom in ecotourism. It has taken concrete measures to limit the number of overseas visitors, however, by putting a ceiling on the construction of tourist resorts. It may be one of the few countries in the world to have adopted such measures.

The president of the Seychelles, France Albert René, explains why the Seychelles has taken the green road. "It is vital for our future to protect what we have," he says, "since our islands are small and our natural resources limited. For this reason, we must plan everything very carefully. Everything new must blend into the environment without harming it. We have even set aside entire islands to conservation" (Hanneberg 1991).

In 1990, the government announced its new Environmental Management Plan for the Seychelles 1990–2000. Ian Collins, a consultant who helped draft the plan, calls it the world's first explicit national sustainable development plan.

Bertrand Rassool, with the Ministry of Planning and External Relations, explains the rationale behind the plan. "Since tourism and fishing make up our economic base," he says, "and since both depend on an unspoiled environment, we are protecting our resources at all costs. Our goal is the total integration of environmental thinking in all segments of society."

Management: The East African Action Plan

In 1985 the countries of East Africa banded together to adopt the Convention for the Protection, Management and Development of the Marine and Coastal Environment of the Eastern African Region, known as the Nairobi Convention, under the auspices of UNEP. The first phase, implemented in the late 1980s, concentrated on the formation of scientific contact points within the region, linking participating institutes and research centers. Baseline pollution studies were carried out in an effort to pinpoint local problem areas and assign priorities.

Under the program, the action plan has five priority projects: (1) coastal zone management; (2) pollution assessment and abatement; (3) contingency planning to combat marine pollution, including oil spills; (4) coastal erosion and siltation control; and (5) the promotion of environmental impact assessments.

As of the spring of 1996, the convention had not entered into force; it requires six ratifications, and only Somalia, Kenya, France (Reunion), the Seychelles, and the Comoros had ratified it. When the plan was launched, most mainland states had little in the way of institutional support or expertise in coastal area management. Building up a qualified network of collaborating agencies took up the first few years of the action plan. UNEP contributed over $1 million to kick-start the research program.

Gradually, institutional capacities to tackle coastal zone problems are being strengthened throughout the region. Some regional NGOs are beginning to take an active interest in coastal zone issues. The East African Wildlife Society, based in Nairobi, has been involved in a national campaign to preserve the extensive wetlands in the Tana River Delta. Educational campaigns have also been launched in an effort to inform the public of the importance of coastal ecosystems such as mangroves, seagrasses, and coral reefs.

In April 1993, representatives from all East African governments met with UN agencies and donor country representatives at a high-level workshop and policy conference held in Arusha, Tanzania. The workshop and conference were organized by an impressive alliance of national and regional institutions, in close collaboration with multilateral donor agencies, including the Swedish Agency for Research Cooperation with Developing Countries (SAREC), the UN Food and Agriculture Organization, the Intergovernmental Oceanographic Commission, the Swedish International Development Cooperation Agency (Sida), the United Nations Environment Programme, and the World Bank (Lindén 1993).

One of the main outcomes of the conference was a pledge by donors and participating governments to strengthen the capacities of national and regional institutions to manage the region's coastal areas in a more sustainable manner. A six-year program of institution strengthening was announced. SAREC, in cooperation with the World Bank, is supporting workshops in the region as a way of fostering national processes for the development and implementation of integrated coastal area management.

As part of the Arusha agreement, the World Conservation Union's Eastern Africa Regional Office in Nairobi has begun to work with the government of Tanzania in an effort to introduce the most comprehensive integrated coastal zone management program in the region. Called the Tanga Coastal Zone Conservation and Development Program, the initial management plan is centered in the coastal province of Tanga. It is currently being implemented by regional and local governing bodies in a progressive partnership with coastal communities and with technical assistance provided by the World Conservation Union.

Conclusions

- Policy makers and planners need better resource assessments in order to make competent judgments about where to invest limited development funds. Management efforts should encompass the three main coastal ecosystems—mangroves, seagrasses, and coral reefs—and should concentrate on those areas that still have viable natural communities.

- Resource management strategies are needed that address population growth and distribution.

- A considerable amount of institution strengthening is needed in order to beef up the management capacity of national institutions charged with coastal resource management and pollution monitoring and control.

- A little investment in training and equipment could help lessen the pressures on near-shore mangrove-, seagrass- and coral reef-based fisheries, while giving artisanal fishing communities a better chance to make a living from the sea.

- The Tanga Coastal Zone Conservation and Development Program needs to be monitored carefully. If successful, it could be used as a model for other countries in the region as they struggle to manage coastal resources sustainably.

CHAPTER 17

West Africa

The "Dismal Coast"

The human landscape of West Africa has changed dramatically since the first Portuguese explorers pushed bravely down the "dismal coast," as they called it, seeking their fortunes. Early traders were confronted by a colossal continent, dark and mysterious. The climate was (and remains) fearsome, with withering humidity and frequent rains. Coastal wetlands were filled with quicksands, man-eating crocodiles, and disease-carrying insects. Malaria was epidemic. The vast coastal jungles seemed impenetrable and foreboding.

During the colonial period, West Africa's coasts became conduits for transporting gold, gemstones, slaves, and other commodities to Europe and the Americas. As coastal trading cities continued to grow and expand, more people were attracted to the coasts. The development patterns established four hundred years ago are still at work, perhaps now more than ever.

The twenty-one countries that stretch out along Africa's west coast, from Mauritania in the north to Namibia in the south, cover 9 million square kilometers. Their collective coastline runs for more than 8,000 kilometers. Coastal shelves extend only 30–40 kilometers offshore before sliding into deep ocean basins. Coastal upwellings make the region's continental shelf a rich fishing ground.

West Africa has sharp economic contrasts: from rapidly industrializing Nigeria and Senegal to the bare subsistence economies of Benin, Sierra Leone, and Guinea-Bissau. Nigeria has built its economy on oil, while Senegal, Democratic Republic of the Congo (formerly Zaire), and Liberia, for example, have evolved advanced mining and processing industries.

Population

West Africa's nemesis—rapid population growth—continues to outpace development efforts, particularly when it comes to badly needed infrastructure improvements. In 1997, the region's collective population was the largest in Africa: over 200 million people, half of them living in Nigeria.

The World Bank estimates that a quarter of the region's population lives and works within 60 kilometers of a coast. This means that around 52 million people occupy less than 10 percent of West Africa's land area (World Bank 1995; see figure 17.1 and table 17.1). The crowded Atlantic corridor has population densities of

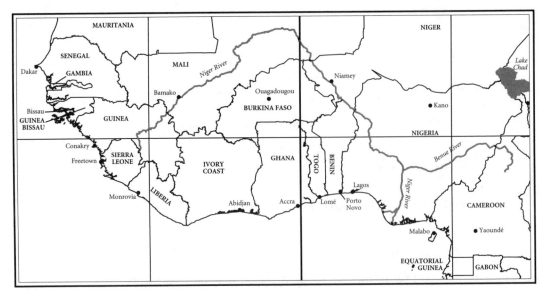

Figure 17.1 *Above:* Geopolitical Map for West Africa. *Opposite:* Estimated Population Densities for West Africa, 1990.

Source: Waldo Tobler, Uwe Deichmann, Jon Gottsegen, and Kelly Maloy, *The Global Demography Project,* National Center for Geographic Information and Analysis, Department of Geography, University of California, Santa Barbara, 1995.

Table 17.1. *Length of Coastline and Demographic Data for Twenty West African Countries*

COUNTRY	COASTLINE (KM)	POPULATION 1997 (MILLIONS)	COASTAL POPULATION (MILLIONS)	POPULATION GROWTH (%/YR)
Angola	1,600	11.6	3.0	3.2
Benin	121	5.9	2.0	3.4
Cameroon	402	13.9	1.7	2.8
Cape Verde	965	0.4	0.4	1.9
Congo	169	2.6	0.36	2.3
Côte d'Ivoire	515	15.0	4.0	2.6
Equatorial Guinea	296	0.4	0.21	2.6
Gabon	885	1.2	0.65	2.0
Gambia	80	1.2	0.7	2.5
Ghana	539	18.1	6.0	2.9
Guinea	346	7.5	1.7	2.4
Guinea-Bissau	274	1.1	0.88	2.1
Liberia	579	2.3	1.5	3.1
Mauritania	754	2.4	0.23	2.5
Namibia	1,489	1.7	0.05	2.6
Nigeria	853	107.1	22.2	3.0
São Tome & Principe	100	0.1	0.1	3.4
Senegal	531	8.8	4.7	2.7
Sierra Leone	402	4.4	2.16	1.9
Togo	56	4.7	1.50	3.5

Sources: World Population Data Sheet, 1997, Population Reference Bureau, Washington, D.C., 1997; *World Resources 1996–97,* Oxford University Press, New York, 1996, p. 268; *Africa: A Framework for Integrated Coastal Zone Management,* World Bank, Washington, D.C., 1995, p. 17.

Note: Democratic Republic of the Congo (formerly Zaire) not included, as it has only 37 kilometers of coastline and minimal population there.

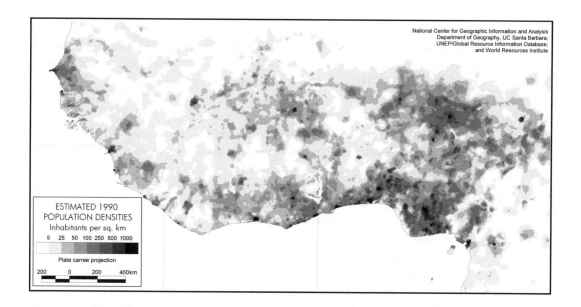

National Center for Geographic Information and Analysis
Department of Geography, UC Santa Barbara;
UNEP/Global Resource Information Database;
and World Resources Institute

ESTIMATED 1990
POPULATION DENSITIES
Inhabitants per sq. km

5   25   50   100   250   500   1000

Plate carree projection

200   0   200   400km

250–500 people per square kilometer. The meteoric growth of coastal towns and cities reflects overall demographic trends, characterized by high birth rates and heavy in-migration from interior provinces.

As a whole, the region's population is growing by close to 3 percent a year. But some coastal cities are growing at nearly twice that rate: In the early 1990s Abidjan was growing at 5 percent a year, Accra-Tema by 6 percent, and Lagos by 5 percent. By the year 2000 the coastal zone around the Gulf of Guinea could contain 40 percent of Benin's population, 25 percent of Côte d'Ivoire's, 20 percent of Togo's, 15 percent of Ghana's, and 10 percent of Nigeria's. Lagos, the capital of Nigeria, already contains around 9 million inhabitants, and population densities are among the highest ever recorded, up to 42,000 people per square kilometer (Moffat and Lindén 1995; *Economist* 1993). The World Bank warns that if current demographic trends do not change, the coastal corridor along the Gulf of Guinea—lying just under West Africa's pregnant belly—will reach the saturation point long before 2025.

Major coastal urban centers in West Africa, as elsewhere, concentrate personnel, skills, industries, and commerce. Over 60 percent of the total industrial infrastructure of those countries bordering the Gulf of Guinea is located in coastal cities and towns. For instance, the capital of the Côte d'Ivoire, Abidjan, contains 20 percent of the country's population and accounts for more than 70 percent of all economic and commercial transactions. Similarly, Lagos accounts for nearly 60 percent of value added in manufacturing and has 40 percent of Nigeria's highly skilled labor force.

The rapid and chaotic growth of the region's coastal urban population has undermined government attempts to build more livable cities. City services everywhere are overburdened. The provision of clean water and sanitation facilities lags far behind the population curve. In Port Harcourt, for instance, over half a million people crowd into waterfront slums that are flooded during the wet season and lack electricity, potable water, and sanitation facilities. "Nearby creeks serve as the solid waste dump, water supply and toilet," states Olof Lindén (Lindén 1990).

217

Epidemics of waterborne diseases such as cholera, hepatitis, typhoid, and malaria are common. Not only do such debilitating diseases erode the health and productivity of the workforce, they also contribute to high infant mortality rates, which in turn feed high fertility levels among the region's women.

Pollution

West Africa, like the rest of the continent, dumps most of its municipal and industrial effluents directly into coastal waters with little or no pretreatment. Raw sewage from nearly 50 million people is channeled into shallow coastal waters, bays, and lagoons. In some areas, high concentrations of bacteria pose a clear threat to human health. The Lagoon of Iddo in Lagos is on the receiving end of some 60 million liters of untreated sewage water a year. Typhoid fever and infectious hepatitis are rampant, especially in the squatter settlements that have sprung up around the city's polluted lagoons (UNEP 1991a).

In Ghana's coastal zone around Accra-Tema, twelve of sixteen lagoons surveyed by the UN were found to be highly contaminated with untreated municipal and industrial wastes. Similarly, in Abidjan, Côte d'Ivoire, the volume of domestic sewage discharged every year into the Ebrie Lagoon was found to be the equivalent of nearly 20 percent of the lagoon's total volume (UNEP 1991a; World Bank 1995).

The coastal waters of virtually every major urban area are clogged with an assortment of municipal and industrial wastes. West African waters are more severely affected by industrial wastes, perhaps, than any other part of the continent. Untreated effluents from petrochemical refineries; mining and metal-smelting operations; and food-processing, chemical, and textile plants are discharged into coastal waters or rivers.

The cement industry in Gabon and Togo contaminates near-shore waters with suspended solids, turning estuaries gray. Wastes from leather and textile industries in Ghana, Benin, and Nigeria go directly into coastal lagoons, creating locally high concentrations of dyes, tannins, and chromium. Heavy metals from mining and metal smelting have ruined shellfish beds along the coasts of Liberia, Senegal, Guinea, and Sierra Leone.

According to UNEP, the wetland zone stretching from the Iddo Lagoon in Lagos to the corridor between Nigeria and Gabon receives enormous quantities of untreated industrial and municipal effluents, which account for the bulk of pollution from the eighteen coastal states situated between Senegal and Angola. Approximately 43 percent of all BOD—nutrient pollution from raw sewage and agricultural chemicals—36 percent of all suspended solids, 83 percent of all oil and grease, and 60 percent of all COD—chemical pollution from industry—are concentrated along this stretch of coast (UNEP 1991a, b).

Runoff from the enormous amounts of pesticides sprayed on agricultural fields every year also degrades water quality in the region. Agricultural exports are still the basis of many of the region's faltering economies, and West African states consume around 10 percent of all pesticides produced in the world every year. "Practically all the fish, crabs and shrimp in the coastal areas are exposed to pesticides," states one UNEP report on the coastal health of the region. Once pesticides infiltrate coastal ecosystems, they tend to accumulate in the fatty tissues of marine organisms. High levels have been detected in crustaceans and shellfish. But their effects on human health are not well understood (UNEP 1991a).

The bulk of West Africa's solid wastes also end up in coastal lagoons, salt marshes, and mangrove swamps. Port Harcourt alone generates close to 100,000 metric tons of solid waste a year; none of it is landfilled. Overall, little of the millions of tons of trash produced in the region are taken to a supervised landfill or buried.

Black Gold

Oil and gas reserves are substantial in several West African states. Collectively, the region holds some 3 billion metric tons of oil and 2.5 trillion cubic meters of natural gas. The bulk of those reserves are in three countries: Nigeria, Angola, and Gabon. During the decade between 1982 and 1992, oil production in Nigeria increased by a factor of three, from 18,500 metric tons to 61,000 metric tons a day. Angola's production increased fivefold to 19,000 metric tons a day, while Gabon's doubled to 14,000 metric tons a day (UNEP 1984b; Moffat and Lindén 1995).

With most oil and gas in the region being produced from offshore platforms in the Gulf of Guinea, particularly off the coast of Nigeria, the waters of the gulf are polluted with oily wastes and tar balls. In addition, there are twenty-seven major oil refineries along West Africa's coast, which also discharge their production wastes into shallow lagoons and coastal waters (Moffat and Lindén 1995).

UNEP reports that between 1976 and 1983, 1,360 oil spills were registered in Nigerian waters alone. In 1983, some 6,400 metric tons of oil leaked from just eleven oil fields operated by one company in Nigeria. Official estimates of the amount of oil spilt every year—around 2,300 cubic meters in three hundred separate incidents—is thought to be a gross underestimation. The real figure is likely ten times that amount (UNEP 1984b).

The entire delta of the Niger River, Africa's third largest, is heavily polluted with petroleum residues from offshore oil and gas platforms. Nigeria has had little success in tightening environmental safeguards for its oil and gas fields.

Resources

West Africa's coasts still harbor some 9 million hectares of mangrove swamps, but nearly everywhere they are under intense development pressures—exploited for timber, tannin, fuelwood, and fodder and bulldozed for urban and industrial expansion. The largest single stand of mangroves on the continent—the 500,000–600,000 hectares found at the mouth of the Niger River in Nigeria—is now under assault from both land and sea. Seaward they are beset by chronic oil pollution and coastal erosion. Landward they are exploited for building materials, fodder, fuel, and other products (Jameson 1995; World Bank 1995).

In tiny Benin, mangrove swamps were replaced by evaporation ponds used for the extraction of salt from seawater. Senegal, Gambia, and Sierra Leone converted thousands of hectares of mangroves into rice paddies. However, the ponds quickly developed acid sulfate conditions, making rice production impossible. Meanwhile, coastal shrimp fisheries declined as a direct result of the loss of their mangrove nursery and feeding areas.

*Table 17.2. Forested Area, Deforestation Rates, and Loss of Mangroves
for Thirteen West African Countries (Data for Year 1992)*

COUNTRY	CLOSED FOREST (IN 1000S OF HA)	DEFORESTATION RATE (%/YR)	LOSS OF MANGROVES (TOTAL %)
Angola	2,900	1.5	50
Benin	47	2.6	—
Cameroon	17,920	0.4	40
Congo	21,340	0.1	—
Côte d'Ivoire	4,458	6.5	60
Gabon	20,500	0.1	50
Ghana	1,718	1.3	70
Guinea	2,050	1.8	60
Guinea-Bissau	660	2.6	70
Liberia	2,000	2.3	70
Nigeria	5,950	5.0	50
Sierra Leone	740	0.8	50
Zaire	105,750	0.2	50

Source: Africa: A Framework for Integrated Coastal Zone Management, World Bank, Washington, D.C., 1995, p. 36.

The region's tropical forests have been reduced to remnants except in remote and underpopulated interior areas. During the early 1990s, both Nigeria and Côte d'Ivoire were losing 5 percent or more of their forest cover every year (see table 17.2).

West Africa's system of coastal lagoons, which run for more than 760 kilometers between Côte d'Ivoire and eastern Nigeria, are in crisis. Highly fecund ecosystems, which cover around 400,000 hectares, the lagoons run parallel to the coast, separated from the sea by a narrow barrier of sand dunes. Traditionally exploited by coastal populations for their fisheries and forest and wildlife resources, these unique wetlands are being used as dumping grounds for municipal and industrial wastes, which has turned many of them into biological graveyards. During the 1980s, many small-scale commercial fishing operations were forced to close down. Fishermen using the Aby Lagoon in the Côte d'Ivoire, for instance, used to haul in 5,000 metric tons of fish a year in the 1970s. By 1981, they were getting no more than 500 metric tons. Today, nothing is caught there except typhoid and cholera.

Coastal Erosion: Losing Ground

"Coastal erosion is probably the most serious environmental problem facing West Africa," maintains Dr. A. Chidi Ibe, head of the Physical and Chemical Oceanography Section of the Nigerian Institute for Oceanography and Marine Research in Lagos. The reasons are due to a complex assortment of ocean-land interactions. In plain terms, the ocean on Africa's west coast acts like a huge scoopbucket, gouging out chunks of the coast and depositing the debris in deep trenches that knife across the continental shelf fairly close to shore.

Coastal erosion is particularly severe along the coasts of Nigeria, Sierra Leone, Liberia, Côte d'Ivoire, Gambia, Ghana, Benin, and Togo. Hundreds of coastal villages have been moved inland as the sea reclaims their land. In the Gulf of Guinea, river deltas had helped build new land by depositing silt from the denuded watershed. Many of these sand spits were occupied by small fishing communities and landless poor, but they were forced off by the advancing sea. In the Niger Delta, riverbank erosion claims 400 hectares of land a year. If present trends continue, 40 percent of the inhabited delta land could be lost in three decades.

Up and down the coast, agricultural lands are being washed out to sea. The twisted trunks of palm trees and mangroves sticking up out of silt-laden coastal waters are a common sight. Sand banks designed to hold back the sea have been battered down. Coastal cities have had to reinforce tidal barriers and jetties at great expense. Victoria Beach in Lagos is now more than 2 kilometers farther inland than it was in 1912. Lagos and Escravos Harbors in Nigeria and Keta Harbor in Ghana have been losing between 20 and 30 meters of beach a year since the late 1970s (Hinrichsen 1990).

Coastal erosion has reached such proportions in land-short countries like Togo and Benin that they have asked the French Ministry for Cooperation for help in combating it. Off the western pier of Lomé Harbor in Togo, the sea has advanced 380 meters in fifteen years. The eastern portion of the harbor is being devoured at the rate of 20 meters a year.

The great mangrove forest in the Niger Delta is being pushed back toward land by the sea. Scientists estimate that it is losing ground at rates reaching tens of meters a year. The huge amount of sediment that used to be brought in by the Niger River every year has been reduced by around 70 percent due to the construction of upstream dams. What sediment does reach the coast is swept away by currents before it can be consolidated by the mangroves and turned into land.

The extent to which human interference has contributed to the processes of coastal erosion is not yet clear. But unplanned coastal development is certainly a factor in some areas. "This problem is receiving a great deal of attention by scientists in the region," observes Dr. Ibe. The problem is that doing something about coastal erosion requires investments far beyond the capacities of most of the region's governments.

Fisheries

Despite a relatively narrow continental shelf, upwellings of nutrient-rich water are found along much of West Africa's extensive coastline. The region has no fringing coral reefs, but mangrove-based fisheries thrive in the north-central regions. The cool Benguela Current, which flows north off the coast of Namibia and Angola, supports an annual catch of around 1.5 million metric tons of seafood, mostly pilchard and hake. In all, somewhere between 2.5 and 3.1 million metric tons of fish and shellfish are harvested every year from West African waters (FAO 1995).

Despite the bounty, over 50 percent of all fish taken is harvested by fishing fleets from sixteen foreign countries, led by Russia. Since most West African states lack the capacity to outfit and maintain deep-water fleets, they sell fishing licenses to foreign fleets, but they are without the monitoring capacity to ensure that those fleets operate within the terms of their fishing agreements.

This lack of enforcement capacity has led to a substantial loss of economic rent from foreign fleets. Most of West Africa's coastal nations do have a licensing system, but the rents they collect do not come close to reflecting the value of fish and shellfish taken by foreign vessels. During the twenty-one-year period from 1969 to 1990, for instance, foreign fishing vessels harvested hake valued at $4.5 billion from Namibia's EEZ, yet they paid a paltry $180,000 to Namibia, 0.004 percent of the total value (Panos Institute 1995).

Faced with collapsing fish stocks in 1990, the Namibian government, in partnership with the private sector, implemented a quota system and banned all foreign fishing fleets from its extensive EEZ until stocks of hake and other fish recovered. Only Russian boats are still permitted to fish, by special agreement, inside Namibian waters. The country also places inspectors on all foreign boats to make sure they adhere to catch quotas and do not throw away large quantities of nontarget species. As a result of those measures, the fisheries contribution to Namibia's GNP is growing by 30 percent a year, despite keeping the total allowable catch for targeted species well within sustainable limits (FAO 1995).

Fishing is still an important source of revenue for West African states. Mauritania's exports of seafood rank second only to its exports of iron ore. Senegal is the region's leading producer of shrimp, which are taken by trawlers operating in the mangrove delta of the Saloum and Sine rivers. About 70 percent of the catch is exported to Europe. Gambia has developed an export industry based on crabs, oysters, and lobsters.

Liberia, Ghana, Togo, and Benin all have thousands of small fishing villages along their coasts. In Ghana, fish consumption exceeds that of meat, and the industry is credited with about 1.2 percent of the country's total GNP. Nigeria's fisheries are also lucrative, bringing in around 200,000 metric tons a year. Much of this catch is taken by commercial operations. Still, around 400,000 artisanal fishermen using simple hook-and-line techniques and gill nets operate in the Niger River Delta. Due largely to chronic overfishing and a lack of modern equipment, demand is so high in Nigeria that three-quarters of the country's fish must be imported (Moffat and Lindén 1995).

The primitive state of West Africa's fishing fleets means that they do not get a fair share of the offshore harvest. The UN Food and Agriculture Organization has long been active in the region in efforts to help upgrade fishermen's skills and equipment. FAO reckons that the commercial fishing potential of offshore waters is on the order of 6 million metric tons a year, 2 million more than is currently harvested (FAO 1995).

As a way of compensating for their lack of capacity, a number of West African states are beginning to follow Namibia's example by taking an active approach to fisheries management within their EEZs. With donor assistance, Mauritania set up a surveillance system worth $7.5 million. The country is now better able to monitor the compliance of foreign fleets with mutually agreed upon catch quotas.

Management: The West African Experience

The action plan for West and Central Africa (WACAF) involves all twenty-one countries bordering the Atlantic (including Democratic Republic of the Congo). Although the WACAF countries adopted the Abidjan Convention for Cooperation in the Protection and Development of the Marine and Coastal Environment of the

West and Central African Region in 1981, it was not ratified until August 1994. The West African states also initialed a Protocol on Controlling Pollution in Cases of Emergency.

A number of important baseline studies have been carried out in an effort to determine what major environmental problems confront the region and to make recommendations for their control. Pollution monitoring is being conducted throughout West Africa by a network of interconnected research institutes. Masses of data have been accumulated on the region's major problem areas. A trust fund was established by UNEP to pay for the program, but as of 1995 it contained only around $400,000.

The priorities of the agreed upon action plan involve the following: monitoring and combating marine pollution; enacting corresponding environmental legislation; adopting protocols on hazardous waste management and land-based pollution control; fighting coastal erosion; conserving critical marine habitats; and setting up a coordinating unit in the region.

In recent years, some bilateral donor agencies, along with the World Bank, have attempted to assist West African nations with the formulation and implementation of coastal zone management plans. But progress is slow since most of those countries lack both a legislative base of supporting regulations and the institutional capacity needed to translate programs on paper into action on the ground. By 1994, only three countries—Guinea-Bissau, Nigeria, and Côte d'Ivoire—had initiated coastal area management plans. Only Guinea-Bissau was moving toward the implementation stage in 1996.

The World Bank puts it most succinctly: "There is clearly an urgent need to broaden both the area and scope of coastal zone management in Africa in order to deal with the increasingly complex problems at hand and to meet the challenges of the future" (World Bank 1995).

Conclusions

- Governing institutions in West Africa need to formulate integrated resource management strategies that take account of rapid population growth and distribution.

- The region's largest cities need to develop the capacity to plan for future infrastructure needs and implement urban management plans.

- Urban areas need to build sewage and industrial wastewater treatment plants in order to reduce pollution and improve the coastal environment. Controlling pollution would also translate into health benefits, since water pollution is one of the main causes of premature death and disability.

- Fisheries management, especially for poor artisanal communities, is sorely needed. Near-shore fisheries are badly affected by pollution, the loss of critical ecosystems, and overexploitation. Attending to the needs of poor fishing communities would help them attain a sustainable system of local, community-based fisheries management.

- The Action Plan for West and Central Africa (WACAF) should be given much higher priority by the region's governments. This is the only vehicle that is capable of focusing *regional* attention on coastal resource management issues. As such, it deserves the continued support of West African states.

- Erosion remains one of the region's top coastal management issues. A region-wide action plan to control coastal erosion could be drafted and implemented under the auspices of WACAF. This would make a good test case in order to see how effective regional coordination can be enhanced.

- By the same token, effective coastal management plans should be adopted and implemented as quickly as possible by each of the region's governments.

CHAPTER 18

A Future for Coastal Seas

Those who cannot remember the past are condemned to repeat it.

—George Santayana

Managing coastal areas requires an interdisciplinary, coordinated effort by governing institutions at all levels, private-sector interests, and community groups working together toward a common set of goals. Such management also requires continued political support.

"Everybody talks about integrated resource management," says one civil servant in Jakarta, "but no one is doing it." In Indonesia, as in most countries, institutions are vertically constructed to handle forestry, for instance, or health care. They are not set up horizontally to handle what UN bureaucrats call "cross-cutting issues," issues that cut across institutional boundaries, disciplinary boundaries, and the boundaries of nation-states. Coastal management is a daunting prospect, even for developed countries with strong institutions, highly educated professionals, and adequate budgets. For developing countries, trying to forge some measure of coastal governance can seem futile.

Over the past three decades, a wealth of experience has been accumulated by international and national institutions, NGOs, regional and local governments, and local communities in managing their coastal areas in a more sustainable fashion for the greater common good. One of the reasons little has been learned from that experience is that programs are often not evaluated in terms of what works and what doesn't and why. Further, lessons are not translated into a language policy makers and planners can understand and use to introduce better coastal management strategies.

Stephen Olsen, director of the Coastal Resources Center at the University of Rhode Island in Narragansett explains, "You have to view coastal area management as a sustained process of governance, not a one-time exercise." Workable coastal management plans are not born as fully implementable programs with all elements in place. Coastal management is a slow and sometimes agonizing process of trial and error in which constant reevaluation leads to adjustments and changes of direction. It is an organic and fluid process, much like coastal areas themselves, constantly under pressure from land-based activities and the ravages of the sea.

Many coastal area managers admit that they need a process of benchmarking, whereby programs and projects can be evaluated and re-evaluated in light of their objectives and constantly changing circumstances. Too many coastal area management plans run aground on the shoals of preconceived notions and rigid institutional constraints, making changes in direction or scope impossible.

All too often the lessons learned from one nation's experiences never get beyond its borders. Similarly, the lessons seldom get out to a broad public constituency, a constituency that might put pressure on its political leaders to press ahead on coastal management plans instead of abandoning them at the first sign of resistance from powerful economic interests.

The constituency problem is particularly vexing. After all, there is now a broad coalition of individuals and institutions around the world that support the conservation and rational management of the world's remaining rainforests. Millions of people care deeply about the fate of marine mammals like dolphins and whales, along with a veritable host of other threatened and endangered species of plants and animals. Yet there has been no groundswell of support to safeguard and manage sustainable coastal areas and the rich and diverse ecosystems they harbor. So far there has been no outcry from an informed public to stop the destruction of the world's coral reefs, an ecosystem containing over 1 million species rivaling tropical rainforests in biodiversity, most of which could be lost within forty years.

We must begin to turn this situation around through well-thought-out and carefully targeted public information campaigns and by incorporating coastal area environment issues into courses in secondary schools, universities, and the informal education sector. The media should be enlisted in this process. Institutions with experience in coastal area management should spend time briefing reporters and the specialized press on coastal area issues. An informed and involved public is essential if governments are to be able to carry out coastal management strategies.

William Matuszeski, director of the EPA's Chesapeake Bay Program Office, underscores the incremental approach to coastal management. "Integrated coastal management is not an effort where you figure it all out at the outset; it is more like peeling an onion," he observes. "As you reach each layer, you learn new things to help you understand the next layer. There is never enough time and money available at one time to act otherwise. In the case of the Chesapeake Bay, it took from 1978 to 1983 to nail down the nutrient issue, from 1983 to 1987 to agree we had to go after nitrogen, until 1992 to begin to understand the interactions and feedback loops of the living resources, and until 1994 to begin to look at air deposition as a major source of nitrogen. If we had tried to figure all this out in 1983, or even taken until 1987, we would have long before lost all public support" (Matuszeski 1996).

An Ideal Coastal Management Process

Coastal management is not just a plan, it's a process. Building a coastal area management plan is akin to building a sturdy house. First a solid foundation must be laid, then the frame added, before the walls and roof can be put in place. Too often, the roof is built before the foundation is in place, so it doesn't fit together. Or the frame is too weak to hold the walls and roof, and it all comes crashing down.

"It is, in essence, an iterative process," explains Peter Thacher, former deputy director of UNEP and one of the key figures behind the launching of the Mediterranean Action Plan, "a process that melds together the natural and social sciences and takes into account the needs and interests of a wide array of stakeholders."

Establishing Leaders

It is usually prudent to give one national institution authority to lead the process of setting up and supervising a coastal area management plan. Too many players with equal authority—often with conflicting sets of interests and priorities as well—can cripple the process of coastal governance before it has a chance to take root and grow. There should also be solid cooperation between national and regional authorities, with clear lines of authority established. And regional authorities need workable links with local leaders in each village or coastal municipality.

This does not necessarily mean that a new institution should be created to oversee the process, though many countries and regional authorities have taken that route. An existing institution, given a clear mandate, can often work just as well. Also, with many government budgets stretched to the limit, trying to create a new institution can result in endless delays. Especially for third world countries, burdened with many pressing environmental problems, it often makes more sense to use whatever relevant institutions are already at hand and can be pressed into service immediately.

Building Constituencies

It is important to take what Stephen Olsen calls the two-track approach in building support for coastal area management plans: work on the national level to create high-level political and administrative support, and at the same time work on the local level to build up grassroots support.

As part of the process of constituency building, it is important to create an open and accountable process that includes all stakeholders. With that in mind, involving local communities in both the planning and the implementation stage is one of the keys to a successful process of coastal governance. It matters not so much which sections of the public get involved, but they must be representative of the local communities and reflect the predominate views of all those who use and live along coastal areas.

Using local NGOs and citizen action groups, or finding individuals who can "make things happen," is a sound way to gain entry into a community. It is equally vital to involve local businesses and industries in the process. Simply put: The more constituent parts, the better the management plan.

Getting Started

Three decades of practical experience in managing coastal areas strongly suggests that the prudent way to begin the process is to start small, with a limited number of concrete and achievable objectives. Once the initial goals have been reached, the process can be broadened to include other issues and concerns or expanded to other areas of the country.

By starting small and working up, coastal managers have a better chance of meeting their objectives. Often governments try to implement overly ambitious plans involving too many levels of government, or too few. The pressure to demonstrate success can warp the whole process, making the outcome more important than the process itself. And that often spells the end for coastal management efforts. With coastal management plans, less is often more.

Increasingly, governments are seeing the benefits of introducing special area management plans that can be applied to specific sections of coastline and designed to deal with a limited number of specified problems. This creates an environment of success rather than failure. For first-time attempts especially, it is crucial that people see results. It reinforces the process and helps muster the momentum needed to keep going and to expand the process to other coastal areas.

Making Institutions Responsive

Institutions charged with the task of formulating and implementing coastal area management plans should be along for the long haul. They should not view the process as having a definite end. Institution building has to go hand in hand with constituency building. Above all, coastal management is a learning process for everyone involved, including institutions. There should be mechanisms in place that provides institutions charged with coastal management functions with a way to evaluate their performance and to fine-tune programs as needed.

Nothing is set in stone when it comes to coastal area management. The best way to proceed is to make the lead institution and its local counterparts as responsive as possible to the needs and concerns of coastal populations. Motivated communities, in partnership with government, can accomplish a great deal. But it is crucial that all parties take a long-term view.

The Trouble with Data

"The trouble with data," as one coastal area manager put it, "is that we often don't have enough of it!" Although he was half joking, his statement cuts to the core of a problem that haunts nearly every facet of natural resource management. Our databases are, for the most part, inadequate for the tasks.

Monitoring coastal resources is particularly difficult. We don't have a very good picture, for instance, of the state of most of the world's mangrove forests and coral reefs. We know we are losing them rapidly, but we don't know enough about where and how bad the damage is. Even where there are data, the analysis is often far behind the information curve.

When it comes to data collection and analysis, any data is better than no data. But scientists, like politicians, often will not put themselves out on a limb unless there is a safety net underneath them. They are loath to advance educated guesses, or educated judgments, about where, when, and how best to tackle resource management problems. The onus of proof may rest with scientists, but they also have a moral obligation to make the best recommendations they can with the facts available. Very likely, waiting until there is "enough" data will condemn many coastal resources to ruination.

As pointed out in the introduction, the World Resources Institute issued an indicator brief in 1995 entitled *Coastlines at Risk: An Index of Potential Development-Related Threats to Coastal Ecosystems.* It contains a promising approach to the assessment of coastal risks from development-related activities. Despite the fact that it considers only five basic indicators—cities, major ports, population density, road density, and pipeline density—it holds great potential as a method for pinpointing coastal areas under severe stress (WRI 1995). Once WRI includes more refined indicators of ecosystem damage, such as conversion of mangrove swamps into fish ponds and dynamiting of coral reefs, its brief should find widespread use with coastal resource managers, city planners, and others involved in caretaking coastal areas and resources.

There is more good news about data. Scientists and resource managers are getting better tools with which to gauge humanity's collective impacts on the biosphere. New GIS—Geographical Information Systems—are being developed with the use of satellite imagery, enabling us to probe the oceans and the earth's crust to sense changes. Comprehensive resource mapping, which gives a detailed view of landscapes and ecosystems, is now possible in many countries.

According to UNESCO, remote sensing can provide reliable data on many of the earth's features: "topography, soil type, near-surface geology, vegetation, surface water, shoreline and coastal resources, the surface temperature and abundance of organic material in the oceans, atmospheric temperature, cloud cover and air and water pollution" (UNESCO 1992b).

GIS systems have numerous applications in integrated coastal area management. The only problem with GIS systems and other new techniques is that they are not being disseminated fast enough in developing countries, where the pace of resource destruction is accelerating in many areas. Many tropical developing countries may lose much of their natural heritage before they even have a chance to properly assess their biological resources.

What Next?

We desperately need to evolve practical and sensible coastal area management plans that have broad public support and can be implemented with a minimum of fuss and bureaucracy. Surely, if we can tackle the problem of ozone depletion and put together, in just a few years, an international effort to phase out ozone-depleting chemicals from hundreds of products, we can collectively tackle the pervasive degradation of coastal areas and near-shore waters.

The following actions could be taken:

1. The UN needs to put teeth into the Agenda 21 section on oceans and coasts as outlined in the Earth Summit. It contains some solid recommendations for integrated coastal zone management that should be given due consideration by national governments, the UN system, regional organizations, bilateral development agencies, and national and international NGOs. But those recommendations need to be translated into a context national governments can understand and act upon: a well-defined set of achievable objectives fashioned within the governing context of each country.

2. Scientific and policy-oriented institutions that have solid hands-on experience in coastal area management—such as the Coastal Resources Center of the University of Rhode Island, the Marine Science Institute of the University of the Philippines (in Metro Manila), the World Conservation Union, the Swedish Agency for Research Cooperation with Developing Countries, the World Bank, and others—need to find a mechanism by which to pool their collective experiences, to disseminate the results of their years of work in coastal management.

3. The intergovernmental organizations already established to deal with the management of regional seas and coastal areas—such as the Mediterranean Action Plan and the East Asian Seas Action Plan—need to be reinforced by the international donor community, as well as national governments in each region. In many cases, solid programs languish because of a lack of funding, which more often than not reflects the ignorance of policy makers concerning coastal issues.

4. UNEP's regional seas programs need to be given proper international and regional support. But UNEP is not an implementing agency. It was meant to function as a catalytic agency on environmental issues, helping other UN agencies take account of environmental concerns in their operational programs. In effect, UNEP operates as the information arm of the UN on environmental matters. That it has failed in many respects is partly due to an uneven, shrinking budget.

In order to revitalize the regional seas program, why not reconstitute it under the joint direction of the United Nations Development Program and the World Bank, with UNEP providing legal expertise and serving in the capacity of an information clearing house? UNDP, the World Bank, and UNEP already manage the Global Environment Facility. The institutional mechanisms for cooperation have already been set up. Gus Speth, UNDP's administrator, has made UNDP the UN agency for sustainable development and has outlined an exciting agenda. UNDP is already involved in the Aral Sea rehabilitation program, the Black and Caspian seas programs, and the Danube River Basin management program (all carried out jointly with the World Bank).

5. The Global Environment Facility, the green fund jointly managed by the World Bank, UNDP, and UNEP, could be asked to provide seed money so that third world coastal states could carry out badly needed biodiversity inventories, identifying areas in urgent need of conservation and protection. A similar process, called Rapid Assessment Procedure, is already being used to evaluate the condition of the world's coral resources. Why not extend this concept to include the entire coastal zone? The GEF, with UNDP, could form a kind of "rapid needs assessment group" consisting of an interdisciplinary team of scientists and resource economists to help developing countries with limited national capacities assess the productive potential of their coastal areas and to draw up special area management plans that could be implemented quickly. The team could draw on the ongoing work of bilaterals and coastal management specialists like the Coastal Resources Center

in Rhode Island and the World Conservation Union. The idea would be to pool resources and expertise, minimize duplication of effort, and use work already done or underway.

6. The International Coral Reef Initiative, launched in 1995 by eight countries—the United States, Australia, France, Japan, Jamaica, the Philippines, Sweden, and the United Kingdom—needs to be given the utmost priority by multilateral and bilateral donors, NGOs, intergovernmental organizations, and the UN system. It constitutes one of the best examples of international and intergovernmental cooperation to tackle the rapid and alarming decline of the world's coral resources. As such, it deserves the international community's undivided attention and support. But one thing seems lacking from the initiative's plans: a coordinated effort to publicize the plight of the world's coral reefs through the media, educational institutions, and NGOs. Why not turn the year of the reef (1997) into the decade of the reef and launch a well-coordinated education and information campaign with a strong media component? The capacity is already in place, what is needed is vision.

7. In 1995, the UN Food and Agriculture Organization, (FAO), based in Rome, managed to push through a voluntary Code of Conduct for Responsible Fisheries. This initiative sets forth principles and standards for the sustainable management of global fisheries. The code addresses six areas of concern: fisheries management, fishing operations, aquaculture and mariculture development, integration of fisheries into coastal area management programs, harvesting and post-harvesting practices (such as discarding bycatch), and trade and fisheries research. FAO's intent is to persuade major fishing nations to observe the Code of Conduct and to enact and enforce laws that contribute to the better caretaking of marine resources. Every major fishing nation should endorse the code and implement it. If observed, it could usher in a new era of fisheries management.

8. In January 1996, UNESCO launched a new endeavor: Environment and Development in Coastal Regions and in Small Islands (CSI). The main objective of CSI is to promote the integrated management of coastal areas. Pilot projects concentrate on four areas: the sustainable development of small islands through the integrated management of freshwater resources; more comprehensive, integrated planning and management to mitigate coastal zone instability; the integrated management of coastal biodiversity; and the management of ecosystem productivity for the sustainable development of coastal communities. Each pilot project includes three elements: (1) cultural and socioeconomic dimensions, including traditional management systems; (2) technical and scientific analyses; and (3) institution building and training of experts. These projects need concerted support from the international community, including both bilateral and multilateral development agencies. Also, efforts should be made to translate the results into new and ongoing management strategies, strategies that can enhance coastal area management by building on community-based efforts.

9. In the longer term, every coastal country should develop a process by which those areas suffering from intense population and development pressures

can be identified and programs crafted specifically for them. Integrated priority action plans could then be put in place that address the most pressing population and resource problems in a timely manner.

This list of priority actions could be expanded on indefinitely. The point to be made is that we have the opportunity to draw on three decades of accumulated experience spread out over fifty-five countries that have launched coastal management plans or are in the process of doing so. What is lacking is a way to weave this wealth of experience into a fabric of new and ongoing initiatives, on both the regional and country levels.

The Kuna Indians of the San Blas in Panama have been managing their resources sustainably for centuries, despite intense development pressures in recent years. Small coastal communities in the Philippines, Indonesia, and Thailand have managed to salvage their resources for the benefit of all. Clearly, it is in everyone's long-term interest to use whatever experience is available to build on, whether of Indian tribes in Central America or of highly paid consultants with the World Bank. There is a very challenging job to be done, and there is no time to lose in doing it.

Introduction

Clark, John. 1996. *Coastal Zone Management Handbook.* New York: Lewis Publishers, pp. 1–10.

Deichmann, Uwe. 1996a. *Asia Population Database Documentation.* National Center for Geographic Information and Analysis, Santa Barbara, California, pp. 2–29.

———. 1996b. *A Review of Spatial Population Database Design and Modeling.* National Center for Geographic Information and Analysis, Technical Report 96–3 (March), Santa Barbara, California, pp. 1–51.

Hinrichsen, Don. 1996. "Coasts in Crisis." *Issues in Science and Technology* 12, no. 4 (summer): 39–47.

———. 1997. "Pushing the Limits." *Amicus Journal* 18, no. 4 (winter): 16–20.

OECD. 1993. *Coastal Zone Management—Integrated Policies.* Paris: OECD, pp. 19–124.

Tobler, Waldo; Deichmann, Uwe; Gottsegen, Jon; and Maloy, Kelly. 1995. *The Global Demography Project.* National Center for Geographic Information and Analysis, Technical Report 95–6 (April), Santa Barbara, California, pp. 4–59.

Chapter 1

Batisse, Michel. 1994. "Mending the Med." *People & the Planet* 3, no. 1: 17–18.

Brown, Lester, and Kane, Hal. 1994. *Full House—Reassessing the Earth's Population Carrying Capacity.* New York: Norton, pp. 75–88.

Chapman, Graham, and Baker, Kathleen, eds. 1992. *The Changing Geography of Asia* London: Routledge, pp. 10–257.

China's Population and Development, Country Report. 1992. Prepared for the Fourth Asian and Pacific Population Conference, Bali, Indonesia, August 19–27, pp. 1–74.

Cruz, Maria Concepcion; Myer, Carrie; Repetto, Robert; and Woodward, Richard. 1992 *Population Growth, Poverty and Environmental Stress: Frontier Migration in the Philippines and Costa Rica,* Washington, D.C.: World Resources Institute (October), pp. 1–69.

Culliton, Thomas; Warren, Maureen; Goodspeed, Timothy; Remer, Davida; Blackwell, Carol; and John McDonough. 1990. *50 Years of Population Change along the Nation's Coasts, 1960–2010.* Rockville, Maryland: National Oceanic and Atmospheric Administration (April), pp. 4–11.

Deichmann, Uwe. 1996a. *Asia Population Database Documentation.* National Center for Geographic Information and Analysis (March), Santa Barbara, California, pp. 2–29.

———. 1996b. *A Review of Spatial Population Database Design and Modeling.* National Center for Geographic Information and Analysis, Technical Report 96–3 (March), Santa Barbara, California, pp. 1–51.

Economist. 1994. "China—Birth of the Instant City" (September 10), pp. 34–35.

Economist. 1994–1995. "Shanghai, City of Glitter and Ghosts" (December 24–January 6), pp. 82–84.

Ehrlich, Paul, and Ehrlich, Anne. 1990. *The Population Explosion.* New York: Simon and Schuster.

Gilbert, Alan. 1990. *Latin America.* Routledge Introductions to Development Series. London: Routledge.

Goldberg, Edward. 1994. *Coastal Zone Space: Prelude to Conflict?* Paris: UNESCO Publishing, pp. 23–30.

Green, Cynthia. 1992. "The Environment and Population Growth: Decade for Action." *Population Reports,* series M, no. 10 (May), pp. 3–25.

Grenon, Michel, and Batisse, Michel, eds. 1989. *Futures for the Mediterranean Basin—The Blue Plan.* Oxford: Oxford University Press.

Hardoy, Jorge, and Satterthwaite, David. 1989. *Squatter Citizen: Life in the Urban Third World.* London: Earthscan Publications.

Hinrichsen, Don. 1994a. "Coasts Under Pressure." *People & the Planet* 3, no. 1: 6–9.

———. 1994b. "Putting the Bite on Planet Earth." *International Wildlife* 24, no. 5 (September–October): 36–45.

———. 1996. "Computing the Risks: A Global Overview of Our Most Pressing Environmental Challenges." *International Wildlife* 26, no. 2 (March–April): 22–35.

Keyfitz, Nathan. 1991. "Population Growth Can Prevent the Development That Would Slow Population Growth." In Jessica Mathews, ed., *Preserving the Global Environment—The Challenge of Shared Leadership.* New York: Norton, pp. 39–77.

Lutz, Wolfgang. 1994. *Population Bulletin: The Future of World Population.* Washington, D.C.: Population Reference Bureau (June), pp. 13–39.

Myers, Norman, and Kent, Jennifer. 1995. *Environmental Exodus—An Emergent Crisis in the Global Arena.* Washington, D.C.: Climate Institute (June), pp. 37–85.

Perez Nieto, Hernan. 1993. "Linking Territorial and Coastal Planning: The Venezuelan Experience." *Ocean & Coastal Management* 21, nos. 1–3: 227–243.

Population Reference Bureau. 1993. "Coastal Living: For a Majority of U.S., It's Home." *Population Today* 21, nos. 7–8 (July–August): 4.

Sadik, Nafis. 1993. *The State of World Population 1993.* New York: United Nations Population Fund, pp. 11–34.

———. 1995. *The State of World Population 1995.* New York: United Nations Population Fund, pp. 9–27.

———. 1996. *The State of World Population 1996.* New York: United Nations Population Fund, pp. 1–62.

Tien, H. Yuan; Zhang, T.; Ping, Y.; Li, J.; and Liang, Z. 1992. *Population Bulletin: China's Demographic Dilemmas.* Washington, D.C.: Population Reference Bureau (June), pp. 2–40.

Tobler, Waldo; Deichmann, Uwe; Gottsegen, Jon; and Maloy, Kelly. 1995. *The Global Demography Project.* National Center for Geographic Information and Analysis, Technical Report 95–6 (April), Santa Barbara, California, pp. 4–59.

Tyler, Patrick. 1994. "Beijing to Impose Huge Fees to Limit Migrants in City." *New York Times,* September 15.

World Bank. 1995. *Africa: A Framework for Integrated Coastal Zone Management.* Washington, D.C.: World Bank, pp. 15–17.

WRI. 1995. *WRI Indicator Brief: Coastlines at Risk: An Index of Potential Development-Related Threats to Coastal Ecosystems.* Washington, D.C.: WRI, pp. 1–8.

WRI. 1996. *World Resources 1996–97,* New York: Oxford University Press, p. 254.

Chapter 2

The data available on the condition of most critical coastal resources are patchy at best; at worst, only rough estimates exist based on a few scientific surveys of one or two coastal areas. I have used what I consider the best estimates available.

The Philippines, Indonesia, and Thailand, have done extensive surveys of their coral resources in an effort to determine their condition and to formulate effective management plans. Much less is known about the fate of the world's vast seagrass meadows. They are also deteriorating, but data on their state and condition are not available, even as gross estimations.

Most of the data on the state of global fisheries comes from the UN Food and Agriculture Organization. An excellent summary can be found in *Review of the State of World Fishery Resources: Marine Fisheries,* published by the FAO in 1995. One of the best sources on the global status of coral reefs is Stephen Jameson et al., *State of the Reefs: Regional and Global Perspectives.* The global state of mangroves can be found in Mark Spalding et al., *World Mangrove Atlas.*

Bateman, Graham, and Egan, Victoria, eds. 1993. *The Encyclopedia of World Geography.* New York: Barnes and Noble Books, p. 493.

Boto, K.; Robertson, A.; and Alongi, D. 1991. "Mangrove and Near-Shore Connections—A Status Report from the Australian Perspective." *Proceedings of the Regional Symposium on Living Resources in Coastal Areas, Manila, Philippines.* Manila: Marine Science Institute, University of the Philippines, pp. 459–467.

Chua, Christopher, and Chou, L. 1992. "Coral Reef Conservation in Singapore—A Case for Integrated Coastal Area Management." In L. Chou and C. Wilkinson, eds., *Third ASEAN Science and Technology Week Conference Proceedings, Marine Science: Living Coastal Resources,* vol. 6 (September 21–23), pp. 437–445.

Clark, John. 1996. *Coastal Zone Management Handbook.* New York: Lewis Publishers, pp. 140, 344.

Cole, Jonathan; Peierls, Benjamin; Caraco, Nina; and Pace, Michael. 1994. "Nitrogen Loading of Rivers as a Human-Driven Process." In M.D. McDonnel and S.T.A. Pickett, eds., *Humans as Components of Ecosystems.* Berlin: Springer-Verlag, pp. 138–154.

Coral Reef Alliance, personal communication, November 1996.

Craft, Lucille. 1994. "$20,000 for One Fish? Bluefin Tuna May Be Worth Too Much for Their Own Good." *International Wildlife* (November–December), pp. 18–21.

Dolar, Louella. 1991. "A Survey on the Fish and Crustacean Fauna of the Seagrass Bed in North Bais Bay, Negros Oriental, Philippines." *Proceedings of the Regional Symposium on Living Resources in Coastal Areas, Manila, Philippines.* Manila: Marine Science Institute, University of the Philippines, pp. 367–377.

Dolar, Louella; Alcala, Angel; and Nuique, J. 1991. "A Survey on the Fish and Crustaceans of the Mangroves of the North Bais Bay, Philippines." *Proceedings of the Regional Symposium on Living Resources in Coastal Areas, Manila, Philippines.* Manila: Marine Science Institute, University of the Philippines, pp. 513–519.

Dugan, Patrick, ed. 1993. *Wetlands in Danger.* A Mitchell Beazley World Conservation Atlas. London: Mitchell Beazley.

Economist. 1994a. "Fish: The Tragedy of the Oceans" (March 19), pp. 21–24.

———. 1994b. "Where Have All the Fishes Gone?" In *World Press Review* (June), p. 42.

FAO. 1995. *The State of World Fisheries and Aquaculture.* Rome: FAO, pp. 3–43.

Fortes, Miguel. 1990. *Seagrasses: A Resource Unknown in the ASEAN Region.* Manila: ICLARM-USAID, pp. 1–31.

Hinrichsen, Don. 1994. "Coasts Under Pressure." *People & the Planet* 3, no. 1: 6–9.

———. 1996. "Reef Revival." *Amicus Journal* 18, no. 2 (summer): 22–25.

Jameson, Stephen; McManus, John; and Spalding, Mark. 1995. *State of the Reefs: Regional and Global Perspectives.* Washington, D.C.: International Coral Reef Initiative Executive Secretariat Background Paper, National Oceanic and Atmospheric Association (May), pp. 5–30.

Jennings, Simon, and Polunin, Nicholas. 1996. "Impacts of Fishing on Tropical Reef Ecosystems." *Ambio* 25, no. 1 (February): 44–49.

King, Michael. 1988. *Coral Reefs in the South Pacific: Handbook.* Noumea, New Caledonia: South Pacific Commission, pp. 1–40.

Lean, Geoffrey, and Hinrichsen, Don. 1994. *Atlas of the Environment.* New York: Harper-Collins, pp. 165–168.

Lindén, Olof. 1990. "Human Impact on Tropical Coastal Zones." *Nature and Resources.* 26, no. 4: 3–11.

Lundin, Carl Gustaf, and Lindén, Olof. 1993. "Coastal Ecosystems: Attempts to Manage a Threatened Resource." *Ambio* 22, no. 7 (November): 468–473.

McNeely, Jeffrey, and Dobias, Robert. 1991. "Economic Incentives for Conserving Biological Diversity in Thailand." *Ambio* 20, no. 2 (April): 86–90.

Miller, Susan. 1993. "Greens Turn Blue to Save the Seas." *New Scientist* (October 16), p. 7.

Ohaus, Tom. 1993. "Fighting Over Fish." *Offshore* (November), 12–19.

Pearce, Fred. 1996. "Only Stern Words Can Save World's Fish." *New Scientist* (February 10), p. 4.

Platt, Anne. 1995. "Reversing Our Path to Destruction." *People & the Planet* 4, no. 2: 6–9.

Porter, G. 1988. *Resources, Population and the Philippines' Future: A Case Study.* Washington, D.C.: World Resources Institute (October).

Safina, Carl. 1995. "The World's Imperiled Fish." *Scientific American* 273, no. 5 (November): 46–53.

Salm, Rodney. 1983. "Coral Reefs of the Indian Ocean: A Threatened Heritage." *Ambio* 8, no. 6: 349–353.

———. 1994. "Corals Hidden Riches." *People & the Planet* 3, no. 1: 19–21.

Schmidt, Carl-Christian. 1993. "The Net Effects of Over-fishing." *OECD Observer,* no. 184 (October–November): 9–12.

Spalding, Mark; Blasco, Francois; and Field, Colin, eds. 1996. *World Mangrove Atlas.* London: International Society for Mangrove Ecosystems, World Conservation Monitoring Centre, International Tropical Timber Organization.

Thorhaug, Anitra. 1981. "Biology and Management of Seagrass in the Caribbean." *Ambio* 10, no. 6: 295–298.

Tolba, Mostafa; El-Kholy, O.; El-Hinnawi, E.; Holdgate, M.; McMichael, D.; and Munn, R. 1992. *The World Environment 1972–1992.* London: Chapman and Hall, pp. 106–130.

Weber, Michael, and Gradwohl, Judith. 1995. *The Wealth of Oceans.* New York: Norton.

Weber, Peter. 1993. *Abandoned Seas: Reversing the Decline of the Oceans.* Worldwatch Paper No. 116. Washington, D.C.: Worldwatch Institute (November), pp. 5–55.

———. 1994. *Net Loss: Fish, Jobs, and the Marine Environment.* Worldwatch Paper No. 120. Washington, D.C.: Worldwatch Institute (July), pp. 5–62.

Wells, Susan, ed. 1988. *Coral Reefs of the World,* vols. 1–3. Nairobi: World Conservation Union and UN Environment Programme.

White, Alan; Hale, Lynne; Renard, Yves; and Cortesi, Lafcadio, eds. 1994. *Collaborative and Community-Based Management of Coral Reefs: Lessons from Experience.* Conn.: Kumarian Press, pp. 1–95.

Wilkinson, Clive. 1992. "Coral Reefs of the World Are Facing Widespread Devastation: Can We Prevent This through Sustainable Management Practices?" In *Proceedings of the Seventh International Coral Reef Symposium,* vol. 1, Guam.

WRI. 1996. *World Resources 1996–97.* New York: Oxford University Press, pp. 296–300.

Chapter 3

The special issue of *Ocean and Coastal Management,* Vol. 21, Nos. 1–3, 1993, is a valuable sourcebook on the entire range of coastal management issues.

Borgese, Elisabeth Mann. 1991. *The Oceans, the Convention and You.* Nairobi: United Nations Environment Programme, OCA/PAC (May), pp. 2–31.

Craik, Wendy. 1992. "The Great Barrier Reef Marine Park: Its Establishment, Development and Current Status." *Marine Pollution Bulletin* 25: 122–133.

Crawford, Brian; Cobb, Stanley; and Friedman, Abigail. 1993. "Building Capacity for Integrated Coastal Management in Developing Countries." *Ocean and Coastal Management* 21: 311–337.

Environment Agency of Japan. 1993. *The Seto Inland Sea—The Largest Enclosed Coastal Sea in Japan.* Tokyo: Environment Agency of Japan (November), pp. 6–8.

Eteuati, Kilifoti. 1991. *The Law of the Sea and the South Pacific.* Nairobi: United Nations Environment Programme, OCA/PAC (November), pp. 2–23.

Great Barrier Reef Marine Park Authority. 1994. *The Great Barrier Reef—Keeping It Great: 25 Year Strategic Plan for the Great Barrier Reef World Heritage Area.* Queensland: Great Barrier Reef Marine Park Authority.

Kelleher, Graeme. 1996. "Case Study 2: The Great Barrier Reef, Australia." In *The Contributions of Science to Integrated Coastal Management.* Rome: FAO, pp. 31–44.

Kenchington, Richard, and Crawford, David. 1993. "On the Meaning of Integration in Coastal Zone Management." *Ocean and Coastal Management* 21, nos. 1–3: 109–127.

Knecht, Robert, and Archer, Jack. 1993. "Integration in the U.S. Coastal Zone Management Program." *Ocean and Coastal Management* 21, nos. 1–3: 183–199.

Kobe City Government. 1989. *A Centennial Tribute to Kobe.* City of Kobe, Japan (July).

———. 1992a. *The City of Kobe.* Kobe, Japan: Mayor's Office.

———. 1992b. *Outline of the Development Bureau Projects.* City of Kobe, Japan.

Kristof, Nicholas. 1995. "The Quake That Hurt Kobe Helps Its Criminals." *New York Times,* June 6.

Levy, Jean-Pierre. 1993. "A National Ocean Policy—An Elusive Quest." *Marine Policy* (March), pp. 75–80.

Lowry, Kem, and Wickremeratne, H.J.M. 1989. "Coastal Area Management in Sri Lanka." In *Ocean Yearbook No. 7.* Chicago: University of Chicago Press, pp. 263–293.

Ness, Gayl, ed. 1992. *Population Dynamics and Port City Development: Comparative Analysis of Ten Asian Port Cities.* Kobe, Japan: Asian Urban Information Center (January), pp. 158–195.

OAS. 1988. *Inventory of Caribbean Marine and Coastal Protected Areas.* Washington, D.C.: Organization of American States.

OECD. 1993. *Coastal Zone Management: Selected Case Studies.* Paris: OECD, pp. 229–243.

Olsen, Stephen. 1993. "Will Integrated Coastal Management Programs Be Sustainable: The Constituency Problem." *Ocean and Coastal Management* 21, nos. 1–3: 201–225.

Olsen, Stephen, and Hale, Lynne. 1994. "Coasts: The Ethical Dimension." *People & the Planet* 3, no. 1: 29–31.

O'Riordan, Brian. 1994. "Tangled Webs of Chaos Adrift." *New Scientist* (August 20), pp. 44–45.

Sorensen, Jens. 1993. "The International Proliferation of Integrated Coastal Zone Management Efforts." *Ocean and Coastal Management* 21: 45–80.

Takayose, Shozo. 1986. "Population Redistribution and the Construction of Port Island and Suma New Town in Kobe." *Population Redistribution in Planned Port Cities.* Kobe, Japan: Nihon University Population Research Institute, pp. 19–74.

Thia-Eng, Chua. 1993. "Essential Elements of Integrated Coastal Zone Management." *Ocean and Coastal Management* 21: 81–108.

Thia-Eng, Chua, and Scura, Louise Fallon, eds. 1992. *Integrative Framework and Methods for Coastal Area Management.* Manila: International Center for Living Aquatic Resources Management, pp. 68–97.

UNESCO. 1993. *Environment and Development Brief No. 6: Coasts.* Paris: UNESCO, p. 3.

United Nations Division for Ocean Affairs and the Law of the Sea. 1993. *Law and Order in the Oceans—United Nations Convention on the Law of the Sea.* New York: United Nations Division for Ocean Affairs and the Law of the Sea (October).

Weber, Peter. 1994. *Net Loss: Fish, Jobs, and the Marine Environment.* Worldwatch Paper No. 120. Washington, D.C.: Worldwatch Institute, pp. 5–62.

WRI. 1996. *World Resources 1996–97.* New York: Oxford University Press, p. 308.

Chapter 4

Instead of calculating coastal populations within 200 kilometers of the coasts of the North and Baltic seas, I have used drainage basin population figures. Europe is so compact and highly developed that it makes sense to look at the coastal impacts from that perspective. Still, wherever I had good data on coastal populations, I used them for illustration.

The best up-to-date material on Baltic pollution and related issues can be obtained from the Helsinki Commission in Helsinki, Finland. Malcolm MacGarvin's excellent book *The North Sea* is still one of the best sources on its condition.

Anderson, Donald. 1994. "Red Tides." *Scientific American* (August), pp. 62–68.

Carter, F.W., and Turnock, D., eds. 1996. *Environmental Problems in Eastern Europe.* London: Routledge, pp. 107–163.

Countryside Commission. 1991. *Europe's Coastal Crisis: A Cooperative Response.* Gloucestershire, U.K.: Countryside Commission, pp. 1–19.

de Jonge, V. N.; et al. 1993. "The Dutch Wadden Sea: A Changed Ecosystem." *Hydrobiologia* 265: 45–71.

Feshbach, Murray, and Friendly, Alfred. 1992. *Ecocide in the USSR.* New York: Basic Books.

Helsinki Commission. 1993. *The Baltic Sea Joint Comprehensive Environmental Action Programme.* Helsinki: Helsinki Commission (March), pp. 1–5.

———. 1994. *20 Years of International Cooperation for the Baltic Marine Environment 1974–1994.* Helsinki: Helsinki Commission, pp. 2–40.

Jepsen, Palle. 1994. "The Implementation of the Ramsar Convention in the Wadden Sea." *Orphelia Supplement* 6 (September), pp. 33–35.

Johnston, Paul; et al. 1993. *The North Sea: Invisible Decline?* London: Greenpeace, pp. 10–70.

Jonsson, Per; et al. 1993. "Pulp-mill Related Polychlorinated Organic Compounds in Baltic Sea Sediments." *Ambio* 22, no. 1 (February): 37–43.

Leppakoski, Erkki, and Mihnea, Pia Elena. 1996. "Enclosed Seas Under Man-Induced Change: A Comparison Between the Baltic and Black Seas." *Ambio* 25, no. 6 (September): 380–389.

MacGarvin, Malcolm. 1990. *The North Sea.* London: Collins and Brown.

MacKenzie, Debora. 1993. "Disease Could Wipe Out Baltic Salmon." *New Scientist* (October 9), p. 8.

Mee, Laurence. 1992. "The Black Sea in Crisis: A Need for Concerted International Action." *Ambio* 21, no. 4 (June): 278–286.

North Sea Task Force, 1993a. *North Sea Quality Status Report, 1993.* London: Oslo and Paris Commissions.

———. 1993b. *North Sea Subregion 1, Assessment Report 1993, Norway and United Kingdom.* Oslo: State Pollution Control Authority, pp. 5–29.

———. 1993c. *North Sea Subregion 6, Assessment Report 1993, Norway.* Oslo: State Pollution Control Authority, pp. 5–54.

OECD. 1993. *OECD Environmental Performance Reviews, Germany.* Paris: OECD.

Pearce, Fred. 1993. "Greenprint for Rescuing the Rhine." *New Scientist* (June 26), pp. 25–29.

———. 1995. "Sea Life Sickened by Urban Pollution." *New Scientist* (June 17), p. 4.

Population Reference Bureau. 1997. *World Population Data Sheet, 1997.* Washington, D.C.: Population Reference Bureau.

Pudlis, Eugeniusz. 1992. "The Baltic Factor." *Warsaw Voice* (January 19), pp. 12–13.

———. 1993. "The Baltic Blues." *The Green Voice.* Special supplement of the Warsaw Voice, July 4, pp. G1–G4.

State Pollution Control Authority. 1993. *Pollution in Norway.* Oslo: State Pollution Control Authority, pp. 33–50.

Stockholm Environment Institute. 1994. *Beauty and the East: An Evaluation of Swedish Environmental Assistance to Eastern Europe.* Stockholm: Stockholm Environment Institute, (September), chapters 1–6.

Surfers Against Sewage. 1993. *Surfers Against Sewage Information Sheet* 1, no. 1 (December): 1–3.

Underdal, Bjaren; et al. 1989. "Disastrous Bloom of *Chrysochromulina polylepis* (*Prymnesiophyceae*) in Norwegian Coastal Waters 1988—Mortality in Marine Biota." *Ambio* 18, no. 5: 265–270.

Vesilind, Priit. 1989. "The Baltic: Arena of Power." *National Geographic Magazine* 175, no. 5 (May): 604–635.

World Wide Fund for Nature. 1991. *The Common Future of the Wadden Sea.* Husum, Germany: World Wide Fund for Nature (November), pp. 9–59.

Wulff, Fredrik; et al. 1990. *Large-Scale Environmental Effects and Ecological Processes in the*

Baltic Sea, Research Programme for the Period 1990–95 and Background Documents. Stockholm: Swedish Environmental Protection Agency.

Chapter 5

One of the best sources of information is Laurence Mee's excellent article in *Ambio,* "The Black Sea in Crisis: A Need for Concerted International Action."

Balkas, T. 1990. *State of the Marine Environment in the Black Sea Region.* UNEP Regional Seas Reports and Studies no. 124. Nairobi: UNEP, OCA/PAC, pp. 1–35.

BSEP. 1994. "The Collapse of the Black Sea Fisheries." *Saving the Black Sea—Official Newsletter of the Global Environment Facility Black Sea Programme.* Istanbul: BSEP (September), p. 7.

FAO. 1993. "Fisheries and Environment Studies in the Black Sea System." *Studies and Reviews: General Fisheries Council for the Mediterranean.* Rome: Food and Agriculture Organization (January).

Griffin, Michael. 1993. "It's Collapsing Completely." *Ceres* 142 (July–August): 28–31.

Hinrichsen, Don. 1991. "Those Danube Blues." *International Wildlife* 21, no. 5 (September–October): 38–47.

———. 1994. "Putting the Blue Back in the Danube: Europeans Try the First Steps of the Watershed Waltz." *Amicus Journal* 16, no. 3 (fall): 41–43.

Kamenova, Tzveta. 1993. *Bulgarian Black Sea Coast Sustainable Development—Problems and Potentialities.* Sofia: Ministry of Regional Development and Construction, unpublished paper, pp. 1–5.

Leppakoski, Erkki, and Mihnea, Pia Elena. 1996. "Enclosed Seas Under Man-Induced Change: A Comparison Between the Baltic and Black Seas." *Ambio* 25, no. 6 (September): 380–389.

Mee, Laurence. 1992. "The Black Sea in Crisis: A Need for Concerted International Action." *Ambio* 21, no. 4 (June): 278–286.

Murray, J. W.; et al. 1989. "Unexpected Changes in the Oxic/Anoxic Interface in the Black Sea." *Nature* 338, no. 6214 (March 30): 411–413.

Pearce, Fred. 1995. "How the Soviet Seas Were Lost." *New Scientist* 148, no. 2003 (November 11): 39–42.

Platt, Anne. 1995. "Dying Seas." *Worldwatch* 8, no. 1 (January–February): 10–12.

Sorensen, Jens; Gable, Fred; Gardner, M; and Hinrichsen, Don. 1997. "The Black Sea: Another Environmental Tragedy in Our Time?" (Paper submitted to the Conference Medcoast, 1997, Qawra, Malta, November 10–14).

UNDP, UNEP, and the World Bank. 1993. *Saving the Black Sea: Programme for the Environmental Management and Protection of the Black Sea.* Washington, D.C.: UNDP, UNEP, World Bank (June), pp. 2–28.

Yesin, N.V.; Kos'yan, R.D.; and Karnaukhova, L.A. 1993. "Ecological Situation of the Black Sea Coastal Zone." In *Coastlines of the Black Sea.* New York: American Society of Civil Engineers, pp. 173–184.

Zaika, V. 1993. *The Drop of Anchovy Stock in the Black Sea: Result of Biological Pollution?* Sevastopol, Ukraine: Institute of Biology of Southern Seas, unpublished paper, pp. 1–3.

Zaitsev, Yuri. 1993. *Biological Aspects of Western Black Sea Coastal Waters.* Odessa: Institute of Biology of Southern Seas, Ukrainian Academy of Sciences, unpublished paper, pp. 1–28.

Chapter 6

Two sources give useful overviews of the sea's problems: The UNEP report by L. Jeftic, *State of the Marine Environment in the Mediterranean Region,* and Fred Pearce's piece in *New Scientist,* "Dead in the Water."

Attenborough, David. 1987. *The First Eden—The Mediterranean World and Man.* London: Collins/BBC Books.

Barberis, Mary. 1994. "Spotlight—Egypt." *Population Today* (Population Reference Bureau, Washington, D.C.) 22, no. 6 (June): 7.

Batisse, Michel. 1990. "Probing the Future of the Mediterranean Basin." *Environment* 32, no. 5 (June): 28–34.

———. 1994. "Mending the Med." *People & the Planet* 3, no. 1: 17–18.

Bohlen, Celestine. 1995. "The Stuff in the Canal Is Not the Stuff of Romance." *New York Times,* May 18.

Buckley, Richard, ed. 1992. *The Mediterranean—Paradise Under Pressure.* Cheltenham, U.K.: European Schoolbooks Publishing, pp. 1–9.

Chircop, Aldo. 1991. *The Law of the Sea and the Mediterranean.* Nairobi: UNEP (November), pp. 4–23.

Cowell, Alan. 1994. "In Turkey's Bleak Cities, Militant Islam on the Rise." *New York Times,* September 15.

Economist. 1995. "A New Crusade" (December 2), pp. 49–50.

FAO. 1995. *Review of the State of World Fishery Resources: Marine Fisheries.* Rome: FAO, p. 35.

Jeftic, L. 1990. *State of the Marine Environment in the Mediterranean Region,* UNEP Regional Seas Reports and Studies, no. 132. Nairobi: UNEP.

MAP Coordinating Unit/UNEP. 1991. "Mr. Aldo Manos Draws a Picture of 15 Years of Cooperation in the Mediterranean." *MedWaves* (winter–spring), pp. 5–8.

Pastor, Xavier, ed. 1991. *The Mediterranean.* London: Collins & Brown, pp. 29–93.

Pearce, Fred. 1995. "Dead in the Water." *New Scientist* 145, no. 1963 (February 4): 26–31.

Ress, Paul. 1986. "The Mediterranean: Surely but Slowly Cleaner." UNEP Press Release, Athens, July.

———. 1987. "Rocking the Cradle of Civilization: Money and Mentality in the Mediterranean." UNEP Press Release, Athens, June.

———. 1988. "Safer Water, Cleaner Beaches and a Sense of Mediterranean Identity." UNEP Press Release, Geneva, August.

———. 1991. "Environmental Harmony in a Sea of Conflicts." UNEP Press Release, Cairo, October.

Thacher, Peter. 1982. "Statement Made at the Meeting of Government Experts on Regional Marine Programmes." Nairobi: UNEP (January), pp. 2–10.

———. 1992. "The Mediterranean: A New Approach to Marine Pollution." In *International Environmental Negotiation.* Vienna: Sage, pp. 110–134.

USAID. 1994. *Environment Project Profile: Alexandria Wastewater Systems Expansion.* Washington, D.C.: Environment and Natural Resources Information Center, Datex, pp. 1–3.

World Bank. 1993–94. "Mediterranean Environment Given Stepped Up Assistance by METAP II." *Environment Bulletin* 6, no. 1 (winter): 7.

———. 1995. *Mainstreaming the Environment: The World Bank Group and the Environment Since the Rio Earth Summit, Fiscal 1995.* Washington, D.C.: World Bank, p. 56.

For a complete EPA survey report, see EPA, *The Quality of Our Nation's Water: 1994.* A good overview of coastal management in the United States is contained in Jack Archer's *Coastal Management in the United States: A Selective Review and Summary.*

Archer, Jack. 1988. *Coastal Management in the United States: A Selective Review and Summary.* Narragansett: Coastal Resources Center, University of Rhode Island, pp. 1–14.

Beatley, Timothy. 1991. "Protecting Biodiversity in Coastal Environments: Introduction and Overview." *Coastal Management* 19, no. 1 (January–March): 1–17.

Chesapeake Bay Foundation. 1991. *Save the Bay, 25th Anniversary, 1966–1991.* Annapolis, Maryland: Chesapeake Bay Foundation, pp. 3–32.

Clark, John. 1996. *Coastal Zone Management Handbook.* New York: Lewis Publishers, pp. 501–507.

Conservation Fund. 1995. *The Sustainable Everglades Initiative Strategic Issues Assessment.* Arlington, Virginia: Conservation Fund (February 20).

Culliton, Thomas; Warren, M.; Goodspeed, T.; Remer, D.; Blackwell, C.; and McDonough, J. 1990. *50 Years of Population Change Along the Nation's Coasts: 1960–2010.* Rockville, Maryland: National Oceanic and Atmospheric Administration (April), pp. 1–27.

Cushman, John. 1996. "Clinton Backing Vast Effort to Restore Florida Swamps." *New York Times,* February 18, pp. 1, 26.

Cushman, John, Jr. 1993. "U.S. Plans Vast Restoration for Everglades." *New York Times,* September 28, p. 10.

Dean, Corey. 1989. "As Beach Erosion Accelerates, Remedies Are Costly and Few." *New York Times,* August 1, p. C1.

Dennison, William C.; Orth, Robert; Moore, Kenneth; Stevenson, Court; Carter, Virginia; Kollar, Stan; Bergstrom, Peter; and Batiuk, Richard. 1993. "Assessing Water Quality with Submersed Aquatic Vegetation—Habitat Requirements as Barometers of Chesapeake Bay Health." *BioScience* 43, no. 2 (February): 86–94.

"The Dirty Seas." 1988. *Time,* August 1, p. 44–50.

Economist. 1993. "Trouble on the Coast 1: The Sands Run Out" (December 18), p. 49.

EPA. 1995. *The Quality of Our Nation's Water: 1994.* Washington, D.C.: EPA (December), p. 209.

Everglades Coalition. 1993. *The Greater Everglades Ecosystem Restoration Plan.* Everglades Coalition (July), pp. 1–35.

Farnsworth, Clyde. 1995. "Pollution in Canada? Believe It." *New York Times,* July 24.

Fraser, C. 1996. *Integrated Coastal Area Management: A Canadian Retrospective and Update.* Paper Presented to the UN Commission on Sustainable Development, New York, April, pp. 1–9.

Godschalk, David. 1992. "Implementing Coastal Zone Management, 1972–1990." *Coastal Management* 20, no. 2 (April–June): 93–112.

Governor's Commission for a Sustainable South Florida. 1994a. *Draft Meeting Summary.* Miami: Crowne Plaza Hotel, June 29–30, pp. 2–14.

———. 1994b. *Draft Meeting Summary.* Naples: Naples Beach Resort and Golf Club, August 4–5, pp. 1–11.

Griffin, Rodman. 1992a. "Marine Mammals vs. Fish." *CQ Researcher,* August 28, pp. 739–755.

———. 1992b. "Threatened Coastlines." *CQ Researcher,* February 7, pp. 99–115.

Hinrichsen, Don. 1995. "Waterworld: A Hundred Years of Plumbing, Plantations and Politics in the Everglades." *Amicus Journal* 17, no. 2 (summer): 23–27.

Holmes, Bob. 1994. "Oil Spill Damages Set at Billions." *New Scientist,* September 24, p. 5.

Horton, Tom. 1993a. "Chesapeake Bay, Hanging in the Balance." *National Geographic* 183, no. 6 (June): 4–35.

———. 1993b. "The Last Skipjack." *New York Times Magazine,* June 13, pp. 32–37.

Horton, Tom, and Eichbaum, William. 1991. *Turning the Tide: Saving the Chesapeake Bay.* Washington, D.C.: Island Press.

Lemonick, Michael. 1987. "Shrinking Shores—Overdevelopment, Poor Planning and Nature Take Their Toll." *Time,* August 10, pp. 38–47.

Lewis, Thomas. 1989. "Tragedy in Alaska." *National Wildlife Magazine,* June–July, pp. 5–9.

Mairson, Alan. 1994. "The Everglades: Dying for Help." *National Geographic* 185, no. 4 (April): 6–35.

"Marine Life Disappearing Off California." 1995. *New York Times,* March 5, p. 10.

McDougall, Walter. 1993. *Let the Sea Make a Noise—A History of the North Pacific from Magellan to MacArthur.* New York: Basic Books, pp. 55–130.

National Oceanic and Atmospheric Administration. 1996. *National Status and Trends Program.* Washington, D.C.: NOAA (spring–summer), pp. 1–4.

National Safety Council. Undated. *Covering the Coasts: A Reporter's Guide to Coastal and Marine Resources.* Washington, D.C.: Environmental Health Center, National Safety Council.

Needham, Brian, ed. 1991. *Case Studies of Coastal Management: Experience from the United States.* Narragansett: Coastal Resources Center, University of Rhode Island, pp. 57–69.

NRDC. 1996. *Testing the Waters: Who Knows What You're Getting Into.* New York: NRDC.

Pain, Stephanie. 1993a. "Species After Species Suffers from Alaska's Spill." *New Scientist,* February 13, p. 5.

———. 1993b. "The Two Faces of the Exxon Disaster." *New Scientist,* May 22, pp. 11–13.

Palm Beach Post. 1989. "Special Report, Lake Okeechobee: Fighting for Its Life," March 12, pp. 1–16.

———. 1993. "Special Report, The Everglades: Reversing Man's Mistakes," April 11, pp. 1–8.

Reiss, Spencer. 1992. "What's a Bay Without Water?" *Newsweek,* November 2, p. 83.

Schneider, Keith. 1994a. "Exxon Ordered to Pay $5 Billion for Spill." *New York Times,* September 17, pp. 1, 10.

———. 1994b. "In Aftermath of Oil Spill, Alaska Waters Languish." *New York Times,* July 7, p. A16.

South Florida Water Management District and Florida Department of Environmental Protection. 1994. *Everglades Program Implementation: Program Management Plan.* West Palm Beach: South Florida Water Management District and Florida Department of Environmental Protection, September 13, pp. 1-1–3-1.

Steiner, Rick. 1995. "Probing an Oil-Stained Legacy." *National Wildlife* 31, no. 3 (April–May): 4–11.

Chapter 8

The definition of the Caribbean used here is UNEP's definition of the *wider Caribbean:* an area that includes the Gulf of Mexico, all of the Caribbean proper, plus the mainland South American states of Columbia, Venezuela, Guyana, and Suriname.

The population data for all countries, including those in Central America that border on two coasts—the Caribbean on the east and the Pacific on the west—are total population figures. Population densities have been calculated by Uwe Deichmann and colleagues at the Center for Geographic Information and Analysis at Santa Barbara, California.

One of the best sources on Central America's coasts is the excellent summary contained in the Coastal Resources Center's *Central America's Coasts: Profiles and Agenda for Action* edited by Gordon Foer and Stephen Olsen. For the Caribbean, see Larry Mosher's chapter, "At Sea in the Caribbean," in *Bordering on Trouble*.

Bilsborrow, Richard. 1991. "Updating Malthus in Latin America." *Earthwatch*, no. 41 (second quarter): 10–12.

Borgese, Elisabeth Mann. 1991. *The Law of the Sea and the Caribbean*. Nairobi: UNEP (November), pp. 2–31.

Bossi, Richard, and Cintron, Gilberto. 1990. *Mangroves of the Wider Caribbean—Toward Sustainable Management*. Washington, D.C.: UNEP, Caribbean Conservation Association, Panos Institute, pp. 1–26.

Broad, William. 1994. "Coral Reefs Endangered in Jamaica." *New York Times*, September 9, p. A12, A21.

Carrasco, Domingo, and Witter, Scott. 1993. "Constraints to Sustainable Soil and Water Conservation: A Dominican Republic Example." *Ambio* 22, no. 6 (September): 347–350.

Clark, John. 1996. *Coastal Zone Management Handbook*. New York: Lewis Publishers, pp. 89, 494–495.

Da Silva, Donatus. 1994. "Harbour View Shows the Way." *Our Planet* 6, no. 1: 16.

Dixon, John. 1993. "Economic Benefits of Marine Protected Areas." *Oceanus* 36, no. 3 (fall): 35–40.

Dixon, John; Scura, Louise Fallon; and van't Hof, Tom. 1993. "Meeting Ecological and Economic Goals: Marine Parks in the Caribbean." *Ambio* 22, nos. 2–3 (May): 117–125.

Elder, Danny, and Pernetta, John, eds. 1991. *Oceans*, A Mitchell Beazley World Conservation Atlas. London: Mitchell Beazley, pp. 124–129.

FAO. 1995. *Review of the State of World Fishery Resources: Marine Fisheries*. Rome: FAO, pp. 14–20.

Foer, Gordon, and Olsen, Stephen, eds. 1992. *Central America's Coasts—Profiles and an Agenda for Action*. Washington, D.C.: U.S. Agency for International Development; and Narragansett: Coastal Resources Center, University of Rhode Island.

Gilbert, Alan. 1990. *Latin America*. London: Routledge, pp. 44–75.

Guzman, Hector, and Jarvis, Kym. 1996. "Vanadium Century Record from Caribbean Reef Corals: A Tracer of Oil Pollution in Panama." *Ambio* 25, no. 8 (December): 523–526.

Hinrichsen, Don, ed. 1981. "The Caribbean." *Ambio Special Issue* 10, no. 6.

Hughes, Terence. 1994. "Catastrophes, Phase Shifts, and Large-Scale Degradation of a Caribbean Coral Reef." *Science* 265 (September 9): 1547–1550.

Jameson, Stephen; McManus, John; and Spalding, Mark. 1995. *State of the Reefs—Regional and Global Perspectives*. Washington, D.C.: NOAA, pp. 6–8.

Lindén, Olof. 1990. "Human Impact on Tropical Coastal Zones." *Nature and Resources* 26, no. 4: 9.

Lundin, Carl Gustaf, and Lindén, Olof. 1993. "Coastal Ecosystems: Attempts to Manage a Threatened Resource." *Ambio* 22, no. 7 (November): 468–473.

Mosher, Larry. 1986. "At Sea in the Caribbean." In *Bordering on Trouble.* Bethesda, Md.: Adler and Adler, pp. 235–269.

———. 1989. "Amazing Boon: Waves of Passengers Set Caribbean Ports Jumping." *Americas* 41, no. 2: 34–40.

Mote Marine Laboratory. 1992. *1992: The Year of the Gulf—America's Sea, Keep it Shining,* Information Pack. Sarasota, Fla.: Mote Marine Laboratory.

Navarro, Mireya. 1994. "Worst Drought in 30 Years Brings Rationing to Half of Puerto Rico." *New York Times,* July 3, pp. 1, 14.

OAS. 1988. *Inventory of Caribbean Marine and Coastal Protected Areas.* Washington, D.C.: Organization of American States.

Pain, Stephanie. 1994. "Living Coastline Suffers Most from Oil Spills." *New Scientist,* January 8, pp. 4–5.

Price, A.R.G., and Heinanen, A.P. 1992. *Guidelines for Developing a Coastal Zone Management Plan for Belize—A Marine Conservation and Development Report.* Gland, Switzerland: IUCN, pp. 1–18.

Southgate, Douglas, and Basterrechea, Manueal. 1992. "Population Growth, Public Policy and Resource Degradation: The Case of Guatemala." *Ambio* 21, no. 7 (November): 460–464.

Thorhaug, Anitra. 1981. "Biology and Management of Seagrass in the Caribbean." *Ambio* 10, no. 6: 295–298.

UNEP (Caribbean Regional Co-ordinating Unit). 1987. *Action Plan for the Caribbean Environment Programme—A Framework for Sustainable Development.* Kingston, Jamaica: UNEP Caribbean Regional Co-ordinating Unit (October), pp. 7–24.

Verhovek, Sam Howe. 1995. "Virus Imperils Texas Shrimp Farms—Mysterious Disease from Ecuador is Moving Up Gulf Coast." *New York Times,* June 14, p. A16.

Weber, Michael, and Gradwohl, Judith. 1995. *The Wealth of Oceans.* New York: Norton, p. 94.

Wille, Chris. 1993. "The Shrimp Trade Boils Over." *International Wildlife* (November–December), pp. 18–23.

Williams, Ted. 1995. "The Turtle Gulf War." *Audubon* 97, no. 5 (September–October): 26–33.

World Bank. 1992–93. "In the Caribbean Sea: Case for Marine Parks Made for Producing Ecological, Economic Benefits." *Environment Bulletin: A Newsletter of the World Bank Environment Community* 5, no. 1 (winter): 10.

WRI. 1994. *World Resources 1994–95—A Guide to the Global Environment.* New York: Oxford University Press, p. 356.

Chapter 9

One of the better sources on the entire west coast of Latin America is UNEP's report *Regional Cooperation on Environmental Protection of the Marine and Coastal Areas of the Pacific Basin* (1991). A good pollution summary can be found in UNEP's 1988 publication *State of the Marine Environment in the South-East Pacific Region.* Those readers who want to read an excellent summary of Ecuador's coastal management efforts can turn to Don Robadue, ed., *Eight Years in Ecuador—The Road to Integrated Coastal Management.*

Bergeron, Lou. 1996. "Will El Niño Become El Hombre?" *New Scientist,* January 20, p. 15.

Boraiko, Allen. 1995. "Bright Hopes for Chile?" *International Wildlife* (July–August), pp. 36–43.

Brooke, Elizabeth Heilman. 1992. "As Forests Fall, Environmental Movement Rises in Brazil." *New York Times,* June 2, p. C4.

Brooke, James. 1994. "Europeans Chase the Sun to Brazil's Northeast." *New York Times,* March 15.

Bryant, Dirk; Rodenburg, E.; Cox, T.; and Nielsen, D. 1995. *WRI Indicator Brief: Coastlines at Risk—An Index of Potential Development-Related Threats to Coastal Ecosystems.* Washington, D.C.: World Resources Institute, pp. 1–8.

Clark, John. 1996. *Coastal Zone Management Handbook.* New York: Lewis Publishers, pp. 511–513.

Couper, Alastair, ed. 1989. *The Times Atlas and Encyclopaedia of the Sea.* London: Time Books, pp. 172–173.

D'Croz, Luis. 1988. *Survey and Monitoring of Marine Pollution in the Bay of Panama,* unpublished paper. Panama City: University of Panama.

Economist. 1993. "A Survey of Latin America: Under Construction" (November 13): 5–28.

———. 1995a. "Brazil's Service Industries: Life's a Beach" (April 8), pp. 93–94.

———. 1995b. "Latin America's Drift to the Cities" (October 14), p. 56.

———. 1995c. "A Survey of Brazil: Half Empty or Half Full?" (April 29), pp. 3–34.

———. 1996. "The Black Hole of São Paulo and Others" (January 20), pp. 41–42.

Elder, Danny, and Pernetta, John. eds. 1991. *Oceans,* A Mitchell Beazley World Conservation Atlas. London: Mitchell Beazley, pp. 166–171.

FAO. 1995. *Review of the State of World Fishery Resources: Marine Fisheries.* Rome: FAO, p. 51

Foer, Gordon, and Olsen, Stephen, eds. 1992. *Central America's Coasts: Profiles and an Agenda for Action.* Washington, D.C.: U.S. Agency for International Development; and Narragansett: Coastal Resources Center, University of Rhode Island.

Gallopin, Gilberto; Gutman, Pablo; and Winograd, Manuel. 1991. *Environment and Development: A Latin American Vision.* Report to UNCED, Rio de Janeiro, June, pp. 1–53.

Herz, Renato. 1990. "Brazil's National Program." *CAMP Network—The International Newsletter of Coastal Area Management and Planning.* Kingston: Coastal Resources Center, University of Rhode Island (May), p. 7.

Hinrichsen, Don. 1989. "Income-Generation for Women on Colombia's Mangrove Coast." *Populi* 16, no. 1: 42–47.

———. 1990. *Our Common Seas: Coasts in Crisis.* London: Earthscan, pp. 82–95.

Ministry of the Environment (Water Resources and Legal Amazon). 1996. *Brazilian Policy for Integrated Management of Coastal and Marine Areas—Report to the UN Commission of Sustainable Development,* fourth session, April, New York, pp. 1–8.

Olsen, Stephen; Robadue, Donald; and Arriaga, Luis. 1994. *A Participatory and Adaptive Approach to Integrated Coastal Management: Ecuador's Eight Years of Experience.* Washington, D.C.: U.S. Agency for International Development; Narragansett: Coastal Resources Center, University of Rhode Island (November), pp. 1–47.

Panos Institute. 1995. "Fish: A Net Loss for the Poor—Developing Countries and the Impact of Dwindling World Fish Stocks." *Panos Media Briefing no. 15.* London: Panos Institute (March), p. 11.

Pearce, Fred. 1993. "El Niño Roars Back Against the Odds." *New Scientist,* May 8, p. 7.

Population Action International. 1990. *Cities—Life in the World's 100 Largest Metropolitan Areas.* Washington, D.C.: Population Action International.

Population Reference Bureau. 1996. *World Population Data Sheet, 1996.* Washington, D.C.: Population Reference Bureau.

Robadue, Donald, ed. 1995. *Eight Years in Ecuador: The Road to Integrated Coastal Management.* Narragansett: Coastal Resources Center, University of Rhode Island (September).

Robadue, Donald, and Arriaga, Luis. 1993a. "Ecuador's Coastal Resources Management Program." *Intercoast Network.* Narragansett: Coastal Resources Center, University of Rhode Island (spring), pp. 2–4.

———. 1993b. *Policies and Programs Toward Sustainable Coastal Development in Ecuador's Special Area Management Zones: Creating Vision, Consensus and Capacity.* Narragansett: Coastal Resources Center, University of Rhode Island, pp. 1–13.

Sadik, Nafis. 1993. *The State of World Population 1993.* New York: United Nations Population Fund, p. 14.

Spalding, Mark; Blasco, François; and Field, Colin, eds. 1996. *World Mangrove Atlas.* Cambridge, England: International Society for Mangrove Ecosystems, World Conservation Monitoring Centre, the International Tropical Timber Organization (summer), pp. 81–103.

UNEP. 1983. *Action Plan for the Protection of the Marine Environment and Coastal Areas of the South-East Pacific,* UNEP Regional Seas Reports and Studies, No. 20. Geneva: UNEP.

———. 1988a. *Implications of Climatic Changes in the South-East Pacific Region.* Bogota: UNEP and CPPS (April).

———. 1988b. *State of the Marine Environment in the South-East Pacific Region.* Bogota: CPPS.

———. 1991. *Regional Cooperation on Environmental Protection of the Marine and Coastal Areas of the Pacific Basin.* Nairobi: UNEP, p. 6.

———. 1992. "South-East Pacific—Looking Back and Forward." *Siren,* no. 47 (September–December): 28–29.

UNFPA. 1994. *Latin America and the Caribbean.* New York: UNFPA, pp. 1–8.

Whelan, Tensie. 1989. "Environmental Contamination in the Gulf of Nicoya, Costa Rica." *Ambio* 18, no. 5: 302–304.

Chapter 10

A good overall source on the state of the South Pacific's environment is *The State of the Environment in the South Pacific,* by Arthur Dahl and L. Baumgart. A more up-to-date overview can be found in Anjali Acharya's article "Small Islands: Awash in a Sea of Troubles."

Acharya, Anjali. 1995. "Small Islands: Awash in a Sea of Troubles." *World Watch* 8, no. 6 (November–December): 24–33.

Anderson, Ian. 1995. "Fallout in the South Pacific." *New Scientist,* September 2, pp. 12–13.

———. 1996a. "Polluter Pays Up in Papua." *New Scientist* 150, no. 2035 (July 22): 9.

———. 1996b. "Return of the Coral Eaters." *New Scientist,* February 3, p. 7.

Baines, Grahm. 1981. *Mangrove Resources and Their Management in the South Pacific,* Topic Review No. 5. Noumea, New Caledonia: SPREP (March), pp. 1–7.

———. 1984. "Environment and Resources—Managing the South Pacific's Future." *Ambio* 8, nos. 5–6: 355–358.

Barnaby, Wendy. 1983. "Australia: Environment Down Under." *Ambio* 7, no. 1: 27–33.

Beyer, Lisa. 1989. "Swimming in Sydney's Swill." *Time,* March 27, p. 53.

Connell, John. 1984. "Islands Under Pressure—Population Growth and Urbanization in the South Pacific." *Ambio* 8, nos. 5–6: 306–312.

Cox, Paul Alan, and Elmqvist, Thomas. 1991. "Indigenous Control of Tropical Rainforest Reserves: An Alternative Strategy for Conservation." *Ambio* 20, no. 7: 317–321.

Crawford, David. 1992. "The Injured Coastline—A Parliamentary Report on Coastal Protection in Australia." *Coastal Management* 20, no. 2 (April–June): 189–198.

Cropper, Angela. 1994. "Small Is Vulnerable." *Our Planet* 6, no. 1: 12.

Dahl, Arthur. 1984. "Biogeographical Aspects of Isolation in the Pacific." *Ambio* 8, nos. 5–6: 302–305.

Dahl, Arthur, and Baumgart, L. 1983. *The State of the Environment in the South Pacific,* UNEP Regional Seas Reports and Studies, No. 31. Geneva: UNEP.

Elder, Danny, and Pernetta, John, eds. 1991. *Oceans,* A Mitchell Beazley World Conservation Atlas. London: Mitchell Beazley, pp. 152–153.

Eteuati, Kilifoti. 1991. *The Law of the Sea and the South Pacific.* Nairobi: UNEP (November), pp. 2–23.

FAO. 1995. *Review of the State of World Fishery Resources: Marine Fisheries.* Rome: FAO, pp. 53–54.

Grigg, Richard, and Birkeland, Charles. 1996. *Status of Coral Reefs in the Pacific.* Sydney, Australia: Scientific Committee on Coral Reefs of the Pacific Science Association, pp. 1–4.

Hinrichsen, Don. 1990. *Our Common Seas: Coasts in Crisis.* London: Earthscan, pp. 63–81.

———, ed. 1984. "The South Pacific." *Ambio Special Issue* 13, nos. 5–6.

Holmes, Bob. 1994. "Special Section: The Other Australia." *New Scientist,* October 29, pp. 34–38.

Jameson, Stephen; McManus, John; and Spalding, Mark. 1995. *State of the Reefs: Regional and Global Perspectives.* Washington, D.C.: NOAA (May), pp. 20–23.

Johannes, Robert. 1982. "Traditional Conservation Methods and Protected Marine Areas in Oceania." *Ambio* 5, no. 2: 258–261.

Lal, Padma Narsey. 1984. "Environmental Implications of Coastal Development in Fiji." *Ambio* 8, nos. 5–6: 316–321.

Osborne, Patrick. 1995. "Biological and Cultural Diversity in Papua New Guinea: Conservation, Conflicts, Constraints and Compromise." *Ambio* 24, no. 4 (June): 231–237.

Patterson, Carolyn Bennett. 1986. "In the Far Pacific: At the Birth of Nations." *National Geographic* (October), pp. 460–499.

Peau, Lelei. 1991. "The American Samoa Coastal Management Program." In Brian Needham, ed., *Case Studies of Coastal Management—Experience from the United States.* Narragansett: Coastal Resources Center, University of Rhode Island, pp. 47–49.

Pernetta, John. 1989. "ASPEI and the South Pacific Action Plan." *Siren,* no. 42 (September): 13–16.

Phillips, David. 1992. "The Marine Environment and Its Management in Australia: An Introduction." *Marine Pollution Bulletin* 25, nos. 5–8: 121.

Shenon, Philip. 1995a. "French Plans for A-Tests Incite Anger." *New York Times,* June 17.

———. 1995b. "Tahiti's Antinuclear Protests Turn Violent." *New York Times,* September 8, p. A8.

SPREP. 1984. *Coastal and Inland Water Quality in the South Pacific,* Topic Review No. 16. Noumea, New Caledonia: South Pacific Commission (December).

———. 1985. *Traditional Environmental Management in New Caledonia: A Review of Existing Knowledge,* Topic Review No. 18. Noumea, New Caledonia: South Pacific Commission (March).

Teisch, Jessica. 1995. "Spotlight: Papua New Guinea." *Population Today* 23, no. 1 (January): 7.

Theroux, Paul. 1992. *The Happy Isles of Oceania, Paddling the Pacific.* London: Hamish Hamilton.

Wells, Susan. 1986. "The Pacific Under Pressure." *Earthwatch,* no. 23: 1–8.

Whitney, Craig. 1996. "France Ending Nuclear Tests That Caused Broad Protests." *New York Times,* January 30, pp. 1, 4.

WRI. 1994. *World Resources 1994–95.* New York: Oxford University Press, pp. 354–357.

WWF. 1990. *South Pacific Conservation Programme.* Gland, Switzerland: WWF, pp. 5–49.

Chapter 11

Most of the Chinese pollution data comes from a special report written by Chinese specialists for UNEP's northwest Pacific regional action plan, which has not yet been approved by the region's countries.

In calculating coastal populations for China, I used the population estimates for each coastal province, an area much larger than the 200-kilometer-wide corridor used elsewhere in this book. Disaggregated population data for that corridor were not available.

One of the best sources on pollution data for the northwest Pacific is UNEP's report *Working Plan of North-West Pacific Region—China National Report.*

Beasley, Conger. 1992. "Two Faces of Japan." *Buzzworm: The Environmental Journal* 4, no. 6 (November–December): 32–37.

Catton, Chris. 1992. *Tears of the Dragon—China's Environmental Crisis.* London: Channel 4 Television, pp. 3–23.

China's Population and Development, Country Report. 1992. Prepared for the Fourth Asian and Pacific Population Conference, Bali, Indonesia, August 19–27, pp. 1–74.

China Population Information and Research Centre. 1992. "China Population Data Sheet, 1992." *China Population Today* 9, no. 5 (October): 6–7.

Chinese Delegation to the Fourth Asian and Pacific Population Conference. 1992. *China's Population and Development—Country Report.* Bali, Indonesia: Fourth Asian and Pacific Population Conference (August 19–27), pp. 1–74.

Clark, John. 1996. *Coastal Zone Management Handbook.* New York: Lewis Publishers, p. 508.

Economist. 1991. "Sons of the Soil" (June 22), p. 68.

———. 1994a. "China's Communists—The Road from Tiananmen" (June 4), pp. 23–25.

———. 1994b. "Hong Pong" (July 16), p. 32.

———. 1995. "China, the Numbers Game" (October 14), pp. 38–39.

Environment Agency of Japan. 1993. *The Seto Inland Sea—The Largest Enclosed Sea in Japan.* Tokyo: Environment Agency of Japan (November), pp. 1–20.

Faison, Seth. 1995. "China to Keep Coast Areas' Special Status—East Shanghai Zone Is Haven from Taxes." *New York Times,* July 7, p. D2.

FAO. 1995. *Review of the State of World Fishery Resources: Marine Fisheries.* Rome: FAO, pp. 43–44.

Guo Jinghui and He Qiang. 1993. *Nearshore Pollution Control of China in Ten Years.* Beijing: Environmental Engineering Department of Tsinghua University, unpublished paper, pp. 1–8.

Hinrichsen, Don. 1993. "War on Wildlife." *Scanorama Magazine* 23, no. 10: 76–80.

Hong, Seoung-Yong. 1991. "Assessment of Coastal Zone Issues in the Republic of Korea." *Coastal Management* 19, no. 4 (October–December): 391–415.

Hunt, Paul. 1993. "Hong Kong Rounds Up Toxic Waste." *New Scientist,* June 5, p. 6.

Kalish, Susan. 1995. "Spotlight China." *Population Today* 23, no. 9 (September): 7.

Kristof, Nicholas, and WuDunn, Sheryl. 1994. *China Wakes—The Struggle for the Soul of a Rising Power.* New York: Vintage Books.

Lei, Xiong. 1993. "China's War on Waste and Pollution." *People & the Planet* 2, no. 2: 19–21.

Linden, Eugene. 1995. "The Tortured Land." *Time,* September 4, pp. 45–51.

National Environmental Protection Agency (China). 1992. *Report on the State of the Environment in China, 1991.* Beijing: NEPA (May), pp. 1–12.

Pitt, David. 1993. "Despite Gaps, Data Leave Little Doubt That Fish Are in Peril." *New York Times,* August 3, p. C4.

Platt, Anne. 1995. "Dying Seas." *Worldwatch* 8, no. 1 (January–February): 10–19.

———. 1995. "Russians Capture 2 Japan Trawlers, Wounding Captain." *International Herald Tribune,* September 28, p. 9.

Smil, Vaclav. 1983. "Deforestation in China." *Ambio* 7, no. 5: 226–231.

Specter, Michael. 1995. "A Too-Free Enterprise Endangers Siberian Tigers." *New York Times,* September 5, pp. A1, A8.

Tian Xueyuan. 1989. "Reform and More Flexible Policies Promote Urbanization." *Chinese Journal of Population Science* 1, no. 3: 275–283.

Tyler, Patrick. 1994a. "Poor on the Move in China—Millions of Peasants Seek Work in Cities." *International Herald Tribune,* June 30, pp. 1, 4.

———. 1994b. "A Tide of Pollution Threatens China's Prosperity." *New York Times,* September 25.

UNEP. 1992. *Working Plan of North-West Pacific Region—China National Report.* Beijing: UNEP (October).

WRI. 1994. *World Resources 1994–95.* New York: Oxford University Press, p. 73.

WuDunn, Sheryl. 1994. "China's Rush to Riches." *New York Times Magazine,* September 4, pp. 38–54.

Xias Duning, Li Xiuzhen, Hu Yuanman, and Wang Xianli. 1996. "Protection of the Littoral Wetland in Northern China: Ecological and Environmental Characteristics." *Ambio* 25, no. 1 (February): 2–5.

Yeung, Yue-man, and Hu, Xu-wei, eds. 1992. *China's Coastal Cities: Catalysts for Modernization.* Honolulu: University of Hawaii Press.

Chapter 12

A good source on overfishing is the *Ambio* article by Daniel Pauly and Chua Thia-Eng, "The Overfishing of Marine Resources: Socioeconomic Background in Southeast Asia." Cyanide fishing is summarized by Leigh Dayton in "The Killing Reefs," *New Scientist,* November 11, 1995.

Alcala, Angel. ed. 1991. *Proceedings of the Regional Symposium on Living Resources in Coastal Areas.* Manila: Marine Science Institute, University of the Philippines.

ASEAN-Australia Marine Science Project. 1992. "The Status of Living Coastal Resources of ASEAN Countries." *ASEAN Marine Science—Newsletter of the ASEAN-Australia Marine Science Project,* no. 19 (April): 6–17.

Centre for Research of Human Resources and the Environment. 1991. *Environmental Profile of Indonesia 1990.* Jakarta: Centre for Research of Human Resources and the Environment, University of Indonesia.

Chiang, K.M. 1993. *Water Pollution Control—The Cleaning Up of Singapore River and Kallang Basin Catchment.* Singapore: Department of Sewerage, Ministry of Environment.

Clark, John. 1996. *Coastal Zone Management Handbook.* New York: Lewis Publishers, pp. 462, 594–595.

Cohen, Margot. 1995. "Indonesia: Cautious Cooperation." *Far Eastern Economic Review.* Special Supplement: Environment in Asia (November 16): 67–68.

College of Social Work and Community Development, University of the Philippines. 1993. *Our Sea, Our Life—Proceedings of the Seminar Workshop on Community-Based Coastal Resources Management.* Silliman University, Dumaguete City, Philippines, February 7–12, pp. 13–19.

Country Report, Philippines. 1993. *Status and Management of Living Coastal Resources: The Philippines,* report presented at the 6th Management Committee Meeting of the ASEAN-Australia Marine Science Project: Living Coastal Resources, Phase II, Manila, February 22–24, pp. F1–F8.

Cruz, Maria Concepcion; Meyer, Carrie; Repetto, Robert; and Woodward, Richard. 1992. *Population Growth, Poverty and Environmental Stress: Frontier Migration in the Philippines and Costa Rica.* Washington, D.C.: World Resources Institute (October), pp. 1–43.

Cruz, Wilfrido, and Cruz, Maria Concepcion. 1990. "Population Pressure and Deforestation in the Philippines." *ASEAN Economic Bulletin* 7, no. 2 (November): 200–212.

Dahuri, Rokhmin. 1994. "The Challenge of Sustainable Coastal Development in East Kalimantan, Indonesia." *Coastal Management in Tropical Asia* (March): 12–15.

Dayton, Leigh. 1995. "The Killing Reefs." *New Scientist,* November 11, pp. 14–15.

Economist. 1993. "Wealth in Its Grasp: A Survey of Indonesia." April 17, pp. 1–18.

Gargan, Edward. 1996. "A Boom in Malaysia Reaches for the Sky." *New York Times,* February 2, pp. D1, D4.

Gomez, Ed, and McManus, Liana. 1996. "Case Study 4: Coastal Management in Bolinao Town and the Lingayen Gulf, the Philippines." In *The Contributions of Science to Integrated Coastal Management.* Rome: FAO, pp. 57–66.

Hale, Lynne, and Olsen, Stephen. 1993. "Coral Reef Management in Thailand—A Step Toward Integrated Coastal Management." *Oceanus* 36, no. 3 (fall): 27–34.

Harger, J.R.E. 1988. "Community Displacement in Stressed Coral Reef Systems and the Implication for a Comprehensive Management Strategy for Coastal and Offshore Productivity Enrichment." *Galaxea* 7: 185–196.

———. 1993. *Organic Pollution in Jakarta Bay.* Jakarta, Indonesia: UNESCO Regional Office for Science and Technology for Southeast Asia and the Pacific, unpublished paper (April), pp. 1–9.

Hinrichsen, Don. 1990. *Our Common Seas: Coasts in Crisis.* London: Earthscan, pp. 96–120.

———. 1991. "The Quest for a Common Future—Nepal, the Philippines and Thailand." *People* 18, no. 3: 16–19.

———. 1994a. "Sea Watchers." *People & the Planet* 3, no. 1: 27.

————. 1994b. "Where Women Take Control." *People & the Planet* 3, no. 1: 22–23.

————. 1997. "Requiem for Reefs?" *International Wildlife* 27, no. 2: 12–20.

Hooten, Anthony, and Hatziolos, Marea, eds. 1995. *Sustainable Financing Mechanisms for Coral Reef Conservation: Proceedings of a Workshop,* Environmentally Sustainable Development Proceedings Series No. 9. Washington, D.C.: World Bank, (June), pp. 60–63.

ICLARM. 1988. "The Coastal Environmental Profile of South Johor, Malaysia." *Tropical Coastal Area Management* 3, no. 3 (December): 1–4.

Indonesia Observer. 1993. "Coast of East Java in Critical Condition," February 11.

Jakarta Post. 1993. "Mangrove Forest in North Jakarta in Critical Condition," April 29.

Jameson, Stephen; McManus, John; and Spalding, Mark. 1995. *State of the Reefs—Regional and Global Perspectives.* Washington, D.C.: NOAA (May), pp. 16–18.

Kano, Ichiro. 1992. "Vietnam—Busily Making Up for Lost Time." *INFOFISH International,* no. 4: 21–26.

Kelleher, Graeme; Bleakley, Chris; and Wells, Sue, eds. 1995. *A Global Representative System of Marine Protected Areas, Volume III: Central Indian Ocean, Arabian Seas, East Africa and East Asian Seas.* Washington, D.C.: World Bank (May), p. 15.

Kemf, Elizabeth. 1988. "The Re-Greening of Vietnam." *New Scientist,* no. 1618 (June 23): 53–57.

————. 1990. *Month of Pure Light—The Regreening of Vietnam.* London: Women's Press.

Kuznik, Frank. 1994. "Healing on the Plain of Reeds." *International Wildlife* (January–February), pp. 14–17.

Lemay, Michele, and Hale, Lynne. 1991. *A National Coral Reef Strategy for Thailand, Vol. 1: Statement of Need.* Narragansett: Thailand Coastal Resources Management Project, University of Rhode Island; Washington, D.C.: USAID (January), pp. 1–33.

————. 1993. *A National Coral Reef Strategy for Thailand, Vol. 2: Policies and Action Plan.* Narragansett: Thailand Coastal Resources Management Project, University of Rhode Island: Washington, D.C.: USAID (February), pp. 1–76.

Lindén, Olof. 1995. "The State of the Coastal and Marine Environment of Vietnam." *Ambio* 24, nos. 7–8: 525–526.

Llana, Ethel. 1991. "Production and Utilisation of Seaweeds in the Philippines." *INFOFISH International,* no. 1: 12–17.

Manila Bulletin. 1993. "Mangrove Plantation Proposed in Manila Bay," April 5.

McManus, John; Nanola, Cleto; Reyes, Rodolfo; and Kesner, Kathleen. 1992. *Resource Ecology of the Bolinao Coral Reef System.* Manila: ICLARM, University of Rhode Island, and University of the Philippines, pp. 1–76.

McNeely, Jeffrey, and Dobias, Robert. 1991. "Economic Incentives for Conserving Biological Diversity in Thailand." *Ambio* 20, no. 2 (April): 86–90.

O'Neill, Thomas. 1993. "Mekong River." *National Geographic* (February), pp. 2–35.

Pauly, Daniel. 1988. "Fisheries Research and the Demersal Fisheries of Southeast Asia." In J.A. Gulland, ed., *Fish Population Dynamics.* London: John Wiley, pp. 329–346.

Pauly, Daniel, and Thia-Eng, Chua. 1988. "The Overfishing of Marine Resources: Socioeconomic Background in Southeast Asia." *Ambio* 17, no. 3: 200–206.

Pauly, Daniel; Silvestre, Geronimo; and Smith, Ian. 1989. "On Development, Fisheries and Dynamite: A Brief Review of Tropical Fisheries Management." *Natural Resource Modeling* 3, No. 3 (summer): 307–329.

Philippines Department of Environment and Natural Resources. 1992. *Philippine Environment and Development 1992,* National Report to the United Nations Conference on

Environment and Development, Rio de Janeiro, Brazil (June). Manila: Department of Environment and Natural Resources, pp. 36–41.

Primavera, Honculada. 1991. "Intensive Prawn Farming in the Philippines: Ecological, Social and Economic Implications." *Ambio* 20, no. 1 (February): 28–33.

Sien, Chia Lin. 1992. *Singapore's Urban Coastal Area: Strategies for Management,* Technical Publications Series 9. Manila: Association of Southeast Asian Nations and U.S. Coastal Resources Management Project.

Sien, Chia Lin, and Ming, Chou Like, eds. 1989. *Urban Coastal Area Management: The Experience of Singapore,* proceedings of the Singapore National Workshop on Urban Coastal Area Management, Singapore, November.

Sloan, N.A., and Sugandhy, A. 1994. "An Overview of Indonesian Coastal Environmental Management." *Coastal Management* 22, no. 3 (July–September), pp. 215–230.

Smith, Russell. 1994. "Coastal Tourism in the Asia Pacific: Environmental Planning Needs." *Coastal Management in Tropical Asia* (March), pp. 9–11.

Spalding, Mark; Blasco, François; and Field, Colin, eds. 1996. *World Mangrove Atlas.* Cambridge, England: International Society for Mangrove Ecosystems, World Conservation Monitoring Centre, International Tropical Timber Organization (summer), pp. 38–64.

Sukardjo, Sukristijono. 1990. "Indonesia's Mangroves Need Study, Care." *Japan Times,* September 30.

Sunday Observer. 1993. "Java's Delicate Mangrove Forests Damaged by Development," April 18.

Tangwisutijit, Nantiya. 1996. "Must the Mekong Die?" *People & the Planet* 5, no. 3: 10–13.

Thia-Eng, Chua. 1993. *Asian Fisheries Towards the Year 2000: A Challenge to Fisheries Scientists.* Manila: ICLARM, pp. 1–30.

UNEP. 1985. *Management and Conservation of Renewable Marine Resources in the East Asian Seas Region,* UNEP Regional Seas Reports and Studies No. 65. Nairobi: UNEP.

———. 1990. *Proceedings of the First ASEAMS Symposium on Southeast Asian Marine Science and Environmental Protection,* UNEP Regional Seas Reports and Studies no. 116. Nairobi: UNEP.

Vusse, Frederick. 1991. "The Use and Abuse of Artificial Reefs: The Philippine Experience." *Tropical Coastal Area Management* (April–August), pp. 8–9.

WALHI (Indonesian Environmental Forum). 1990. "Evicted from the Sea." *Environesia* 4, no. 1 (March): 13–15.

Warfvinge, Hans, and Pham Dinh Huan. 1991. *Environmental Profile for Vietnam.* Ministry of Forestry, Socialist Republic of Vietnam, unpublished paper, November, pp. 1–11.

Wilkinson, Clive; Chou, L.M.; Gomez, E.; Mohammed, I.; Soekarno, S.; and Sudara, S. 1992. "A Regional Approach to Monitoring Coral Reefs: Studies in Southeast Asia by the ASEAN-Australia Living Coastal Resources Project." *Proceedings of the 7th International Coral Reef Symposium, June 1992, Guam.*

Chapter 13

Population data come from the UN and the Population Reference Bureau. Coastal populations were estimated using mostly coastal urban data and disintegrated data from the latest censuses as presented by Uwe Deichmann in his population density map (p. 179). India has population data for districts as well as states, and both have been used in this analysis. However, India's centralized data collection system means that the states, in most instances, have to appeal to New Delhi in order to get population totals. And as of late 1996, population data from the 1991 census were still being processed by the office of the Registrar General of India in Delhi. The populations of Bangladesh, Sri Lanka, and the Maldives are all essentially coastal.

A good overview of pollution and development problems in South Asia can be found in Paul Harrison's award-winning book, *The Third Revolution—Environment, Population and a Sustainable World*.

Clark, John. 1996. *Coastal Zone Management Handbook*. New York: Lewis Publishers, pp. 144–146, 545–546, 580–586.

Coast Conservation Department. 1990. *Coastal Zone Management Plan—Sri Lanka*. Colombo: Coast Conservation Department.

Conly, Shanti, and Camp, Sharon. 1992. *India's Family Planning Challenge: From Rhetoric to Action*. Washington, D.C.: Population Action International.

D'Monte, Darryl. 1996. "Filthy Flows the Ganga." *People & the Planet* 5, no. 3: 20–22.

Elder, Danny, and Pernetta, John, eds. 1991. *Oceans*, A Mitchell Beazley World Conservation Atlas. London: Mitchell Beazley, p. 142.

FAO. 1995. *Review of the State of World Fishery Resources: Marine Fisheries*. Rome: FAO, pp. 41–42.

Gordon, Ann. 1991. *The By-Catch from Indian Shrimp Trawlers in the Bay of Bengal: The Potential for Its Improved Utilization*. Madras, India: Bay of Bengal Programme, p. 5.

Govind, Har. 1989. "Recent Developments in Environmental Protection in India: Pollution Control." *Ambio* 18, no. 8: 429–433.

Harrison, Paul. 1991. "Living Dangerously." *People* 18, no. 3: 6–10.

———. 1992. *The Third Revolution—Environment, Population and a Sustainable World*. London: Penguin Books, pp. 202–203, 234–235.

Hasan, Samiul, and Mulamootti, George. 1994. "Natural Resource Management in Bangladesh." *Ambio* 23, no. 2 (March): 141–145.

Haub, Carl. 1996. "Spotlight: India." *Population Today* (Population Reference Bureau, Washington, D.C.) 24, no. 5 (May): 7.

Hinrichsen, Don. 1990. *Our Common Seas: Coasts in Crisis*. London: Earthscan, pp. 121–131.

Jameson, Stephen; McManus, John; and Spalding, Mark. 1995. *State of the Reefs—Regional and Global Perspectives*. Washington, D.C.: NOAA, ICLARM, World Conservation Monitoring Centre (May), pp. 13–14.

Kelleher, Graeme; Bleakley, Chris; and Wells, Sue. 1995. *A Global Representative System of Marine Protected Areas, Vol. III: Central Indian Ocean, Arabian Seas, East Africa and East Asian Seas*. Washington, D.C.: Great Barrier Reef Marine Park Authority, World Bank, World Conservation Union (May), pp. 15–36.

Lean, Geoffrey, and Hinrichsen, Don. 1994. *Atlas of the Environment*. New York: Harper-Collins, pp. 109–114.

Lowry, Kem, and Sadacharan, Dianeetha. 1993. "Coastal Management in Sri Lanka." *Coastal Management in Tropical Asia*, no. 1 (September): 1–7.

Mitchison, Amanda. 1986. "Theirs Is to Catch, But Not Consume—Small Scale Fishing Communities in India." *Food and Nutrition* 12, no. 2: 11–18.

Montgomery, Sy. 1995. *Spell of the Tiger: The Man-Eaters of Sundarbans*. Boston: Houghton Mifflin.

Nayak, B. U.; Chandramohan P.; and Desai, B.N. 1992. "Planning and Management of the Coastal Zone in India: A Perspective." *Coastal Management* 20, no. 4 (October–December): 365–374.

Nielsen, Henrik. 1991. "Shrimp Farming in West Bengal." *Bay of Bengal News,* no. 43 (September): 6–9.

Olsen, Stephen; Sadacharan, D.; Samarakoon, J.; White, A.; Wickremeratne, H.; and Wijeratne, M. 1992. *Coastal 2000: Recommendations for a Resource Management Strategy for Sri Lanka's Coastal Region,* volumes 1 and 2. Narragansett: Coastal Resources Center, University of Rhode Island, pp. 15–79.

Panos Institute. 1995. *Fish: A Net Loss for the Poor: Developing Countries and the Impact of Dwindling World Fish Stocks,* Panos Media Briefing no. 15. London: Panos Institute (March), pp. 5–6.

Pearce, Fred. 1996. "Squatters Take Control." *New Scientist* (June 1), pp. 39–42.

Population Action International. 1990. *Cities—Life in the World's 100 Largest Metropolitan Areas.* Washington, D.C.: Population Action International.

Population Reference Bureau. 1994. *India Health and Family Welfare: Data from the National Family Health Survey, 1992–93,* wall chart. Washington, D.C.: Population Reference Bureau.

Rajasuriya, Arjan; Ranjith, M.W.; De Silva, N.; and Ohman, Marcus. 1995. "Coral Reefs of Sri Lanka: Human Disturbance and Management Issues." *Ambio* 24, nos. 7–8: 428–437.

Repetto, Robert. 1996. "The Second India Revisited: Population Growth, Poverty, and Environment over Two Decades." *Proceedings of the Conference on Population, Environment and Development, March 13–14, 1996.* Washington, D.C.: Tata Energy and Resources Institute, pp. 2–31.

Sampat, Payal. 1996. "The River Ganges' Long Decline." *World Watch* 9, no. 4 (July–August): 25–32.

Shankar, Uday. 1992. "Purifying the Ganga." *Down to Earth* 1, no. 9 (September 30): 25–32.

Sherk, Kirsten. 1993. "Spotlight: Bangladesh." *Population Today* 21, no. 10 (October): 11.

Spalding, Mark; Blasco, François; and Field, Colin, eds. 1996. *World Mangrove Atlas.* Cambridge, England: International Society for Mangrove Ecosystems, World Conservation Monitoring Centre, and International Tropical Timber Organization (summer), pp. 38–64.

UNEP. 1985. *Environmental Problems of the Marine and Coastal Area of India: National Report,* UNEP Regional Seas Reports and Studies No. 59. Nairobi: UNEP, pp. 1–18.

———. 1986. *Environmental Problems of the Marine and Coastal Area of Maldives: National Report,* UNEP Regional Seas Reports and Studies No. 76. Nairobi: UNEP, pp. 1–25.

———. 1987. *Environmental Problems of the South Asian Seas Region: An Overview,* UNEP Regional Seas Reports and Studies No. 82. Nairobi: UNEP, pp. 1–31.

Visaria, Leela, and Visaria, Pravin. 1995. "India's Fertility Declines, But It Still Leads World in Population Growth." *Population Today* 23, no. 10 (October): 1–2.

WRI. 1994. *World Resources 1994–95.* New York: Oxford University Press, pp. 83–103.

Chapter 14

Fairly good population data exist on coastal populations in the Arabian Gulf; both the UN and the Arab countries of the gulf have useful data sets.

A good source on gulf environmental problems is Michael Renner's article, "Military Victory, Ecological Defeat."

Campbell, Meredith. 1987. "Save Our Scenery." *Oman Daily Observer,* March 8, p. 7.

Canby, Thomas. 1991. "After the Storm." *National Geographic* 180, no. 2 (August): 2–35.

Cava, Francesca. 1993. "The Environmental Response to the Gulf War." *Oceanus* 36, no. 3 (fall): 60–62.

FAO. 1995. *Review of the State of World Fisheries: Marine Fisheries.* Rome: FAO, p. 40.

Hinrichsen, Don. 1990. *Our Common Seas: Coasts in Crisis.* London: Earthscan, pp. 40–48.

Jameson, Stephen; McManus, John; and Spalding, Mark. 1995. *State of the Reefs—Regional and Global Perspectives.* Washington, D.C.: NOAA, ICLARM, World Conservation Monitoring Centre (May), pp. 10–12.

Kelleher, Graeme; Bleakley, Chris; and Wells, Sue, eds. 1995. *A Global Representative System of Marine Protected Areas, Vol. III: Central Indian Ocean, Arabian Seas, East Africa and East Asian Seas.* Washington, D.C.: Great Barrier Reef Marine Park Authority, World Bank, World Conservation Union (May), pp. 39–70.

Omran, Abdel, and Roudi, Farzaneh. 1993a. "The Middle East Population Puzzle." *Population Bulletin* 48, no. 1 (July): 2–35.

———. 1993b. "Oil Jobs Have Big Impact on Heavily Populated Middle East." *Population Today* 21, no. 9 (September): 1–2.

Pearce, Fred. 1993. "Gulf Fisheries Still Suffering in Wake of War." *New Scientist* (March 20), p. 10.

Population Reference Bureau. 1997. *World Population Data Sheet, 1997.* Washington, D.C.: Population Reference Bureau.

Renner, Michael. 1991. "Military Victory, Ecological Defeat." *World Watch* (July–August): 27–33.

Salm, Rodney. 1988. "Mud, Mud, Glorious Mud." *PDO News,* no. 4: 15–19.

———. 1991. "Oman's Gold Coast." *PDO News,* no. 3: 20–26.

———. 1992. "Facing Change in Oman." *IUCN Bulletin,* no. 4: 16–17.

Salm, Rodney, and Dobbin, James. 1987. "A Coastal Zone Management Strategy for the Sultanate of Oman." Reprinted from proceedings of conference *Coastal Zone 1987,* Seattle, Washington, May 26–29, pp. 97–106.

Shuaib, H.A. 1998. "Oil, Development and the Environment in Kuwait." *Environment* 30, no. 6 (July–August), pp. 18–20, 39–44.

UNEP. 1992a. "Assessing the Impact of Kuwait Oil Fires," *Siren,* no. 47 (September–December): 4.

———. 1992b. "Lessons of War." *Siren,* no. 45 (March): 1–4.

WMO. 1992. *Report of the Second WMO Meeting of Experts to Assess the Response to and Atmospheric Effects of the Kuwait Oil Fires,* Report no. 81. Geneva: World Meteorological Organization (May), pp. 1–36.

World Conservation Monitoring Centre. 1991. *Gulf War Environmental Information Project,* press release. Cambridge, England: World Conservation Monitoring Centre, pp. 1–3.

World Conservation Union. 1989. "Oman's Coasts Mapped." *IUCN Bulletin* 20, nos. 7–9 (July–September): 13.

———. 1992. "Gulf Environment Revisited." *IUCN Bulletin* 23, no. 1 (March): 11.

WWF. 1991. *Lessons to Be Learned from the Consequences of the Arabian Gulf War,* a WWF Discussion Paper. Gland, Switzerland: WWF (August), pp. 2–18.

Chapter 15

There is little in the way of reliable demographic data for the Red Sea and the Gulf of Aden. Population has been estimated based on the data for urban centers, including growth rates.

Berkeley, Bill. 1996. "The Longest War in the World." *New York Times Magazine*, March 3, pp. 59–61.

Doubilet, David. 1993. "The Desert Sea." *National Geographic* 184, no. 5 (November): 60–87.

Economist. 1996. "Aqua-Tourism—Paradise Retained?" (February 3), p. 72.

Elder, Danny, and Pernetta, John, eds. 1991. *Oceans*, A Mitchell Beazley World Conservation Atlas. London: Mitchell Beazley, pp. 140–141.

FAO. 1995. *Review of the State of World Fishery Resources: Marine Fisheries*. Rome: FAO, p. 40.

Government of Eritrea. 1995. *National Environmental Management Plan for Eritrea*. Asmara: Government of Eritrea.

Hillman, Jesse. 1993. "In Eritrea, Africa's Newest Nation, Scientists Find That Wildlife Has Survived." *African Wildlife Update* 2, no. 3 (May–June): 1–2.

Hinrichsen, Don. 1990. *Our Common Seas: Coasts in Crisis*. London: Earthscan, pp. 159–165.

———. 1992 "Egypt at the Crossroads." *People & the Planet* 1, no. 4: 10–19.

Jameson, Stephen; McManus, John; and Spalding, Mark. 1995. *State of the Reefs—Regional and Global Perspectives*. Washington, D.C.: NOAA, ICLARM, World Conservation Monitoring Centre (May), p. 10.

Kelleher, Graeme; Bleakley, Chris; and Wells, Sue, eds. 1995. *A Global Representative System of Marine Protected Areas, Vol. III: Central Indian Ocean, Arabian Seas, East Africa and East Asian Seas*. Washington, D.C.: Great Barrier Reef Marine Park Authority, World Bank, World Conservation Union (May), pp. 39–62.

Population Reference Bureau. 1995. "Philippine Fertility Down, Remains High in Yemen." *Population Today* 23, no. 3 (March): 8.

———. 1997. *World Population Data Sheet, 1997*. Washington, D.C.: Population Reference Bureau.

Rowntree, John. 1993. *Final Report—Eritrea Coastal and Marine Resources Assessment and Project Identification*. Asmara, Eritrea: USAID (October).

Sadik, Nafis, ed. 1994. *Making a Difference: 25 Years of UNFPA Experience*. New York: United Nations Population Fund, pp. 78–91.

UNEP. 1988a. *Siren*, no. 36 (April): 4–36.

———. 1988b. *Siren*, no. 38 (October): 1–34.

———. 1988c. *Siren*, no. 39 (December): 1–30.

———. 1989. *Siren*, no. 41 (July): 17–20.

———. 1990. *Africa's Seas—Challenge of the 1990s*. Nairobi: UNEP, pp. 15–16.

UNICEF. 1994. *Children and Women in Eritrea*. Asmara, Eritrea: UNICEF, pp. 1–20.

Watzman, Haim. 1995. "Red Sea Pays the Price of Peace." *New Scientist*, February 18, p. 9.

Welwig, Tony. 1995. "Coral Grief: Red Sea Reefs Crumble." *New Internationalist*, no. 270 (August): 5.

World Bank. 1995. *Africa: A Framework for Integrated Coastal Zone Management*. Washington, D.C.: World Bank, p. 17.

Chapter 16

East Africa's population data comes from the World Bank, which calculated the entire coastal population within 60 kilometers of the coast (see World Bank, *Africa: A Framework for Integrated Coastal Zone Management*).

Bryson, Bill. 1993. "Paradise Found." *Observer Magazine: On the Waterfront, 1993 Travel Special* (January 3), pp. 11–17.

Carter, Marion. 1996. "Spotlight Kenya." *Population Today* 24, no. 1 (January): 7.

Clark, John. 1996. *Coastal Zone Management Handbook.* New York: Lewis Publishers, pp. 587–589.

Coughanowr, Christine; Ngoile, Magnus; and Lindén, Olof. 1995. "Coastal Zone Management in Eastern Africa Including the Island States: A Review of Issues and Initiatives." *Ambio* 24, nos. 7–8: 448–457.

East African Wildlife Society. 1993. *Ramsar and Tana River Delta Wetlands.* Nairobi: East African Wildlife Society (March), pp. 1–5.

FAO. 1995. *Review of the State of World Fishery Resources: Marine Fisheries.* Rome: FAO, pp. 39–41.

Girardet, Edward. 1993. "Conservation Amid a Civil War." *WWF Features,* no. 20: 1–4.

Hanneberg, Peter. 1991. "The Seychelles—Conservation Keeps the Tourists Coming." *Tomorrow* 1, no. 1: 25–37.

Hanssen, Nina. 1992. "Experts Warn over Ecological Disaster." *Daily Nation,* December 2.

Hinrichsen, Don. 1990. *Our Common Seas: Coasts in Crisis.* London: Earthscan, pp. 132–149.

————, ed., 1983. *Ambio Special Issue:* "The Indian Ocean." Vol. 12, no. 6.

Iqbal, M.S. 1992. *Assessment of the Implementation of the Eastern African Action Plan and the Effectiveness of Its Legal Instruments,* UNEP Regional Seas Reports and Studies No. 150. Nairobi: UNEP, pp. 1–21.

Jameson, Stephen; McManus, John; and Spalding, Mark. 1995. *State of the Reefs—Regional and Global Perspectives,* Washington, D.C.: NOAA, ICLARM, World Conservation Monitoring Centre, p. 13.

Kelleher, Graeme; Bleakley, Chris; and Wells, Sue, eds. 1995. *A Global Representative System of Marine Protected Areas, Vol. III: Central Indian Ocean, Arabian Seas, East Africa and East Asian Seas.* Washington, D.C.: World Bank, pp. 71–105.

Lindén, Olof. 1993. "Resolution on Integrated Coastal Zone Management in East Africa, Signed in Arusha, Tanzania." *Ambio* 22, no. 6 (September): 408–409.

Lorch, Donatella. 1995. "As Population Swells, Nairobi Plunges into Poverty." *New York Times,* December 12, p. A14.

Lutz, Wolfgang. 1992. "Population-Development-Environment Interactions: A Case Study on Mauritius." *Popnet,* no. 21 (Spring): 1–12.

Muchena, F.N., and Ndaraiya, F.M. 1990. *The Impact of Agriculture in the Coastal Areas of East Africa.* Nairobi: National Agricultural Laboratories, unpublished paper, pp. 1–17.

Opala, Ken. 1992. "Game Reserve Land Given Out." *Daily Nation,* August 11.

————. 1993a. "Sh. 5.4 Billion Tourism Grant at Risk over Tana Delta Deal." *Daily Nation,* February 1.

————. 1993b. "Tana Delta Row Drags on with No End in Sight." *Daily Nation,* January 1.

Population Reference Bureau. 1989. "A Gray Land in a Red Ocean?" *Population Today* (July–August), pp. 3–4.

Republic of Kenya. 1991. *Tana River Delta Wetlands Survey.* Nairobi: Ministry of Reclamation and Development of Arid, Semi-arid Areas and Wastelands (May), pp. 5–47.

Salm, Rodney. 1996. "The Status of Coral Reefs in the Western Indian Ocean with Notes on the Related Ecosystems," working paper prepared for the International Coral Reef Initiative Workshop, Seychelles, March, pp. 1–23.

Semesi, Adelaida, and Howell, Kim. 1992. *The Mangroves of the Eastern African Region.* Nairobi: UNEP, pp. 1–45.

UNEP. 1982a. *Conservation of Coastal and Marine Ecosystems and Living Resources of the East African Region,* UNEP Regional Seas Reports and Studies No. 11. Nairobi: UNEP, pp. 1–64.

———. 1982b. *Environmental Problems of the East African Region,* UNEP Regional Seas Reports and Studies No. 12. Nairobi: UNEP, pp. 1–65.

———. 1985. *Management and Conservation of Renewable Marine Resources in the Eastern African Region,* UNEP Regional Seas Reports and Studies No. 66. Nairobi: UNEP, pp. 1–93.

———. 1989. *A Coast in Common—An Introduction to the Eastern African Action Plan.* Nairobi: UNEP, p. 3–40.

———. 1990. *Africa's Seas—Challenge of the 1990s.* Nairobi: UNEP, pp. 1–19.

Wickremeratne, Shanti. 1991. *The Law of the Sea and the Indian Ocean.* Nairobi: UNEP, (November), pp. 2–32.

World Bank. 1995. *Africa: A Framework for Integrated Coastal Zone Management.* Washington, D.C.: World Bank.

Chapter 17

West Africa's coastal population estimates come from the World Bank and are based on demographic data for coastal zones up to 60 kilometers inland. The main source for population data for both East and West Africa is the World Bank's *Africa: A Framework for Integrated Coastal Zone Management.*

Economist. 1993. "A Survey of Nigeria" (August 21), pp. 1–14.

Elder, Danny, and Pernetta, John, eds. 1991. *Oceans,* A Mitchell Beazley World Conservation Atlas. London: Mitchell Beazley, pp. 134–136.

FAO. 1995. *Review of the State of World Fishery Resources: Marine Fisheries.* Rome: FAO, pp. 20–23.

Hinrichsen, Don. 1990. *Our Common Seas: Coasts in Crisis.* London: Earthscan, pp. 150–158.

Jameson, Stephen; McManus, John; and Spaulding, Mark. "State of the Reefs: Regional and Global Perspectives." Washington, D.C., International Coral Reef Initiative Executive Secretariat Background Paper, NOAA (May).

Lindén, Olof. 1990. "Human Impact on Tropical Coastal Zones." *Nature and Resources* 26, no. 4: 8.

Moffat, David, and Lindén, Olof. 1995. "Perception and Reality: Assessing Priorities for Sustainable Development in the Niger River Delta." *Ambio* 24, nos. 7–8: 527–538.

Panos Institute. 1995. *Panos Briefing—Fish: A Net Loss for the Poor.* London: Panos Institute (March), pp. 1–13.

Price, A.R.G. 1992. *Coastal Assessment of Parc National du Banc d'Arguin, Mauritania.* Gland, Switzerland: IUCN, pp. 1–40.

UNEP. 1984a. *Environmental Management Problems in Resource Utilization and Survey of Resources in the West and Central African Region,* UNEP Regional Seas Reports and Studies no. 37. Geneva: UNEP.

———. 1984b. *Onshore Impact of Offshore Oil and Natural Gas Development in the West and Central African Region,* UNEP Regional Seas Reports and Studies No. 33. Geneva: UNEP.

———. 1991a. *Africa's Seas—Challenge of the 1990s.* Nairobi: UNEP, pp. 1–19.

———. 1991b. *Change and Challenge—An Introduction to the West and Central African Action Plan.* Nairobi: UNEP, pp. 1–20.

World Bank. 1995. *Africa: A Framework for Integrated Coastal Zone Management.* Washington, D.C.: World Bank.

WRI. 1994. *World Resources 1994–95.* New York: Oxford University Press, pp. 336, 354, 356.

Chapter 18

The ideas in this chapter are not my own, though I agree with them. Numerous institutions and thoughtful individuals concerned about coastal management and how to make it sustainable have shared their knowledge and perspectives with me. I am especially indebted to the staff of the Coastal Resources Center of the University of Rhode Island for their insightful comments and for helping me understand some of the complex pitfalls that await individuals and institutions as they try to grapple with integrated coastal area management.

One of the best sources on coastal zone management and what is required to make it successful are the case studies found in *The Contribution of Science to Integrated Coastal Management* (1996).

Cicin-Sain, Biliana. 1993. "Sustainable Development and Integrated Coastal Management." *Ocean & Coastal Management* 21, nos. 1–3: 11–40.

Earle, Sylvia. *Sea Change: A Message of the Oceans.* New York: Ballentine Books.

Hinrichsen, Don. 1996a. "Coasts in Crisis." *Issues in Science and Technology* 12, no. 4 (summer): 39–47.

———. 1996b. "Computing the Risks." *International Wildlife* 26, no. 2 (March–April): 22–35.

Loftas, Tony, ed. 1995. *Dimensions of Need—An Atlas of Food and Agriculture.* Rome: FAO, pp. 54–55.

Matuszeski, William. 1996. "Case Study 1: The Chesapeake Bay Programme, USA." In *The Contributions of Science to Integrated Coastal Management.* Rome: FAO, pp. 25–30.

OECD. 1993a. *Coastal Zone Management—Integrated Policies.* Paris: OECD.

———. 1993b. *Coastal Zone Management—Selected Case Studies.* Paris: OECD.

Olsen, Stephen. 1996. "Case Study 3—Ecuador's Coastal Resources Management Programme." In *The Contributions of Science to Integrated Coastal Management.* Rome: FAO, pp. 45–56.

Pain, Stephanie. 1996. "Treasures Lost in Reef Madness." *New Scientist,* February 17, p. 9.

Pearce, Fred. 1995. "Call for Action to Save the Oceans." *New Scientist,* January 28, p. 7.

UNEP. 1996. *The Diversity of the Seas: A Regional Approach.* London: UNEP, World Conservation Monitoring Centre.

UNESCO. 1992a. *Coastal Systems Studies and Sustainable Development: Proceedings of the COMAR Interregional Scientific Conference.* Paris: UNESCO.

———. 1992b. *New Technologies—Remote Sensing and Geographic Information Systems,* Environment and Development Briefs No. 3. Paris: UNESCO, pp. 1–16.

———. 1995. *Coasts and Small Islands—Targets for Integrated Efforts.* Paris: UNESCO, p. 3.

Vallega, Adalberto. 1993. "A Conceptual Approach to Integrated Coastal Management." *Ocean & Coastal Management* 21, nos. 1–3: 149–161.

WRI. 1995. *Coastlines at Risk: An Index of Potential Development-Related Threats to Coastal Ecosystems,* WRI Indicator Brief. Washington, D.C.: World Resources Institute, pp. 1–8.

———. 1996. *World Resources, 1996–97.* New York: Oxford University Press, p. 300.

INDEX